T0191790

Earth and Environmental Sciences Library

Earth and Environmental Sciences Library (EESL) is a multidisciplinary book series focusing on innovative approaches and solid reviews to strengthen the role of the Earth and Environmental Sciences communities, while also providing sound guidance for stakeholders, decision-makers, policymakers, international organizations, and NGOs.

Topics of interest include oceanography, the marine environment, atmospheric sciences, hydrology and soil sciences, geophysics and geology, agriculture, environmental pollution, remote sensing, climate change, water resources, and natural resources management. In pursuit of these topics, the Earth Sciences and Environmental Sciences communities are invited to share their knowledge and expertise in the form of edited books, monographs, and conference proceedings.

More information about this series at http://www.springer.com/series/16637

Mansour Ghorbani

The Geology of Iran:
Tectonic, Magmatism
and Metamorphism

 Springer

Mansour Ghorbani
Faculty member of Shahid Beheshti University
Tehran, Iran

Management of Pars Geological Research Center
(Arian Zamin)
Tehran, Iran

ISSN 2730-6674   ISSN 2730-6682 (electronic)
Earth and Environmental Sciences Library
ISBN 978-3-030-71111-5   ISBN 978-3-030-71109-2 (eBook)
https://doi.org/10.1007/978-3-030-71109-2

This Springer imprint is published by the registered company Springer Nature Switzerland AG
The registered company address is: Gewerbestrasse 11, 6330 Cham, Switzerland

Preface

From the fierce wind of vicissitude, one cannot see,
That, in this sward, hath been even a red rose, or a wild white rose

—Persian Poet Hafiz (1315–1390)

The beginning of any scientific, economic, investigational, and executive achievement depends on the proper knowledge of basic information. The application of geosciences to meet the needs of each country is related to the level of knowledge of the geologists. Accordingly, the comprehensive use of geoscience resources, including energy resources, dimension stones, metallic, and non-metallic minerals, as well as infrastructural projects, such as roads, dams, bridges, tunnels, large structures, and reducing and preventing the natural hazards are based on the following two principles:

1. Geological knowledge of the geologists.
2. Having the necessary types of equipment and capabilities, and managing policy to use them.

The book "Geology of Iran," written in conjunction with the first principle, is the result of three decades of fieldworks and laboratory works, intended to raise the knowledge of scholars, researchers, and students about the geological knowledge of Iran that is one of the goals and missions of the "Pars Geological Research Center" (Arianzamin). In addition to personal information and available resources, this book is intended to provide a wealth of reports from reputable geological organizations, academic dissertations, and reports from companies and organizations related to geology, including the Geological Survey of Iran, National Company Iran Oil, Iran Steel Co., Iran Copper Co. and also the latest findings and achievements of this science have been used.

This book tries to describe the geological setting of Iran in the world throughout geological history, referring to paleogeography and general geodynamics. Also, all structural units, faults, tectonic phases, and orogeny occurred in the geology of Iran have been evaluated.

Magmatic and metamorphic rocks along with ophiolitic complexes have extensive outcrops in Iran and these rocks with Precambrian age constitute its basement. Study and identification of such rocks not only throws light on the geodynamic issues of Iran but also helps in recognition of the mode of formation and evolution of the sedimentary basins located within various structural divisions of the country. Moreover, the majority of metallic and non-metallic mineral deposits are associated either directly or indirectly with magmatic, and at time metamorphic, rocks.

In the Magmatism and Metamorphism parts, it is tried to thoroughly consider the various aspects of the igneous rocks, whether intrusive, extrusive or young volcanoes, from the point of view of petrography, geochemistry, and geodynamics. In addition, the major intrusive bodies of Iran have been presented along with their petrologic and chronologic specifications in tables, mentioning the bibliographic resources.

The ophiolite complexes from all over the country have been spatially and temporally classified and described from the point of view of petrology, genesis and geodynamics. The results of the research by young scientists on the ophiolite rocks of Iran have been included in Chapter 7.

Chapter 8 presents an attempt to describe the metamorphic rocks on the basis of their petrography and the protolithic metamorphic facies, so as to reveal their spatio-temporal distribution in each of the geological zones of Iran.

When writing all parts of the book, I have been consulted by friends with the contents of certain sections. I am very grateful to all of them who encouraged and cooperated with me, including Drs. Reza Kohansal, Anoshiravan Kani, Azin Ahifer, Mozhgan Salehi Yazdi, Masoud Ovissi, Keyvan Zand Karimi, and Mohsen Ghorbani. I am particularly indebted to Prof. Dr. Clayton for his linguistic and structural tectonic edition on the final draft of the book.

Tehran, Iran   Mansour Ghorbani

Contents

About the Author

Dr. Mansour Ghorbani was born in Nanaj village, a rural district of Malayer County in the west of Iran on 1961. He completed his primary and secondary education in his hometown by 1979.

He graduated from high school in 1983. He studied geology at the University of Shahid Beheshti and concurrently Chemistry at Islamic Azad University. He continued his academic studies in geology at the University of Shahid Beheshti and received his masters (M.Sc.) and Doctor of Philosophy (Ph.D.) degrees in 1993 and 1999, respectively.

Following his academic accomplishments, he joined the geology faculty at the Shahid Beheshti University and has been teaching undergraduates, postgraduates and Ph.D. students till now. He currently holds the associate professor position at the University.

From 1991 to 1996, he was involved in the treatise on the geology at geological survey of Iran. He wrote and compiled a lot of literatures on the geology and mineral deposits, such as economic deposits, soils, iron, antimony, arsenic, mercury, copper, lead, and zinc in Iran.

Aside from teaching, he has been working on international and national research projects with mining, oil and gas companies.

The rewards and outcome of these years of studying and working are 39 books, more than 200 academic papers, supervising 7 research projects, advising and supervising of 75 theses in M.Sc. and Ph.D. degrees, over 120 scientific and technical reports in reference to natural and mineral resources in Iran and Iranology, as well as the compilation of international metallogeny and gem distribution maps of the Middle East.

He enjoys traveling around the country and abroad; he maintains that while visiting and working in various regions, he meets different ethnic groups with different cultures and traditions in Iran. He has learned how the habitat and the natural surroundings have a greater effect on the people's socio-economic aspects of life in some places than others. So, he wrote a comprehensive atlas of Iran, from tourism, ecotourism, and geotourism point of view in 10 volumes in persian.

Years of working experiences and personal beliefs in private research work compelled him to establish his own research centre called Pars Geological Research Centre (Arianzamin) in 2002. website: http://arianzamin.com.

He, from a sociocultural standpoint, endeavors to help countries and people who speak the same language and have had the same or similar cultures, to establish a long-lasting sociocultural bond with one another.

Chapter 1
Geological Setting and Crustal Structure of Iran

Abstract This chapter is an attempt to describe the geological setting of Iran within the global tectonic framework as a part of the Alpine-Himalayan belt and to provide related evidence from the Tethyan realm. In this chapter, the origins of Proto-Tethys, Paleo-Tethys and Neo-Tethys and their role in geology of Iran are mentioned. Another issue mentioned in this chapter is the Basement of Iran as well as the mode of formation and thickness of the Iranian crust in different regions and structural zones. Paleogeography of Iran from the Late Precambrian to the Quaternary is described in this chapter by mentioning the events that occurred during these times, and at each time the position of Iran in the whole planet is described.

Keywords Geological setting · Alpine-Himalayan belt · Iranian crust · Basement · Gondwana · Paleogeography

1.1 Geological Setting of Iran Within the Alpine-Himalayan Belt

Based on stratigraphic, paleontological and lithofacies evidence as well as chronometric, metallogenic and tectonic evaluations (on igneous bodies such as the Doran Granite), it can be deduced that the Iran Plateau was a part of northern Gondwana during Neoproterozoic–Cambrian (probably all of the Paleozoic) times. The northern parts of Iran and the neighboring countries record the collision of the Gondwana and Laurasia supercontinents during the late Paleozoic.

Throughout the Early and early Late Paleozoic (Cambrian to Carboniferous), Gondwana was the largest supercontinent on Earth and comprised Arabia, Africa, Antarctica, Australia, India, Madagascar, most parts of South America and some minor terranes surrounding these (Stampfli et al. 2001).

The Laurasia supercontinent, which was located to the north of the Gondwana from late Neoproterozoic to (early) Paleozoic, comprised three major terranes viz. Avalonia, Baltica and Lorenzia, and several volcanic island arcs. As Stampfli et al. (2001) and Scotese (2005) pointed out, there are indications that the Gondwana and Laurasia supercontinents were initially united as "Pangea" but later separated to form

© The Author(s), under exclusive license to Springer Nature Switzerland AG 2021
M. Ghorbani, *The Geology of Iran: Tectonic, Magmatism and Metamorphism*,
Earth and Environmental Sciences Library,
https://doi.org/10.1007/978-3-030-71109-2_1

the Tethys, before rejoining later to form Pangea in the late Paleozoic (Torsvik and Cocks 2004). Nonetheless, that reintegration was not as complete as that of the late Neoproterozoic and some parts of the Gondwanan supercontinent gradually joined Laurasia during Mesozoic and Cenozoic time, which will be described in the next section about the Tethys.

As explained by the Plate Tectonic Theory, the Tethyan Ocean extended from east to west dividing Pangea supercontinent into two portions viz. Gondwana in the south and Laurasia in the north, which converged, closing the Tethys between them, and forming the Alpine-Himalayan orogenic belt. A large number of reports have been published on the time of opening and closure of Tethys that are not in agreement with one another; the estimates for the opening and closure events of the Tethys, ranging from Late Proterozoic to Neogene.

All through the Late Neoproterozoic to Middle Paleozoic, Iran was a part of the Gondwana supercontinent. However, the closure of the Paleo-Tethys (Tethys II) leads to the integration of Iran to Laurasia. During Late Paleozoic time, the opening of the Neo-Tethys (Tethys III) resulted in the detachment of the southern part of Iran (Zagros) and its integration into the northern extension of the Arabian Plate.

There have been mentions of various terms such as "Paleo-Tethys I", "Paleo-Tethys II" and "Neo-Tethys" (Tethys III) in the geological literature of Iran, which is confusing to both novice geologists and researchers. A few researchers are of the opinion that during the Early Paleozoic an extension of "Paleo-Tethys" that covered Tianxian, Pamir, Northern Hindukush, and Turkmenistan separated Iran from Laurasia and was closed by the Hercynian event. Berberian (1973), Stöcklin (1977) and Majidi (1981) considered the Mashhad and Asalem-Shanderman ophiolite suites as an indication of the collision of Iran with Laurasia. Stöcklin (1977) believed that the southern Caspian basin and its oceanic crust are the remnants of "Paleo-Tethys", while Eftekharnezhad (1991) considered rifting in the northeastern Afro-Arabian platform during Carboniferous-Permian time was responsible for the formation of another ocean named "Paleo-Tethys II" which was closed in Triassic times.

The Tethys Ocean should never be compared with any of the contemporary oceans such as the Atlantic or the Red Sea because it consisted of a number of interconnected sedimentary basins separated from one another by continental or oceanic crusts (like present-day oceans). In fact, the microcontinents along with the intervening oceans formed a continental complex with relatively narrow elongate oceanic basins. These basins were zigzag in shape and their depths never reached that of a true ocean (Dercourt et al. 1986, 1993, 2000).

It should be stated that the various Tethyan basins did not form simultaneously, but were successive, with the closure of one phase corresponding to the opening of another. The geological evidence within Asia and the Middle East indicates that the trend of formation and closure was from north to south (perpendicular to the E-W trend of the Tethys). As a consequence, they become younger in a northward direction. In fact, during this process, segments of the Gondwana super-continent were separated from it and after traveling in a northward direction, collided with Eurasia. This view of the Tethys is more in accordance with the geological observations made

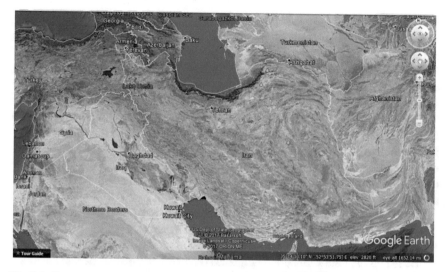

Fig. 1.1 Geographical setting of Iran within the Middle East (Google Earth 2017)

in the Middle East, and the terms "Tethys I", "Tethys II" and "Tethys III" refer to such phases.

Iran is considered as part of Alpine-Himalayan Mountain Belt from a global tectonic point of view. This Mountain Range formed as a result of various cycles of the Alpine Orogeny (Fig. 1.1). European and Asian geologists believe that this belt was formed due to the closure of a large east-west trending ocean (the Tethys) between the two huge continents viz. Gondwana and Laurasia. The mode of formation and evolution of this belt throws light on the Paleozoic, Mesozoic and Cenozoic sedimentary basins and magmatic zones of Iran. The characteristics of the Paleo-Tethys reflect the Paleozoic settings and geological phenomena, while those of the Neo-Tethys are indicative of the Mesozoic and the Cenozoic situations. The various phases of the Tethys Ocean with respect to the geology of Iran are discussed below.

1.1.1 Tethys I (Proto-Tethys)

This ocean formed between the two supercontinents of Gondwana and Laurasia in the northern part of the Iranian Plateau during the Ediacaran or early Cambrian and closed during Silurian time. Based on the Infracambrian-Cambrian sedimentary facies, Lasemi (2001) suggested that the Proto-Tethys existed before the Paleo-Tethys in northern regions of Gondwana.

According to Ghorbani (2007): "Tethys I probably appeared in Late Proterozoic and closed in Ordovician. There are no indications of the inception of this phase of Tethys in Iran and it perhaps covered the regions to the north of Iran, i.e. Tianxiang, north of Pamir, and possibly Touran."

As mentioned above, the Proto-Tethys Ocean spread out during Cambrian to Ordovician time. However, with the development of an extensional basin towards the south, it began shrinking until the Silurian when it closed completely.

The geological evidence indicates that "Tethys I" extended over southern regions of present-day Eurasia, correlating with the Caledonian phase, and came to an end at a time corresponding to the Caledonian Orogeny. The reason for the absence of the Caledonian phase in Iran is the attachment of the Iranian terrane to another continent (north of Gondwana) during this time interval, while the Caledonian orogeny acted on southern Eurasia (Berberian and King 1981; Ghorbani 2007). The ophiolitic suites of Greater Caucasus and the ophiolitic complex of Sultan Uzdag in Aral can be considered as the remnants of "Tethys I". The relevant facts in regard to "Tethys I" are as follows:

There have not been any reports of orogenic activity encompassing folding, metamorphism, and plutonism in Iran during the entire Paleozoic (especially in Early and Middle Paleozoic). The Paleozoic rocks of Iran are in fact metamorphosed due to the Early and the Late Cimmerian orogenies (Aghanabati 2004; Zanchi et al. 2006).

No indications of deep marine facies exist in Neoproterozoic-late Cambrian sedimentary successions of Iran (Lotfali Kani and Ghorbani 2016).

The mafic and ultramafic rocks of the Mashhad and Asalem areas that Berberian (1973), Stöcklin (1977) and Majidi (1981) regarded as early to early late Paleozoic in age and considered to be the indications of "Paleo-Tethys" actually formed in an intra-continental rift which is attributed to middle Paleozoic (Eftekharnezhad 1991; Ghorbani 1999, 2007).

During the early Paleozoic, China, India, and Siberia were separated from Iran and Arabia (Stampfli 1978).

As mentioned earlier, the closure of "Tethys I" commenced in Ordovician and was completed by the Silurian or early Devonian. The reason for extensive impacts of the Caledonian orogeny in the regions to the north of Iran, and its near absence in Iran is the distant position of Iran from it.

The stratigraphic record of Cambrian and Ordovician rocks (i.e. Soltanieh, Barut, Zaigun, Lalun, Mila and Kuhbanan formations) and some of the Neoproterozoic successions of Iran and Arabia, as well as the hiatuses during Silurian and Devonian (e.g. Berberian and King 1981), correlate well with the sedimentary gap of the "Tethys I" in these regions.

1.1.2 Tethys II

This phase of development of the Tethyan Ocean was more extensive than the first and occurred to the south of "Tethys I" within the northern areas of Gondwana. Tethys II can be traced from Afghanistan to Turkey, and in fact, refers to what has been described as Paleo-Tethys in the literature. It occurred as a rift in the northern areas of Gondwana from Silurian to Carboniferous (or even Late Devonian) and developed into an ocean named "Tethys II" in the Late Paleozoic (with intervening continental fragments all along its length). This ocean closed at the end of the Paleozoic and beginning of the Mesozoic i.e. the onset of the Triassic in response to the early phase of the Early Cimmerian orogeny (Zanchi et al. 2006). The Paleo-Tethys suture can be traced in northeastern Iran (Binaloud Mountains; Motaghi et al. 2012b).

"Tethys II" extended from the east and north of Iran, through Turkey to Western Europe in the form of intra-continental rifts that progressively developed into an ocean. The ophiolites to the south of Mashhad are the remnants of the Paleo-Tethys that closed during Early Triassic time. The metamorphic complex of Asalem-Shanderman (with its initial volcano-sedimentary and mafic-ultramafic parent rocks), and the Gorgan Schists (with an acidic volcanic, mafic-ultramafic, sedimentary or pyroclastic origin) can be explained by considering the "Tethys II" coverage on north and east of Iran. Moreover, Baikal Lake, Caspian Sea, and the Black Sea can be taken as remnants of "Tethys II."

The relevant facts concerning "Tethys II" are as follows:

Mafic-ultramafic rocks of Fariman are komatiitic lavas (Dakhili 1995) that were formed in an intra-continental rift setting (Eftekharnezhad and Stocklin 1992; Dakhili 1995); this rift later developed into an ocean. The age assigned to these rocks, according to their fossil content, is Late Paleozoic.

The Asalem-Shanderman Complex includes actinolite schists, garnet-biotite-muscovite schists, and fine-grained gneisses along with serpentinized ultramafics. This complex originally formed from a mafic-ultramafic parent rock along with a sedimentary sequence that might have been emplaced in response to rifting that resulted in the formation of an ocean.

Many researchers (e.g. Stampfli 1978) believe that an intra-continental rift caused the separation of Iran and Afghanistan from Gondwana in the Carboniferous and Permian; this rift could be the precursor of "Tethys II".

The ophiolitic complexes of Allahyarloo-Gharebagh and Zangzoor of Armenia are extensions of the Asalem-Shanderman complex and are thus related to "Tethys II".

The Dareh Anjir ophiolites in the southeastern corner of Aghdarband inlier are colored melanges along with diabasic tuffs, radiolarites and chert-bearing radiolarites, ultrabasic rocks, phillitic shales and minor amounts of limestones that contain Late Permian fossil fragments (Kashani and Bozorgnia cited in Eftekharnezhad and Behroozi 1991).

Geodynamic analyses have revealed the presence of an oceanic trench that separated Eurasia from Iran in the Permian; this trench did not last longer than Triassic times (Ruttner 1984).

Extension of Baikal Lake, Caspian Sea and the Black Sea on an east-west trend is an indication of "Tethys II".

Floral similarities of Iran, Afghanistan, Touran, and China in the Middle Triassic point to the closure of "Tethys II" (Darvishzadeh 1992). The absence of floral similarities during the Carboniferous-Permian indicates the separation of these areas by "Tethys II" times.

The presence of basaltic rocks (continental tholeiitic or alkaline basalts) in Silurian and Devonian successions of Northern Iran, Central Iran, and Alborz points to the opening of a rift and the formation of the Tethys to the north.

1.1.3 Tethys III

Simultaneous with the closure of "Tethys II", several intra-continental rifts were developed in northern areas of Gondwana (to the south of "Tethys II") parallel with the major ancient faults of Zagros. The Zagros Rift (between the Arabian Plate and the Sanandaj-Sirjan Zone) later developed into an extensive ocean (Neo-Tethys).

In addition, the East Iran Rift, parallel to Nehbandan fault (that finally culminated in a sea similar to the present-day Red Sea) and the rift along the Darouneh Fault, might have also resulted in the formation of oceanic crust; although their formation was not contemporaneous with the Neo-Tethys, though they were by-products of the same process that gave rise to the latter. The geographic distribution of these rifts and basins is shown in Fig. 1.2.

The formation of "Tethys III" took place during Triassic and Early Jurassic times with some parts forming as late as the Lower to Middle Cretaceous. However, the majority of these basins closed by Late Cretaceous or Paleocene times but their remnants existed into the Early Neogene. The closure of "Tethys III" was probably due to either subduction that started in Early Jurassic, or opening of Atlantic Rift, as well as opening of the Red Sea in Late Jurassic and Early Cretaceous.

The trend of the Atlantic Rift was N-S, while that of the "Tethys II" was E-W. The central Atlantic cut across Neo-Tethys at the location of the Caribbean Sea.

The Atlantic Rift expanded towards south during the Early and Middle Cretaceous, while in Late Cretaceous and Early Tertiary it moved to the north (Moein-Vaziri 1996). The presence of thrusts that trend from north to south in northern Iran is dominantly due to the development of the Atlantic rift and its push towards the south. The closure of East Iran (Zabol-Balouch) Rift at the current position of the flysch-zone was probably due to the northward movement of India. This resulted in compression of the Afghan Block, pushing it towards the west, thus closing the basin and deforming (folding) the sediments between the Lut and Helmand (Afghan) Blocks.

1.2 Basement Rocks of Iran

Considering the geological record of Iran and the Middle East and comparing the geology of Iran with Arabia, it seems that stabilization of the basement rocks of Iran was attained by the Neoproterozoic to early Cambrian. Various characteristics of the Neoproterozoic and lower Cambrian successions of Iran and Arabia point towards the Gondwanan origin of both of these terranes (Berberian and King 1981).

a

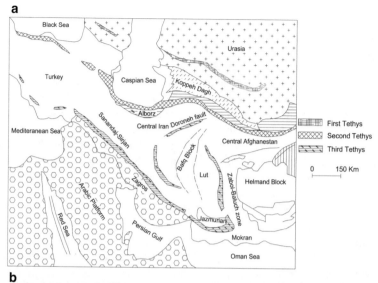

b

Fig. 1.2 a. Tethys in Iran (Ghorbani 2007). **b.** Plate tectonic reconstruction of the Tethys II and III (Si: Siberia, KZ: Kazakhestan, KR: Karakum, TM: Tarim Block, NC: North China, SC: South China, IC: Indochina, NT: North Tibet, SS: Sanandaj-Sirjan, AL: Alborz, CI: Central Iran, AF: Afghanistan, ST: South Tibet), (modified after Dèzes 1999)

Studies carried out by BRGM and USGS scientists and German researchers on the Arabian Plate indicate that the Arabian Shield was placed adjacent to Mozambique on the east of Africa before the formation of the Red Sea. Continental accretion has gradually taken place from Africa towards Iran because Central Africa possesses the Late Archaean-Early Proterozoic basement; moreover as one moves towards the north, the age of the basement decreases so that both Iran and Arabia lack Archaean basement rocks. Also, the extent of Precambrian outcrops in the northern terranes of Gondwana reduces as compared to the central regions.

Some geologists relate the stabilization of Iran's basement rocks to the Baikalian, Assyntic and Pan-African orogenies. According to the studies carried out in the Arabian Shield (by BRGM and USGS), the basement rocks of the Arabian Plate were formed during the Proterozoic. The absolute age of this basement has been determined at around 800–900 Ma by radiometric methods. However, it should be kept in mind that this age relates only to surface exposures.

The oldest sedimentary rocks of Iran are those of the Kahar and Tashk formations exposed in Alborz, Azarbaijan and Central Iran. Lithologically, these rocks are composed of clastic sediments including shales, dolomitic sandstone, and tuffs that have been metamorphosed to slate and phyllite grade. The Kahar Formation is estimated to be 650–850 Ma. The chronostratigraphic age of the overlying strata of the Kahar Formation is considered to be Neoproterozoic on paleontological evidence (Ghavidel-Syooki 1995). However, the relative age and the thickness of its lower parts are not known, due to the lack of any fossils.

A metamorphosed massif has been described in the Central Iran and Takab regions, which is older than Kahar Formation and is metamorphosed to amphibolite facies. These have been attributed to very old ages in the literature but the observations of the author in Takab, as well as the information provided by Etemad-Saeed et al. (2016) concerning Central Iran both, infer that the metamorphism and the protoliths of these rocks are not much older than Kahar (Neoproterozoic). In fact, the oldest metamorphosed or unmetamorphosed igneous rocks of Iran are those having the same age as the Doran Granite and Qaradash Series.

It should be further emphasized that the radiometric dating of rocks of Iran has previously never yielded an age older than 900 Ma (Crawford 1977; Seger 1977; Hamdi 1995; Ramezani and Tucker 2003).

If the geological facts about the Arabian Plate are compared with those of Iran, and if we consider the intrusion of granitoid bodies necessary for the formation of continental crust, it can be concluded that the Iran continental shelf must have formed in Neoproterozoic-Early Cambrian and before the deposition of Lalun Sandstone.

Therefore, what constituted the basement underneath Kahar Formation is a point to ponder. The geological evidence, including the lithological and chemical composition of the Kahar Formation, shows that this formation lacks any characteristic features of crustal origin. Having been formed during the Neoproterozoic-early Cambrian, the rift that occurs in Central Iran (see Chap. 5) indicates that an intermediate crust must have preceded it, to explain the formation of the rift. Therefore, it is justified to consider the Kahar Formation as an equivalent to the flysch sediments of Eastern Iran and the Makran Zone, with only minor differences. Since the origin and development of the Flysch and Makran Zone are more or less understood, and the existence of oceanic crust underneath them has been ascertained, the same mode of origin and presence of oceanic crust below Kahar Formation is also possible.

1.3 Early-Middle Proterozoic/Archean Basement Rocks

The old metamorphic and igneous basement rocks of Iran are not those that outcrop at the surface but are the rocks that underlie the oldest formations and successions which belong to Early-Middle Proterozoic or even Archean. The basement rocks of Iran are older than Neoproterozoic with no surface outcrops. In other words, the oldest rocks that outcrop in Iran belongs to the Late Proterozoic. The possibility of the existence of Early and Middle Proterozoic and perhaps Archaean basement is important from the seismo-tectonic and seismologic, fundamental exploration and geodynamics points of view as well as for interpreting the geological and subsequent events, especially in the Phanerozoic.

There are many methods to evaluate the presence of Early-Middle Proterozoic/Archean basement rocks:

– Deep drilling (thousands of kilometers) on the Neoproterozoic terrane, which is currently not feasible.
– Seismic and gravimetric investigations and comparison with similar terranes, which has been used to some extent to estimate the thickness of crustal rocks of Iran.
– Geological correlation with neighboring countries, especially those in the North Gondwana realm.

As mentioned before, Iran was located on the northern extension of Gondwana, which becomes younger in northward direction; Archean and Early-Middle Proterozoic basement rocks outcrop in the central parts of Gondwana, i.e. Arabia, Afghanistan, and Pakistan. Consequently, Iran must also possess basement rocks of similar ages since it adjoins these countries.

We can use the mode of formation of old magma associated with the ancient continental crust to evaluate the existence of basement rocks older than those currently considered. For example, if the petrological characteristics of Doran and Zarigan granites are indicative of their intrusion into the continental crust, then there must have been some older rocks than these granites into which they were intruded.

There are some volcanic rocks in Azarbaijan, Central Iran and Zagros which were not formed by differentiation of basic magmas, but were created by partial melting due to injection of basic magma into the crust; e.g. the Qaradash Rhyolites of Azarbaijan, rhyolites of the Hormoz Series in Zagros, and rhyolites and acidic tuffs of the Rizu Series in Central Iran. These are all indicative of the existence of a relatively thick continental crust which produced them through partial melting.

In Saudi Arabia and the late Proterozoic terranes of Iran (refer to magmatism of Iran), many intra-continental rifts occur, which needed to have a relatively thick crustal setting to develop on.

The ancient formations (late Proterozoic) of Iran like the Kahar in Alborz-Azarbaijan, as well as the Tashk I, II, and Kalmard in Central Iran are clastic sedimentary deposits with continental provenance. If these formations are Late Proterozoic in age, then their source rocks should be older and of a continental nature.

Age determination of samples from igneous rocks (especially zircon grains) has shown varied results, for example, 1.5 Ga, 800 Ma, etc. probably due to having inner zones belonging to much older continental sources (recycled). Taking into account all the above-mentioned facts, a prior Late Proterozoic basement can be considered to occur underneath Iran which may be of Early-Middle Proterozoic age. However, the reduction in the age of the basement rocks observed in a northward direction rules out the possibility of an Archean age for the underlying basement of Iran (Omrani pers. comm. Jan. 2016).

1.4   Seismological Estimation of Crustal Thickness of Iran

To measure the thickness of the crustal rocks of the Middle East, a series of studies have been carried out under the Alp-Himalayan International Geodynamic Project. This project has been termed as "Geotraverse Research Project 1" performed through the collaboration of the Geological Survey of Iran and the German National Institute (DFG) in the year 1979.

Using the Deep Seismic Survey (DSS) method, most of the structural zones of Iran (including Folded Zagros, Zagros Thrust, Sirjan metamorphics, and Central Iran) have been investigated. To carry out seismic profiling, explosions in the Sarcheshmeh Copper Mine and Bafgh Iron Mine were used as the sources of seismic energy. The following summarises the outcome of seismic profiling:

The maximum thickness of crustal rocks in Iran occurs underneath the Zagros Thrust Zone (more than 50–55 km), while the minimum is that of Makran on the coast of Oman Sea (less than 25 km) (Dehghani and Makris 1983).

The thickness of crustal rocks in Central Iran varies from 40 to 45 km and that of the southern coasts of the Caspian Sea is around 35 km. It is estimated at around 40 km under the Lut Block, gradually increasing to 60 (or even 70 km) when approaching the metamorphics of Sirjan and the Zagros Suture Zone. Moving southwards to the Persian Gulf, the thickness of the crust decreases gradually so that on the coasts of the Persian Gulf it is around 30 km thick. These results are supported by gravimetric surveys carried out in Zagros. It is worth mentioning that these figures indicate the depth of the boundary between the crust and the mantle i.e. Moho's Discontinuity (Fig. 1.3).

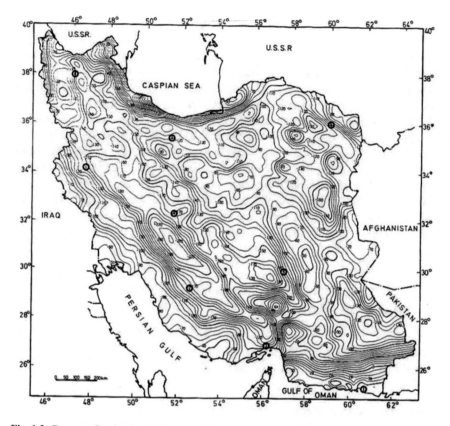

Fig. 1.3 Bouguer Gravity Anomaly map of Iran (Dehghani and Makris 1983)

The thickness of the crust under the southern shores of the Caspian Sea is about 35 km. The gravity field varies between 100 and 120 mgal which matches the estimated thickness. The Alborz Mountains seem to have no roots (Dehghani and Makris 1983).

Overall, it can be deduced from the above-mentioned studies that the continental crust of the Arabian Plate extends not only under Zagros but up to the Zagros Thrust Zone and parts of the Sirjan metamorphics, so that in the extreme northeast, the thickness of the crust increases; the overlying part is that of Central Iran Microcontinent while the underlying section is that of the eastern margin of the Arabian Plate. Most researchers believe that the collision of Iran and the Arabian Plate and further subduction of the latter are responsible for the extraordinary thickness of the crust under Iran, especially within the Sanandaj-Sirjan Zone. Of course, it should be kept in mind that these two crustal components had been parts of one plate up until Triassic time; later (in the Late Triassic), formation of a rift that gave rise to Tethys III (Neo-Tethys) separated the two from each other but they once again joined in the Late Cretaceous-Paleocene to form the present structure.

Based on recent studies by the receiver function apparatus, valuable data about Iranian crust has been obtained (Paul et al. 2006, 2010; Kaviani et al. 2007; Motaghi et al. 2012a). In addition to the thickness of the crustal rocks, these investigations have revealed other characteristics such as the geometry of the subduction zone as well as the location of the Neo-Tethys suture zone. It is obvious that the Main Zagros Thrust (MZT) superposes the location of the Arabian-Eurasian collision line (Paul et al. 2006), which is currently referred to as the Zagros Suture Zone. According to Omrani et al. (2008) the thickness of the crust along the line joining Bushehr and Yazd is initially 25 km, increasing gently to about 45 km at the southern boundary of Main Zagros Thrust, and then suddenly to 70 km under Sanandaj-Sirjan. This sharp increase is due to the underthrusting of the Arabian Plate below Central Iran's active margin (Sanandaj-Sirjan) which is located at a distance of about 50–90 km from the northeastern border of the boundary. Further to the northeast, the depth of Moho reduces to 42 km under the Central Iran and Sanandaj-Sirjan Zones.

1.5 Paleogeography of Iran from Neoproterozoic to Recent—An Overview

We are unaware of the beginning and the end of the world.
The first and last of this old book has fallen

Persian Poet Kaleem Kashani (1581–1651)
The oldest formations that crop out in Iran are composed of metamorphosed shallow-marine sedimentary rocks which are referred to as the "Kahar Formation" in Alborz and Azarbaijan or their equivalents in Central Iran that are of middle Neoproterozoic (Cryogenian) age (Ghavidel-Syooki 1995). The oldest metamorphic rocks which underlie these sediments have scattered exposures in Alborz-Azarbaijan (Takab area) and Central Iran in the form of metamorphic complexes with amphibolite facies. Taking into account the geological evidence available in Iran and other Gondwanan terranes, these complexes can only be as late as early Neoproterozoic (Tonian), i.e. younger than 1 Ga.

1.5.1 Neoproterozoic

Following the deposition of the Kahar Fm. in northern Iran, and Tashk, Morad and Kalmard formations in Central Iran, a shallow sea covered the whole of Northern Gondwana and Iran, laying down a succession of marine sediments that began with Bayandor facies (Stöcklin et al. 1964) and continued with Soltaniyeh, Barut, Zaigun and Lalun formations, which can be attributed to Late Neoproterozoic (Ediacaran)-Early Cambrian (Stöcklin and Setudehnia 1991). The equivalents of these rock formations in Central Iran are Kushk, Rizu, Desu, Heshem, Aghda and Dahu (Zaigun and Lalun), while those of Zagros basin comprises of Hormoz series (Neoproterozoic), Barut, Zaigun and Lalun (Early Cambrian).

It seems that during the middle Neoproterozoic (Cryogenian) and the beginning of the Late Neoproterozoic (Ediacaran), an intra-continental rift developed in Northern Gondwana that led to the formation of oceanic crust in some places. This rift closed by the end of the Early Neoproterozoic (Tonian) and its activities gradually diminished.

The presence of acidic-tuffaceous volcanic rocks (sometimes basaltic), as well as their intrusive equivalents such as the Doran Granite and Narigan-Zarigan in Neoproterozoic terranes of Iran (e.g. within Alborz, Azarbaijan, Central Iran and Zagros), are attributed to the development of this rift (Berberian and King 1981). This rift fully closed by the end of the Neoproterozoic and was replaced by a shallow sea that covered northern Gondwana and lasted until the end of the Middle Paleozoic (Devonian).

1.5.2 Paleozoic

During Early Paleozoic, the development of platformal conditions over the region comprising Iran and its Gondwanan neighboring countries (established sometimes earlier) led to the deposition of shallow marine sediments including dolomites, sandstones and red shales (Soltanieh, Barut, Zaigun, Lalun, and Mila formations) and their equivalents. During this time, Iran was a part of Gondwana positioned at 35°-53° in the southern hemisphere (Fig. 1.4).

The average depth of the basin gradually increased in some parts during the middle and late Cambrian, providing the conditions for laying down the dolomites, limestones, and shales of the Mila Fm. Nonetheless, environmental conditions similar to those of the Zaigun and Lalun facies continued in some places in Central Iran throughout the middle and late Cambrian, giving rise to the rocks of the Kuhbanan Formation (Hamedi and Alavi-Naeini 1984).

Ordovician times witnessed deepening of the sea over most parts of Iran (especially the southern regions) and in the northern Gondwanan terranes. Consequently, the Ilbeyk, Zardkuh and Seyahoo formations (Zagros; Setudehnia 1975; Harrison 1930) along with the Shirgesht Formation (Central Iran; Ruttner et al. 1968) and the Lashkarak Formation (Alborz; Gansser and Huber 1962) were deposited.

The climatic changes over the northern Gondwana towards the end of Ordovician resulted in the formation of land-locked water bodies and thus lowered sea-levels. This phenomenon caused the regression of sea and the emergence of platformal areas (Berberian and King 1981), which in turn left breaks in the stratigraphic column of various parts of Iran. The lowered sea-levels continued well into the Silurian Period (e.g. Ghavidel-Syooki et al. 2011).

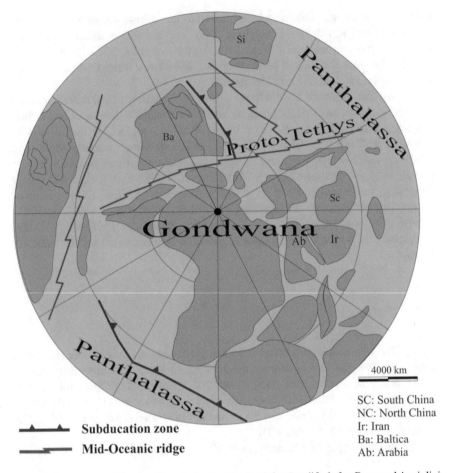

Fig. 1.4 Paleogeographic map of Iran during the late Cambrian (modified after Berra and Angiolini 2014)

Platformal conditions persisted in some parts of Iran into the middle Paleozoic (Silurian and Devonian). Clastic successions, dolomites, and evaporites (Niur, Padeha and Zakeen formations) are the evidence of such environments. The tectonic conditions prevailing towards the north of the Gondwana Supercontinent (Caledonian Orogeny) perhaps resulted in pulses of epeirogenic movements (Berberian and King 1981) in the southern territories (i.e. northern Gondwana), being reflected in form of minor unconformities within the Silurian (mostly) and Devonian successions due to transgressive-regressive phases which at times even led to complete emergence. It should be noted that, in spite of the prevailing shallow water platformal conditions, scattered areas of high land existed in the form of islands within this vast shallow ocean.

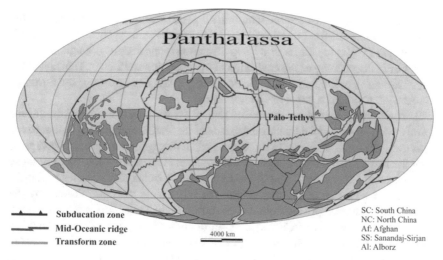

Fig. 1.5 Paleogeographic map of Iran during the early Devonian (modified after Domeier and Torsvik 2014)

As previously suggested, tectonic conditions on the northern edge of Gondwana Supercontinent were not as calm as in the regions located further south (southwest Central Iran and Zagros) during the Middle Paleozoic; in fact, the ancient Tethyan Ocean, or Paleo-Tethys, was being formed and the tensile forces acting on the northern edge of Gondwana and the southern edge of Laurasia caused extensive fracturing which led to extrusion of basaltic flows associated with the Silurian and Devonian formations of Iran. Figure 1.5 shows the position of Iran during the early Devonian. During this time, the Paleo-Tethys Ocean was opening in northern Alborz.

In Late Paleozoic time (Carboniferous-Permian), the transgression of the sea which began in the Late Devonian, covered the northern and central parts of Iran. Prevalence of such conditions until the Early Carboniferous led to the deposition of the diachronous Khoshyeilagh (Bozorgnia 1973), Geirud, and Shishtu formations in Alborz and Central Iran. However, since the Paleo-Tethys extended over Northern Gondwana, no such evidence is found in southern Iran (Zagros) where the Carboniferous System is either missing or underdeveloped.

During the Late Carboniferous, climatic conditions deteriorated and ended in glaciation. As a result of the continental glacier accumulation, and a slowing of ocean-ridge activity (beneath the Paleo-Tethys), the low-lying land areas emerged from under the sea. This led to the omission of the Late Carboniferous succession in Iran (e.g. Gaetani et al. 2009; Leven and Gorgij 2011; Zandkarimi et al. 2016, 2019).

Changing climatic conditions during Early Permian time witnessed the melting of continental glaciers, while further evolution of the Paleo-Tethys and development of a convergent continental margin on the southern edge of Laurasia caused sea-levels to rise. This phenomenon led to further deepening of the marine environments as well as extensive transgression over the northern margin of the Gondwana Supercontinent, which in turn caused vast sedimentation in all parts of Iran. At the beginning of the Permian, similar clastic formations were laid down in Alborz (Dorud Formation), Central Iran (Vazhnan and Bagh-e Vang formations) and Zagros (Faraghan Formation). Further deepening of the environment of deposition resulted in a change to the carbonate facies of the Jamal Fm. (Central Iran), Ruteh Fm. (Alborz), and Surmagh and Abadeh formations (west of Central Iran).

It seems that the commencement of separation of the Zagros basin during the Permian was accompanied by the quiescence of the tectonic conditions in that basin. Moreover, the trend of events in the Zagros basin diverged ever since its separation from the rest of Iran, so that the Permian facies of Zagros are different from those of Alborz and Central Iran. The deposition of the Permian clastic facies in Zagros starts with a slight delay compared with the other parts of Iran, followed by the non-clastic facies comprising dolomite-evaporite in the lower part and limestone in the upper part. Alborz and Central Iran, on the other hand, experienced evaporitic conditions in the end-Permian, which shifted towards continental conditions later on. The marine environments of Zagros continued almost unchanged into Triassic times. By the late Permian, parts of the Paleo-Tethys Ocean started to be closed forming the supercontinent Pangea. Then, the Cimmerian terranes broke off from the northern margin of Gondwana and drifted northward across the Paleo-Tethys as a result of the Neo-Tethys opening to collide finally with the Eurasian margin in the Late Triassic, giving rise to the Eo-Cimmerian Orogen (Fig. 1.6).

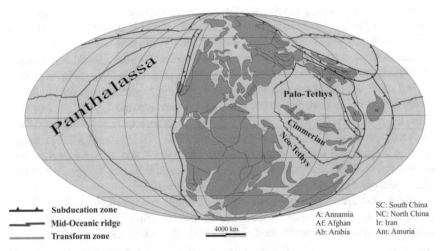

Fig. 1.6 Paleogeographic map of Iran around the Permian-Triassic boundary (modified after Domeier and Torsvik 2014)

1.5.3 Mesozoic-Cenozoic

It can be argued that Iran experienced extensive basin development and facies changes during the Mesozoic. Towards the north of present-day Iran, the Paleo-Tethys Ocean closed during the Paleozoic-Mesozoic transition due to continental collision (Stampfli and Borel 2002). Consequently, the terrances of Iran and Turan, which were separate entities during Palaeozoic, joined together. Conversely, the new Tethyan Ocean, or Neo-Tethys, was being created in southern Iran, resulting in the separation of Iran and the Arabian terranes (Alavi 1980), which were joined together during the Paleozoic (until late Permian).

Though, the collision in the north (between Iran and Turan) and rifting in the south (Iran from Arabia) gave rise to drastic changes in facies, in some places such as Abadeh, Shahreza, Julfa and Amol (northern Iran) and some parts of Zagros, sedimentation continued without a break during the late Paleozoic-early Mesozoic transition (Stepanov et al. 1969; Teichert et al. 1973; Partoazar 1995; Gaetani et al. 2009; Zandkarimi et al. 2014). This is because the Paleo-Tethys had not yet closed completely in the north, the newly-developed Neo-Tethys was opening to the south, and the two oceans were connected together through shallow platformal or epicontinental seas.

At the end of the Middle Triassic, most of the Paleo-Tethys closed, and Iran and Turan terranes coalesced (Alavi 1991). Simultaneously, Iran and Arabia were completely separated owing to the development of Neo-Tethys.

The rifting in the south and the collision in the north during early Permian and Middle Triassic times, respectively, caused the formation of magmatic and metamorphic rocks which demonstrates characteristics of extensional settings in the south (Delavari et al. 2016) and compressional settings in the north (Ghorbani 2013). These events represent the Cimmerian Orogeny over all of Iran (Zanchi et al. 2009).

The outcomes of collision and rifting led to the development of a shallow marine environment in northern and central Iran during the Early and Middle Triassic (Gaetani et al. 2009). The consequence of the Early Cimmerian event was activation of old basement faults as well as the development of new ones during the Late Triassic and Early Jurassic that resulted in the formation of an extensive deeper marine environment over many areas of the present-day Iran (northern and Central Iran, as well as Sanandaj-Sirjan). On the one hand, sediments equivalent to the Shemshak Formation sometimes reaching a thickness of 4000 m were deposited in many localities, and on the other hand, a new basin of deposition, called Kopeh-Dagh, was created in northeast Iran. The newly-formed Kopeh-Dagh basin maintained its independence until the Late Cenozoic, hosting a continuous sedimentary succession spanning Jurassic to Pliocene (Afshar-Harb 1994).

During the same interval of time, the Zagros basin of Southern Iran, located on the northern edge of the Arabian shield, developed sedimentary facies similar to those of Arabia and neighboring countries and persisted from end-Permian to Cenozoic. As mentioned earlier, this was a result of the opening of Neo-Tethys between Iran and Arabia. During Triassic and Early Jurassic time, the Neo-Tethys had developed all

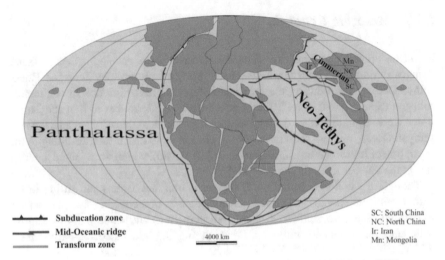

Fig. 1.7 Paleogeographic map of Iran during late Jurassic (modified after Golonka 2007)

the features of an ocean similar to that of the present-day Atlantic. The subduction of this ocean beneath Central Iran began during the latest Jurassic and continued until the Eocene (e.g. Alavi 2007), eventually resulting in a collision of the Arabian plate with Iran. Closure of the Neo-Tethys, suturing and continued pressing of the two plates led to the subduction of the continental portion of the Arabian shield under Iran.

During Jurassic and Cretaceous time, continued subduction under Iran in the Sanandaj-Sirjan Zone resulted in extensive emplacement of intrusive igneous bodies, even as far as the west of Central Iran. Consequently, a magmatic arc developed in southwestern Central Iran forming the Urmia-Dokhtar belt (Fig. 1.7).

According to Omrani (pers. comm. 2017), "With the commencement of subduction under the active margin of Central Iran (Sanandaj-Sirjan) the magmatic arc is created. This trend continues from Triassic to Late Cretaceous; ceases for a period of about 20 million years; and restarts towards Central Iran in Eocene and continues up to Quaternary. The lengthy magmatic arc known as Urmia-Dokhtar with a NW-SE trend parallels the Mesozoic magmatic arc in Central Iran."

The collision of the northern edge of the Arabian continent with Iran shaped the current morphology of mountain ranges in south-southwest of the country (Zagros Mountains). The other parts of the oceanic plate which were obducted during the collision constitute the ophiolitic complexes on the northern edge of the sutured zone. Nonetheless, it should be kept in mind that many ophiolites in the Zagros suture zone are the result of obduction over the Arabian plate during Late Cretaceous (Alavi 2004).

The consequences of formation of new basins during the late Permian-Early Triassic in southern Iran (Zagros basin) and Late Triassic-Early Jurassic in northeastern (Kopeh-Dagh basin), central, eastern and northern Iran were dynamism

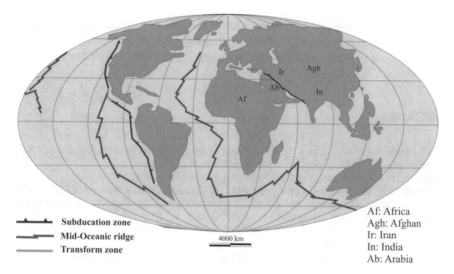

Subducation zone
Mid-Oceanic ridge
Transform zone

4000 km

Af: Africa
Agh: Afghan
Ir: Iran
In: India
Ab: Arabia

Fig. 1.8 Paleogeographic map of Iran during the late Eocene (based on the data herein and Omrani et al. 2007)

and continuous upheavals in most parts of Iran during Cretaceous, so this period constitutes major shifts in all marine basins of Iran.

A new oceanic basin similar to the present-day Red Sea opened up between the Lut and Afghan blocks to the east of Iran in Cretaceous times and continued until the Early Paleogene. The East Iran flysch facies accumulated in this basin. Since the extension of this basin into the southeastern part of Iran (Makran region) is connected with Neo-Tethys, it can be considered as a part of the Neo-Tethys Ocean. The eastern Iran basin closed due to the convergence of the Afghan and Lut blocks, probably due to the movement of the Indian plate towards the north. The resultant folding gave rise to the flyschoid subzone of eastern Iran (Fig. 1.8).

The Laramide Orogeny in the Late Cretaceous induced shrinkage of internal basins, whether oceanic or shallow continental (except Zagros and Kopeh-Dagh). As a result, in Alborz, Central Iran and Sanandaj-Sirjan, the Cenozoic succession overlies the earlier sediments often with an angular unconformity and conglomeratic lithology (Fajan Conglomerate in Alborz; Kerman Conglomerate in Central Iran) (e.g. Aghanabati 2004). The Paleogene is either not represented (in Sanandaj-Sirjan and northern Alborz) or vaguely exemplified due to strong manifestation and diversification of volcano-sedimentary rocks, with the exception of Zagros, Kopeh-Dagh and southeastern Iran (Makran). Simultaneous with the Pyrenean Orogeny, magmatic activity intensified during Eocene-Oligocene times. Consequently, most of the exposures of Central Iran (Urmia-Dokhtar), eastern Iran and southern Alborz are the results of these events (Emami et al. 1993; Hassanzadeh 1994). During this time Iran was a part of Asia and experienced the final stages of Neo-Tethyan subduction.

During the late Oligocene-Neogene, magmatic activity in Central Iran and southern Alborz gradually reduced. Concurrently, Kopeh-Dagh and Zagros began to be folded and became emergent. Almost all parts of Iran emerged during the younger Alpine Orogeny and are folded (Sanandaj-Sirjan and northern Alborz were above sea-level for most of Cenozoic) assuming their present morphology.

References

Afshar-Harb A (1994) Geology of Kopet Dagh, in treatise on the geology of Iran. In: Hush-mandzadeh A (ed) Geological Survey of Iran, Tehran, 275 p

Aghanabati A (2004) Geology of Iran. Geological Survey of Iran, Tehran

Alavi M (1980) Tectonostratigraphic evolution of the Zagrosides of Iran. Geology 8(3):144–149

Alavi M (1991) Sedimentary and structural characteristics of the Paleo-Tethys remnants in northeastern Iran. Geol Soc Am Bull 103(8):983–992

Alavi M (2004) Regional stratigraphy of the Zagros fold-thrust belt of Iran, and itsproforeland evolution. Am J Sci 304:1–20

Alavi M (2007) Structures of the Zagros fold-thrust belt in Iran. Am J Sci 307:1064–1095. https://doi.org/10.2475/09.2007.02

Berberian M (1973) The seismicity of Iran preliminary map of epicentres and focal depth 1: 2 500 000. Geol Surv Iran (Seismotectonic Group)

Berberian M, King GCP (1981) Towards a paleogeography and tectonic evolution of Iran. Can J Earth Sci 18(2):210–265

Berra F, Angiolini L (2014) The evolution of the Tethys region throughout the Phanerozoic: a brief tectonic reconstruction. In: Marlow L, Kendall C, Yose L (eds) Petroleum systems of the Tethyan region: AAPG Memoir, vol 106, pp 1–27

Bozorgnia F (1973) Paleozoic foraminiferal biostratigraphy of central and east Alborz Mountains, Iran. Nat Iranian Oil Company Geol Laboratories, Publ 4:1–185

Crawford MA (1977) A summary of isotopic age data for Iran, Pakistan and India. In: Libre a la memoire del A.F. de Lapparent. Mémoire hors-serie 8. Societé Géologique de France, pp 251–260

Dakhili MT (1995) Petrology of magmatic and metamorphic rocks of the northeastern Fariman. Unpublished M.Sc. thesis, Faculty of Earth Sciences, Shahid Beheshti University

Darvishzadeh A (1992) Geology of Iran. Amirkabir Publication Company, Tehran

Dehghani G, Makris J (1983) The gravity field and crustal structure of Iran. Geol Surv Iran Rep 51:51–68

Delavari M, Dolati A, Marroni M, Pandolfi L, Saccani E (2016) Association of MORB and SSZ ophiolites along the shear zone between Coloured Mélamge and Bajgan Complexes (North Maran, Iran): evidence from the Sorkhband area. Ofioliti 41:21–34. https://doi.org/10.4454/ofioliti.v41 i1.440

Dercourt J, Gaetani M, Vrielynck G, Barrier B, Biju-Duval B, Brunet MF, Cadet JP, Sandulescu M (2000) Atlas Peri-Tethys, Palaeogeographical maps, CCGM/CGMW, Paris: maps and explanatory notes, volume 24

Dercourt J, Ricou LE, Vrielynck B (eds) (1993) Atlas Tethys Palaeoenvironmental Maps, Gauthier-Villars, 260 p

Dercourt J, Zonenshain LP, Ricou LE, Kazmin VG, Le Pichon X, Knipper AL, Grandjacquet C, Sbortshikov IM, Geyssant J, Lepvrier C, Pechersky DH (1986) Geological evolution of the Tethys belt from the Atlantic to the Pamirs since the Lias. Tectonophysics 123(1–4):241–315

Dèzes P (1999) Tectonic and metamorphic evolution of the central Himalayan domain in southeast Zanskar (Kashmir, India). Doctoral dissertation, Université de Lausanne, Faculté des sciences

Domeier M, Torsvik TH (2014) Plate tectonics in the late Paleozoic. Geosci Front 5(3):303–350

Eftekharnezhad J, Behroozi A (1991) Geodynamic significance of recent discoveries of ophiolites and Late Paleozoic rocks in NE-Iran (including Kopet Dagh). Abh Geol B-A 38:89–100

Eftekharnezhad J, Stocklin J (1992) Geological maps of Iran, sheet K8. Geological Survey of Iran, Birjand

Eftekharnezhad J (1991) Geodynamic significance of recent discoveries of ophiolites and late Paleozoic rocks in NE. Iran (including Kopet-Dagh). Abh Geol B.A. Wien, 89–110

Emami MH, Sadeghi MM, Omrani SJ (1993) Magmatic map of Iran. Scale 1(1):000

Etemad-Saeed N, Hosseini-Barzi M, Adabi MH, Miller NR, Sadeghi A, Houshmandzadeh A, Stockli DF (2016) Evidence for ca. 560 Ma Ediacaran glaciation in the Kahar Formation, central Alborz Mountains, northern Iran. Gondwana Res 31:164–183

Gaetani M, Angiolini L, Ueno K, Nicora A, Stephenson MH, Sciunnach D, Rettori R, Price GD, Sabouri J (2009) Pennsylvanian-Early Triassic stratigraphy in the Alborz Mountains (Iran). Geol Soc Lond Spec Publ 312(1):79–128

Gansser A, Huber H (1962) Geological Observations in the Central Elburz, Iran. Schweiz Mineral Petrogr Mitt 42(2):583–630

Ghavidel-Syooki M (1995) Palynostratigraphy and palaeogeography of a Palaeozoic sequence in the Hassanakdar area, Central Alborz Range, northern Iran. Rev Palaeobot Palynol 86(1–2):91–109

Ghavidel-Syooki M, Álvaro JJ, Popov L, Pour MG, Ehsani MH, Suyarkova A (2011) Stratigraphic evidence for the Hirnantian (latest Ordovician) glaciation in the Zagros Mountains, Iran. Palaeogeogr Palaeoclimatol Palaeoecol 307(1–4):1–16

Ghorbani M (1999) Petrological investigations of Tertiary-Quaternary magmatic rocks and their metallogeny in Takab area. PhD thesis, Shahid Beheshti University

Ghorbani M (2007) Economic geology of natural and mineral resources of Iran, Pars Geological Research Center (arianzamin), 492 p

Ghorbani M (2013) Economic geology of Iran, vol 581. Mineral deposits and natural resources. Springer

Golonka J (2007) Phanerozoic paleoenvironment and paleolithofacies maps: late Palezoic. Geologia/Akademia Górniczo-Hutnicza im. Stanisława Staszica w Krakowie 33(2):145–209

Hamdi B (1995) Precambrian and Cambrian sedimentary rocks of Iran. Treatise on the Geology of Iran, Ministry of Mines and Metals

Hamedi M, Alavi-Naeini M (1984) A revision of the Dezu and Rizu 'Series', and introduction of a new Banestan formation of the Kuhbanan area, Kerman

Harrison JV (1930) The geology of some salt-plugs in Laristan, southern Persia. Q J Geol Soc 86(1–4):463–522

Hassanzadeh J (1994) Consequence of the Zagros continental collision on the evolution of the central Iranian plateau. J Earth Space Phys 21:27–38 (in Persian, with English abstract)

Kaviani A, Paul A, Bourova E, Hatzfeld D, Pedersen H, Mokhtari M (2007) A strong seismic velocity contrast in the shallow mantle across the Zagros collision zone (Iran). Geophys J Int 171(1):399–410

Lasemi Y (2001) Facies analysis, depositional environments and sequence stratigraphy of the Upper Precambrian and Paleozoic rocks of Iran. Geological Survey of Iran

Leven EJ, Gorgij MN (2011) First record of Gzhelian and Asselian fusulinids from the Vazhnan Formation (Sanandaj-Sirjan zone of Iran). J Stratigr Geol Correl 19:486

Lotfali kani A, Ghorbani M (2016) Late Precambrian-Early Paleozoic Stratigraphy of Northern Gondwana Region with special emphasis on Iran. In: International congress on Palaeozoic stratigraphy of Gondwana

Majidi B (1981) The ultrabasic lava flows of Mashhad, North East Iran. Geol Mag 118(1):49–58

Moein-Vaziri H (1996) An introduction of geology of magmatism in Iran. Tehran University of Teacher Education, Tehran, p 440

Motaghi K, Tatar M, Priestley K (2012a) Crustal thickness variation across the northeast Iran continental collision zone from teleseismic converted waves. J Seismol 16(2):253–260

Motaghi K, Tatar M, Shomali ZH, Kaviani A, Priestley K (2012b) High resolution image of uppermost mantle beneath NE Iran continental collision zone. Phys Earth Planet Inter 208:38–49

Omrani J, Agard P, Whitechurch H, Benoit M., Prouteau G, Jolivet L (2007) Arc-magmatism and subduction history beneath the Zagros Mountains. A new report of adakites and geodynamic consequences, Iran

Omrani J, Agard P, Whitechurch H, Benoit M, Prouteau G, Jolivet L (2008) Arc-magmatism and subduction history beneath the Zagros Mountains, Iran: a new report of adakites and geodynamic consequences. Lithos 106:380–398

Partoazar H (1995) Permian deposits in Iran. In: Hushmandzadeh A (ed) Treatise on the geology of Iran, vol 22, pp 1–340. Geological Survey of Iran, Tehran, Iran (In Persian with English abstract)

Paul A, Kaviani A, Hatzfeld D, Vergne J, Mokhtari M (2006) Seismological evidence for crustal-scale thrusting in the Zagros mountain belt (Iran). Geophys J Int 166(1):227–237

Paul A, Hatzfeld D, Kaviani A, Tatar M, Péquegnat C (2010) Seismic imaging of the lithospheric structure of the Zagros mountain belt (Iran). Geol Soc Lond Spec Publ 330(1):5–18

Ramezani J, Tucker RD (2003) The Saghand region, central Iran: U-Pb geochronology, petrogenesis and implications for Gondwana tectonics. Am J Sci 303:622–665

Ruttner AW (1984) The pre-Liassic basement of the eastern Kopet Dagh Range. Neues Jahrbuch für Geologie und Paläontologie-Abhandlungen, pp 256–268

Ruttner A, Nabavi MH, Hajian J (1968) Geology of the Shirgesht Area (Tabas Area, East Iran). Geol Surv Iran.

Scotese CR (2005) Plate Tectonic and paleogeographic maps and animations. PALEOMAP Project (www.scotese.com)

Seger FE (1977) Zur Geologie des Nord-Alamut Gebeites (Zentral-Elburz, Iran): Eidgenossische Technischule. Zurich Thesis (6093)

Setudehnia A (1975) The Paleozoic sequence at Zard Kuh and Kuh-e-Dinar. Bull Iran Pet Inst 60:16–33

Stampfli GM, Borel GD (2002) A plate tectonic model for the Paleozoic and Mesozoic constrained by dynamic plate boundaries and restored synthetic oceanic isochrones. Earth Planet Sci Lett 196(1–2):17–33

Stampfli GM, Borel GD, Cavazza W, Mosar J, Ziegler PA (2001) Palaeotectonic and palaeogeographic evolution of the western Tethys and PeriTethyan domain (IGCP Project 369). In Episodes (p. 222)

Stampfli GM (1978) Etude géologique generale de l'Elbourz oriental au sud de Gonbad-e-Qabus (Iran NE). Doctoral dissertation, Université de Genève

Stepanov DL, Golshani F, Stocklin J, with contributions by Hamzepour B, Seyed-Enami K, Mehrmush M, Nikravesh-Rosen R (1969) Upper Permian and Permian-Triasslo Boundary in North Iran, Geological Survey of Iran, Report No. 12, 1-72, Fig. 1-6, PI. 1-3, 1-15, Tehran

Stöcklin J (1977) Structural correlation of the Alpine ranges between Iran and Central Asia

Stöcklin J, Setudehnia A (1991) Stratigraphic lexicon of Iran. Geol Surv Iran Rep 18:1–376

Stöcklin J, Ruttner A, Nabavi MN (1964) New data on the lower Palaeozoic and Precambrian of North Iran. Geol Surv Iran Rep 1:1–29

Teichert C, Kummel B, Sweet W (1973) Permian- Triassicstrata, kuh-e-Ali Bashi, northwestern Iran. Bull Mus Com Zool 145(8):359–471

Torsvik TH, Cocks LRM (2004) Earth geography from 400 to 250 Ma: a palaeomagnetic, faunal and facies review. J Geol Soc 161(4):555–572

Zanchi A, Berra F, Mattei M, Ghassemi MR, Sabouri J (2006) Inversion tectonics in central Alborz, Iran. J Struct Geol 28(11):2023–2037

Zanchi A, Zanchetta S, Berra F, Mattei M, Garzanti E, Molyneux S, Nawab A, Sabouri J (2009) The Eo-Cimmerian (Late? Triassic) orogeny in North Iran. Geol Soc Lond Spec Publ 312(1):31–55

Zandkarimi K, Najafian B, Bahrammanesh M, Vachard D (2014) Permian foraminiferal biozonation in the Alborz Mountains at Valiabad section (Iran). Permophiles 60:10–16

Zandkarimi K, Najafian B, Vachard D, Bahrammanesh M, Vaziri SH (2016) Latest Tournaisian–late Viséan foraminiferal biozonation (MFZ8–MFZ14) of the Valiabad area, northwestern Alborz (Iran): geological implications. Geol J 51(1):125–142

Zandkarimi K, Vachard D, Najafian B, Mosaddegh H, Ehteshami-Moinabadi M (2019) Mississippian lithofacies and foraminiferal biozonation of the Alborz Mountains, Iran: implications for regional geology. Geol J 54(3):1480–1504

Chapter 2
Structural Units of Iran

Abstract In this chapter, the structural units of Iran are described. It begins with the structural divisions of Iran based on the opinions of researchers who have worked on structural units and sedimentary basins of Iran since 1968. After that the literature concerning studies on the structural units of Iran associated with structural maps are listed and illustrated, and then the structural units are stratigraphically, magmatically and metamorphically outlined. In this chapter, the structural units of Iran are described. It begins with the structural divisions of Iran based on the opinions of researchers who have worked on structural units and sedimentary basins of Iran since 1968. After that the literature concerning studies on the structural units of Iran associated with structural maps are listed and illustrated, and then the structural units are stratigraphically, magmatically and metamorphically outlined. For each geological zone, items such as the boundaries of structural units, differences and similarities of structural zones with other adjacent structural units are mentioned. Each structural zone has been discussed in terms of lithostratigraphy, magmatic and metamorphic phases as well as the evolution of sedimentary facies and their mode of formation. The structural zones include:

Zagros Zone, Kopeh-Dagh Zone, Sanandaj-Sirjan Zone, Lut block, Ophiolitic zone, Central Iran Zone, Paleogene volcanic zone, Makran and Eastern Iran zone, Alborz Zone.

Keywords Structural units · Alborz · Azerbaijan · Central Iran Microcontinent · Urmia-Dokhtar · Sanandaj-Sirjan · Zagros · Kopeh-Dagh · Eastern Iran · Makran

2.1 Literature and Structural Subdivisions

The structural units are the areas representing the same geological history, lithostratigraphy, magmatic and metamorphic activities, tectonic movement and fold trend. Since 1957, with the establishment of the Geological Survey of Iran by the United Nations, considerable efforts have been made with the collaboration of Iranian and International geologists. During the same period, there have been important discussions on the structure and the geological position of Iran in the Middle East, as well as in comparison with Alpine-Himalayan structural patterns. Stöcklin and Nabavi (1973) summarized the updated information, and using maps and findings of the National Iranian Oil Company, published the tectonic map of Iran. In this map Iran was tectonically subdivided into 9 structural units as follows (Fig. 2.1):

1.   Zagros Zone

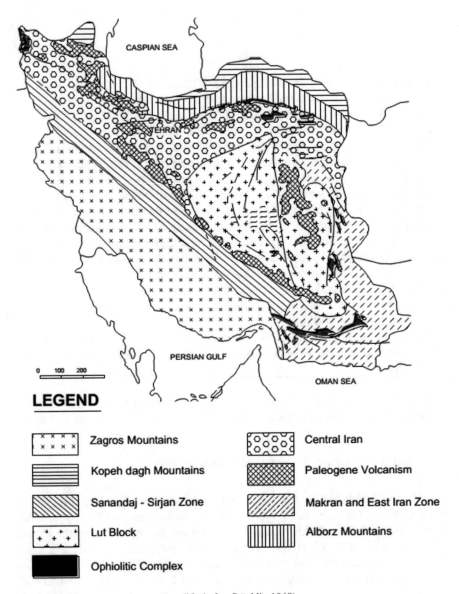

Fig. 2.1 Iranian structural zones (modified after Stöcklin 1968)

2. Kopeh-Dagh Zone
3. Sanandaj-Sirjan Zone
4. Lut block
5. Ophiolitic zone
6. Central Iran Zone
7. Paleogene volcanic zone
8. Makran and Eastern Iran zone
9. Alborz Zone

During the last 3 decades, most researchers have referred to this map. Geological works carried out in different fields have resulted in many reports and have identified aspects of the geology of Iran that need serious revision. However, after the first effort of Stöcklin and Nabavi (1973,) researchers such as Nabavi (1976) Eftekharnezhad (1980), Nogol-e Sadat (1993), Alavi (1991), Stampfli (1978), and Aghanabati (2000, 2004) presented relatively different structural analyses of the geology of Iran; each of which presents some special facts concerning Iranian geology (Figs. 2.2, 2.3, and 2.4). Here, the structural units of Iran are briefly discussed.

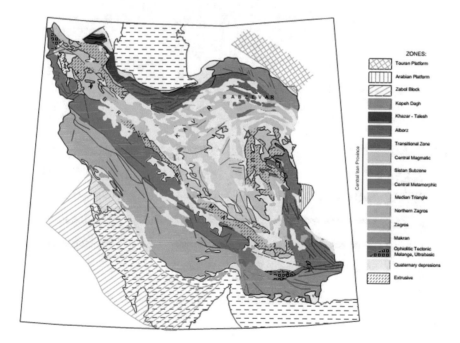

Fig. 2.2 structural subdivision of Iran (modified after Nogol-e Sadat 1993)

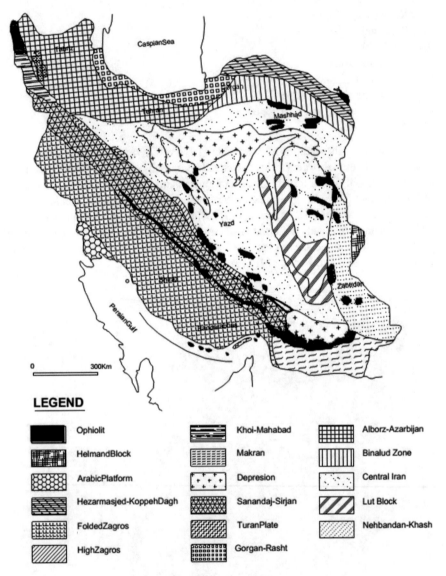

Fig. 2.3 Structural zones of Iran (modified after Nabavi 1976)

2.2 Alborz

Consisting mainly of platformal-type lithofacies of Neoproterozoic to Quaternary age, the Alborz range is limited by the Caspian depression in the north and Central Iran in the south. However, its eastern and western boundaries are not completely defined. Though the Binaloud Range is the continuation of Alborz, its lithofacies resemble

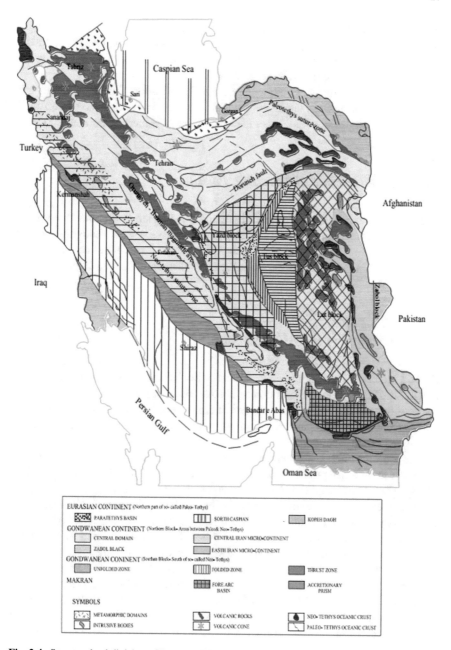

Fig. 2.4 Structural subdivision of Iran (modified after Aghanabati 2004)

those of Central Iran (Nabavi 1976). Azerbaijan, northwest Iran, was considered by Nabavi (1976) to be a part of Alborz (Fig. 2.3). Stöcklin believed that Alborz, based on its structural condition, is an anticlinorium margin of Central Iran, that stratigraphically and structurally resembles it.

According to the present author, Azerbaijan has stratigraphically affinities with Central Iran, especially in the Paleozoic and Mesozoic but during the Cenozoic, its geology is much different and is comparable to that of Alborz (Ghorbani 2005).

2.2.1   Location and Geographic Position

The Alborz is more than 1000 km length and 50–100 km width and extends in a sinuous manner from Gorgan in the east to Astara in the west. Geographically, it trends E-W and is located in northern Iran, the southern Caspian Sea extending from Azerbaijan in the west to Khorasan in the east, terminating in the Caucasus in the west and Afghanistan in the east.

2.2.2   Geology of the Alborz Zone

From a geological point of view, Alborz with its special sedimentological and magmatic characters is mostly a structural zone containing Neoproterozoic to Recent rocks, though it does not show a record of continuous sedimentation and there are associated stratigraphic gaps. Sedimentary rocks make up more of the succession than igneous and metamorphic rocks.

The southern boundary is defined by roughly parallel faults and associated depressions, the latter having most probably been formed by faulting. These faults, such as the Semnan, Attari, Garmsar, Kahrizak and north Alborz faults cannot be observed at surface owing to the alluvial cover. In the northern part, the North Alborz and Khazar faults form the boundary between Alborz and the Caspian Sea. As already mentioned, the northern and southern boundaries of Alborz are completely known but the eastern and western boundaries have always been disputed. Some researchers (e.g. Nabavi 1976) regarded Azerbaijan as a part of Alborz; hence the western boundary is extended to the western limit of Iran (Fig. 2.3). Stöcklin (1968) considered northeastern Iran, i.e. the Binaloud zone of eastern Iran, as a part of the Alborz, whereas Nogol-e Sadat (1993) included the Binaloud zone as a transitional zone between the Alborz and Central Iran. Aghanabati (2004) believed that there are no major differences between Central Iran and Alborz.

According to the present author, except for the Gorgan-Talesh zone, the segregation of Alborz, Azerbaijan, Central Iran, and Binaloud in Paleozoic and Mesozoic was not so clear. Lithostratigraphic, magmatic and metamorphic features of above areas are so similar that they can be considered to be the same unit. However, in the Cenozoic, differences are such that it is possible to separate the Binaloud ranges in

Alborz and Azerbaijan, the Tarom-Hashtjin magmatic belt, western-middle Sabalan and Ahar as a distinct structural unit. For more explanation, the Tarom and Talesh subzones are compared below.

2.2.3 Structural Subdivisions of the Alborz

Generally, the Alborz is divided into 6 structural subzones (based on Stöcklin 1974a):

1. Uplifted Gorgan Zone

This consists of recrystallized basement rocks and is mostly covered by Mesozoic strata. This zone is quite similar to the Talesh area (Ghorbani 1999, 2002).

2. North Neogene Zone

This zone includes a folded belt of Mesozoic and especially Neogene rocks which is overlain by mollase facies. This zone is separated from the North-Central zone to the south (see below) by a thrust fault. These features can be observed in the area located between the Talesh and Tarom mountains (Ghorbani 2002).

3. North-Central Zone

This is characterized by shallow water sediments ranging in age from the Infracambrian to the Late Cretaceous. Short-termed periods of volcanic activity, as well as Cenozoic tectonic deformation of the area also characterize this zone. These features are rarely found in the Talesh Mountains.

4. South-Central Zone

Like the latter zone, the succession comprises pre-Cenozoic shallow-water sediments, covered by a very thick sequence of Cenozoic volcanic rocks, with the Eocene especially well developed. Important Eocene thrusts also can be traced here. The rocks that crop out in this area are more or less similar to those of the Tarom volcanics; though there are some differences that will be mentioned in this chapter.

5. South "Tertiary" Zone

The succession consists of a very thick sequence of Eocene volcanic rocks and continental Neogene sediments. It is characterized by South-dipping thrusts.

6. Uplifted Southern Front

This zone contains shallow water sediments and volcanic rocks. The folding stages of the Early Cretaceous and later on and normal faulting are obviously present.

The general geology of the Talesh and Tarom is presented below. From a tectono-magmatic point of view, these areas described as the Talesh and Tarom subzones (sometimes referred to together as 'Tarom-Hashtjin').

These two subzones are of the different geological character, especially in Cenozoic.

2.2.4　Geological Comparison of the Tarom and Talesh Subzones

Considering the structural unit subdivisions, both zones are located in the western Alborz zone. This zone is mostly refereed in the zonation of Nabavi (1976).

2.2.4.1　Talesh Subzone

This is generally composed of Paleozoic and Mesozoic rocks that are slightly metamorphosed (especially the Paleozoic), and is characterized by its sedimentary provenance. Its southern edge, i.e. the Sefidrud margin, contains Cenozoic volcano-sedimentary rocks. Regarding its tectonic position, it should be mentioned that this subzone is located along the northern Sefidrud River and southwestern Khazar Plain. This subzone is limited to the north by Sefidrud thrust, having the same trend as the Zagros thrust, and the Tarom subzone to the south.

2.2.4.2　Tarom Subzone

This subzone generally lacks Paleozoic and Mesozoic rocks or, more accurately, the Tarom basement cannot be observed and all the outcrops are composed of volcano-sedimentary rocks and Cenozoic intrusions (Ghorbani 2002). Here, the general geology of those areas is presented.

Geology

These oldest known rocks in this subzone are believed to be those of the Gasht meta-morphic complex, consisting of gneiss, mica-schist, and magmatics of Devonian-Carboniferous age. Crawford (1972) and Clark et al. (1975) dated the gneiss and mica-schists as 375±2 Ma based on the Rb-Sr, i.e. pre-Hercynian. The rocks of the Gasht and Shanderman areas are assigned by researchers such as Eftekharnezhad and Behroozi (1991) to the Late Paleozoic. Ghorbani (2002) is of the opinion that these rocks belongs to the middle to late Paleozoic and were metamorphosed by the early Cimmerian event. The Lower Paleozoic deposits do not crop out and the only exposures are of middle to lower Paleozoic andesites and basic volcanic rocks with sandy intercalations. The slates and phyllites interfinger with upper carbonate. The mentioned rocks are interbedding with conglomerates and quartz-arenites. This unit is mostly found in the south and southeast Talesh area with a few records from the northern part.

　　The Triassic rocks are of limited distribution. The Jurassic strata show affinities to the Shemshak Formation, consisting of grey to black shale accompanied by brown sandstone and conglomerate with plant remains, bivalves, ammonites. Volcanic tuffs occur in the north western part of southern Befrodagh Mountains.

Other rocks of the area are as following.

The Jurassic deposits, known as the Shal Formation, are lithologically mainly composed of glauconitic sandy to silty limestones with limey sandstones containing ammonites. The younger deposits are of more restricted distribution, are often recrystallized and are composed of pinkish crème to light grey sandstone, being similar to the Lar Formation. Neocomian deposits consisting of ammonite bearing light grey fine-grained limestone can also be observed (Aghanabati 2004).

A. Maastrichtian silty and sandy limestone

The Berriasian to Maastrichtian succession consists of vacuolar limestone associated with sandstones and volcanic intercalations which crop out near the arcuate faults of Sefidrud in the western part.

Cenomanian to Maastrichtian reddish polygenic conglomerate, black green volcanic tuffs and gross limestone associated with shale, siltstone and sandstone followed by basalt and andesite intercalations are exposed in the western part of the area.

Towards the eastern part, the Cenomanian to Maastrichtian volcanoclastic and black tuffite associated with basalt, andesite and thick sequence of clastic rocks also containing thin limestone and occasionally trachyte in the upper part.

The Cenozoic rocks mostly crop out in the Talesh subzone, along the northern part of the Sefidrud River, and are lithologically divided into following units:

1. Volcanic rocks; andesite to basalt and associated pyroclastic rocks
2. Mixed continental to marine rocks of the Upper and Lower Red formations which mostly formed the Tarom plain, southern part of the Talesh subzone and the Manjil dome. In the northern part of this subzone, Neogene-Quaternary rocks consisting of Pleistocene sediments Khazar plain, coastal sand, dune (being older than trassess and old sandbars) and Quaternary deposits

B. Geology of the Tarom-Hashtjin subzone

In contrast to the Talesh subzone, Paleozoic and Mesozoic rocks cannot be observed in this subzone which is composed of magmatic rocks.

Considering the above, the Alborz zone can be characterized as follows.

The eastern-western boundary of the Alborz during Paleozoic and Mesozoic time was the same as the geographical boundary that enter Afghanistan from Khorasan and Azerbaijan to the Caucasus and Turkey (Fig. 2.5).

The eastern and western boundaries of Alborz during the Cenozoic are limited to the Gorgan and Astara areas, respectively. Consequently, it extends in a sinusoidal manner to the south Caspian Sea. Its maximum width is 150 km. The northern and southern boundaries are well defined and faulted (Figs. 2.6 and 2.7).

The northern wedge is characterized by oceanic crust mixed with sedimentary and volcano-sedimentary rocks metamorphosed to greenschist facies such, as the Gorgan schist and metamorphic rocks of Gasht-Shanderman (Fig. 2.6).

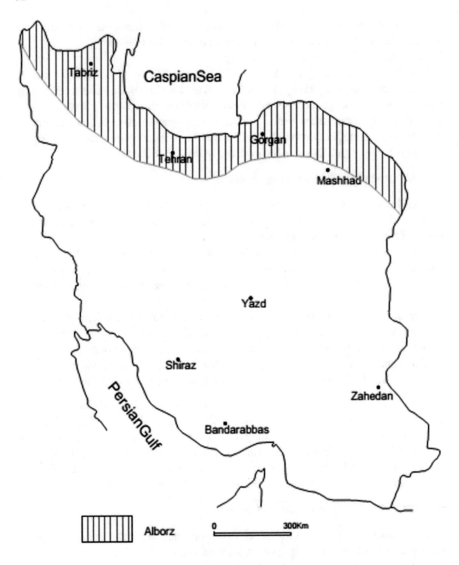

Fig. 2.5 Geographical position of the Alborz (Ghorbani 2013)

Except for the northern part which was previously described, most of the Neoproterozoic, Paleozoic, and Mesozoic rocks are unmetamorphosed sediments.

The Cenozoic succession is mostly composed of volcanic and pyroclastic rocks cropping out from the southern slope to the central area. Towards the west, they increase in alkalinity and thickness. In Azerbaijan, the thickness of these rocks is so high as to form a distinct zone.

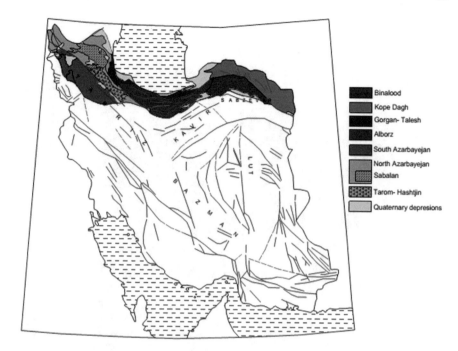

Fig. 2.6 Structural subdivision of Alborz and adjacent areas (Ghorbani 2013)

In the central part the Alborz-Taleghan, Azadkuh, large- to medium-sized, late Eocene-Oligocene acidic masses have intruded the younger rocks.

In the northern Alborz, Masuleh-Astara, basic to acidic dikes is intruded into the Mesozoic strata.

2.2.5 *Lithostratigraphy of the Alborz Zone*

The oldest sedimentary rocks, referred to as the "Kahar Formation", of Iran are observed in this area (Aghanabati 2004). The base of this formation does not crop out anywhere and its true thickness is therefore unknown.

The Kahar formation is composed of slate, sandy shale, sandstone and thin-bedded limestone (Dedual 1967). Upwards, associated with a color change to red, it tends to show a gradual increase in grain size. The Kahar Formation conformably overlies the Soltanieh Formation in most places.

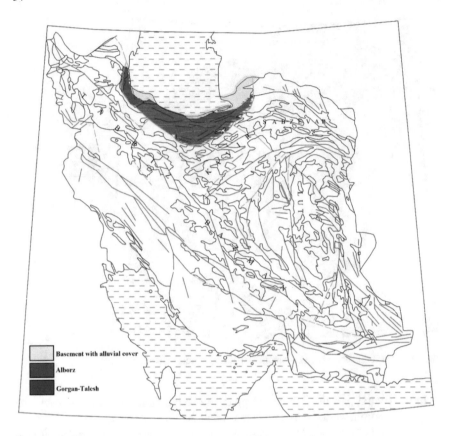

Fig. 2.7 Structural subdivision of Alborz (Ghorbani 2013)

2.2.5.1   Paleozoic

The Paleozoic rocks of Alborz are associated with sedimentary gaps and unconformities. As a result, two major gaps were recorded from this area; the first in Late Ordovician-Silurian to Devonian, the second Late Carboniferous-Early Permian. Paleozoic formations of Alborz indicate shallow water environments and mostly consist of sandstone and limestone (Assereto 1963; Berberian and King 1981; Gaetani et al. 2009; Zandkarimi et al. 2014, 2016, 2017a, b).

2.2.5.2   Mesozoic

Like the Paleozoic, the Mesozoic of Alborz is mostly composed of sedimentary rocks including dolomite and limestone, often representing shallow to relatively deep water depositional environments. However, some Cretaceous rocks have been reported as indicating deep marine sedimentary environments.

2.2.5.3 Cenozoic

As already mentioned, according to the present author, the Cenozoic rocks of the Alborz have a less extensive distribution than those of the Paleozoic and Mesozoic. It is of note that the lithostratigraphic units of this area completely differ between the northern and southern slopes. As a result, Paleogene sediments are not observed in the northern areas, while on the southern areas, these sediments include pyroclastic rocks with lava and broad and thick volcanoclastic rocks.

The stratigraphical features of Alborz were described by Ghorbani (2019).

2.2.6 Magmatism

Though, it has already been stated that the Alborz Mountains are mainly composed of sedimentary rocks, volcanic and intrusive rocks occur sporadically and can be classified as follows.

2.2.6.1 Volcanic Rocks of Paleozoic Age

These are mostly observed in the eastern Alborz and include basalts of Ordovician-Silurian age (Jenny 1977) in the Bastam- Khosh-Yeiylagh belt. They are associated with the Geirud Formation (Assereto 1963), indicating the opening of the Pale-Tethys in northern Alborz.

2.2.6.2 Mesozoic Basaltic to Andesite Volcanic Rocks

Rocks of Triassic-Jurassic age appear in the limited area such as the Upper Jurassic melaphyre-basaltic rocks of the Damavand which can be related to deep basement faults that occasionally experienced extensional movement (Kheirkhah et al. 2015). The northern Alborz magmatism is characterized by Cretaceous to Paleocene gabbro to basic rocks that have been less studied. It is believed that the study of these rocks will resolve some of the geological problems of Alborz.

2.2.6.3 Cenozoic Magmatic Rocks

Although magmatic activity in the Alborz was less significant than in the central Alborz and Eastern Alborz, it was much more pronounced than other during earlier eras. It should be noted that, Azerbaijan is not included in the Alborz in the following description. The Cenozoic magmatism of Alborz can be divided into following classes:

- Pyroclastic-volcanic rocks.
- Plutonic masses, dikes and sills.

The volcanic rocks mostly occur in the southern Alborz and are not observed in the north and central parts (Ghorbani 2002). The southern Alborz to northern Takestan volcanic rocks show many affinities with those of northern central Iran. Eastwards of the Takestan to Astara area of western Alborz, affinities with rocks in Azerbaijan is evident. In fact, on the western slope, the Cenozoic rocks are expanded further.

Cenozoic plutonic bodies in Alborz are relatively abundant in the border and western parts of which the main masses are as follows:

- Granite-diorite of Alamkuh of Klardasht and Akapel.
- Syenite to gabbro small masses which extend from northeastern to northwestern Tehran (Ghorbani 2002).

2.3   Central Iran Zone

Central Iran is one of the main units of Iran with a triangular shape. It is the most complex and geographically-largest unit of Iran, encompassing Neoproterozoic to Quaternary rocks and recording several phases of metamorphism, orogeny, and magmatism are differed by Various researchers have published different views on the geographical boundaries of this unit, with some researchers such as Stöcklin (1968) considering Central Iran to be restricted to the area from the Alborz Mountains in the North, southwest to Sanandaj-Sirjan and east to the Lut block, whereas Nabavi (1976) includes the northern Lut block in Central Iran.

Nogol-e-Sadat (1993) further extended its area and added Azerbaijan and east Iran to Central Iran. However, he subdivided it to several zones. According to Boulin (1991), Central Iran is located between the orogenic belt of the Paleo-Tethys in the north and the Neo-Tethys in the south. The northern limit is assumed to be Alborz Mountains and the south-southwestern limit, the Urmia-Dokhtar volcanic belt. In its south and southeast part, it is separated from Sanandaj-Sirjan by a belt of high-dipping, right lateral faults, that were active until the Mesozoic (Sengör 1990). The eastern boundary is not so clear, because some geologists considered the Lut block as the eastern part of Central Iran, whereas some regarded it as a different zone. Central Iran records several magmatic, orogenic and metamorphism events.

2.3.1   Lithostratigraphy

In the recent studies, the oldest rock units of Central Iran are assumed to be Tashk, Kalmard, Morad, Kushk, Rizu, Desu formations, of which the latter is a volcanoclastic unit (e.g. Huckriede et al. 1962; Haghipour and Pelissier 1968; Ruttner et al.

1968). The Neoproterozoic volcanic rocks are mostly acidic to intermediate Alkaline -calc-alkaline in composition.

The intrusion was initiated after waning volcanic activity, forming Zarigan type granites with a 530–540 Ma. This granitzation resulted in the stabilization of the basement and its subsequent metamorphism. High-grade metamorphism is not recorded, although, it was previously assumed that high-grade metamorphics were widespread. However, recent studies have revealed that they were metamorphosed during Middle or late? Triassic Cimmerian event (Zanchi et al. 2009).

The Paleozoic sediments of Central Iran start with continental facies overlain by white quartzarenite, referred to as the "Dahu Formation". Lake of sedimentation during the Late Ordovician to Early Devonian may be linked to epeirogenic movements of the Caledonian movements (e.g. Berberian and King 1981) and subsequent erosion and climate change. According to Haghipour et al. (1979), these movements were solely epeirogenic and should not be considered as an orogenic phase. Additionally, it seems that another movement, Hercynian or Variscan tectonic movement, occurred in the late Carboniferous in some parts, and resulted in cessation of sedimentation in most of Iran, especially southern Alborz and Azerbaijan (Zandkarimi et al. 2019).

Though Triassic rocks can be observed in the eastern, central and western parts of middle-central Iran, the eastern part shows wider distribution, probably due to the activity of the Kalmard, Kuhbanan and Anar faults. The Middle and Upper Triassic rocks are disconformably overlain by younger rocks. Magmatic activities, such as Esmaeilabad and Dehbid granites, also formed as a result of the early Cimmerian orogeny. Evidence of metamorphism is observed along the central part, especially in the Saghand area.

During the Jurassic period, sandy, shaly and marly facies were deposited in most of central Iran, which were subsequently metamorphosed by the late Cimmerian orogeny (Aghanabati 1977). The Late Cimmerian orogeny was a compressional phase associated with magmatism and metamorphism (Zanchi et al. 2009), with the Shirkuh Granite in Yazd Province an example. This granite intruded the Upper Jurassic rocks and is overlain by Lower Jurassic rocks. The late Cretaceous marine transgression was accompanied by more extension, with deposition of conglomerate, sandstone, and clastic calcareous rocks. Towards the Late Cretaceous-Paleocene, Maastrichtian-Paleocene, most parts of the middle triangle underwent severe folding that resulted in a disconformity between the Upper Cretaceous and Lower Paleocene sequences. As a consequence, the Paleocene succession starts with basal conglomerate and sandstone and overlies older rocks with an angular unconformity. Cenozoic magmatism in these areas was both plutonic and volcanic. Following the Late Cretaceous compressional phase, the Laramide orogeny resulted in metamorphism, folding, uplifting and ophiolite emplacement and formed granodiorite intrusions with calc-alkaline attitude.

During Eocene time, volcanic activity caused by extension resulted in high-volume flows with andesitic-dacitic rocks extruded along faults, such as the Dehshir-Baft fault (Crawford 1972). These volcanic activities have been widespread and continued to the Quaternary as travertine. During the Quaternary period, a marine regression associated with final formation of uplifts, most basins were isolated from the open sea and became land masses, on which halite, gypsum, and clay and marly sediments were deposited. Deserts formed during Quaternary time in this area include Ardakan, Abarkuh, Bafgh and Biabanak.

2.3.2   Central Iran Microcontinent

Stöcklin (1968) in his structural zonation scheme assigned this area to the "Lut zone" subsequently re-named by Aghanabati (2004), the "Central Iran Microcontinent". This area is surrounded by the ophiolite sutures of Sistan, Naein, Baft, Doruneh, and Kashmar-Sabzevar. and can be subdivided by long strike-slip faults into the Lut block, the Shotori horst, the Tabas graben, the Kalmard horst, and the Poshtbadam and Yazd Blocks (e.g. Alavi 1991).

If one considers the Lut block as a part of Central, as we do, northeast Central Iran comprises the Tabas, Bafgh and Yazd blocks during the Paleozoic and Mesozoic exhibit different behavior (Fig. 2.8).

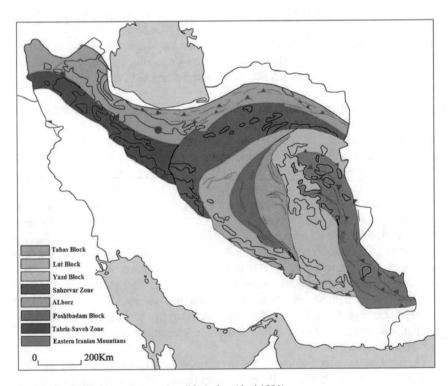

Fig. 2.8   Central Iranian subzones (modified after Alavi 1991)

2.4 "Urmia-Dokhtar" Magmatic Belt

This extends from south of Central Iran to the northern Sanandaj-Sirjan Zone. The Urmia-Dokhtar belt is composed of an extensive and thick sequence of volcanic, volcaniclastic and related rocks extending from Sahand to Bazman. This belt is not limited geographically to Iran, but actually extends from Serbia to Pakistan. The Iranian part of this belt is geologically characterized by:

- A NW-SE trending belt;
- Basaltic to rhyolitic volcanic and volcanoclastic strata
- The volcanic strata
- There are numerous types of Paleocene to Pliocene intrusions of gabbroic to granitic composition, cutting the old rocks of this belt.

In contrast to the northern part, in the southern part, Precambrian to Cambrian basement is rarely observed.

During Cenozoic time, volcanic activity occurred unevenly and was of the highest severity in the middle Eocene, late Eocene-early Eocene, Late Oligocene-early Miocene, and Mio-Pliocene.

The intrusive rocks of the Urmia-Dokhtar zone have distinctive dioritic to granitic composition whereas the extrusives are basaltic to rhyolitic. Their ages range from Eocene to Pliocene or occasionally Quaternary. However, their provenance is disputed, with some researchers considering them to be subduction-related calc-alkaline (e.g., Forster et al. 1972; Jung et al. 1976; Berberian et al. 1982), some as rift-related (Amidi et al. 1984) and others as of island arc affinity (e.g. Shahabpour and Doorandish 2008).

The volcanic-plutonic activities of this belt are characterized by:

1. There is no exact regular pattern in rifting and subduction environments; so that the age of lava in NE direction should ideally decrease in age and compositional variation from calc-alkaline to shoshonitic, but this is not observed. In the rift model, it should show alkaline to tholeiitic composition. Studies carried out till now (e.g. Pearce et al. 1984; Hassanzadeh 1993; Shahabpour 2007) did not support the rift model of Amidi et al. (1984). Accordingly, the magmatism of the Urmia-Dokhtar zone should be interpreted with consideration of the accepted petrological roles and general geological characters of Iran.
2. Regarding the compositional affinity and age of the magmatic belts of the Eocene-Oligocene of western Alborz- Azerbaijan, and Urmia-Dokhtar as well as the position of these two belts on the continental crust, assignment of either belt to an island arc setting (e.g. Shahabpour 2007) is questionable.

It seems that after the end of the Neo-Tethys subduction and during the collision of the Iranian and Arabian plates in the Late Cretaceous and subsequent subduction, uplift occurred on the margin of the Iranian microcontinent, central Iran and Alborz (Ghasemi and Talbot 2006).

The Pyrenean movement was initiated during the Early Eocene, with an extensional episode followed by compression in the late Eocene-Oligocene. The former episode resulted in extrusion of acidic to basic lava flows and related tuffs. With the onset of the later episode, the extrusive process rarely performed which increased the importance of amalgamation, simulation and amalgamation events during the last movements of Pyrenean. This has caused a shift toward a more acidic composition and the emplacement of intrusions having the same sources as the volcanic rocks. Accordingly, the belt should be considered as an after-collision magmatic belt, reflecting rifting and subduction characters.

Many studies have been conducted on the Urmia-Dokhtar zone, though most of them were carried out on only small areas. The works of Emami (1981) and Amidi et al. (1984) in the Natanaz-Naein area showed that volcanic activity started with an alkaline basalt flow associated with low acidic content during the early Lutetian.

Forster et al. (1972) also conducted studies on a 9000 km^3 area and concluded that the middle Eocene volcanic rocks that they studied are composed of olivine andesite, latite-andesite, latite-trachyte, alkali trachyte and lucite-phenolite. Eocene rocks are also accompanied by quartz andesite-latite. The Pleistocene rocks comprise ignimbrite, rhiodacite and dacite.

Volcanic eruption associated with the Lutetian marine transgression deposited the volcanoclastic "green series" succession. This activity was mostly phreatomagmatic and extensively explosive. The volcanic succession of this area is composed of the following.

Late Paleocene-early Eocene rocks of alkaline to intermediate composition, including trachyandesite, trachybasalt and shoshonite; and rhyolitic flow encompasses pumice and breccia tuff and breccia lava flow.

The Middle Eocene succession includes green sedimentary rocks, acidic tuff, tuffite, recrystallized limestone, calcareous rocks and conglomerate. It is worth mentioning that this succession is more complete in the Alborz area (Amidi et al. 1984).

The Urmia-Dokhtar magmatism was evaluated in detail in Ghorbani (2014).

2.5 Azerbaijan Zone

2.5.1 Geographical Setting

Distributed in northwestern Iran with an area of ca.107 km, the Azerbaijan zone is a mountainous area politically limited by Aras River and Talesh mountains, respectively, of which the latter separates Guilan from Azerbaijan.

The Sabalan Mountain with a 4811 m height is the highest and interferes of Aghchai and Aras River as well as the Zanganeh plain, western Poldasht, is the lowest point of this area. It consists of a chain of well-defined mountains with a mainly E-W trend joining two north-south trending mountains.

The eastern and western elevations from east to west direction are as follows:

Gharedagh elevations (Arasbaran)

220 km in length joining Talesh Mountains; highest peak is 3660.

Sabalan

This is actually a line dividing the drainage basins of the Aras and Urmia rivers. It encompasses the well-known Sabalan peak (4811 m) in the western Ardabil. This volcanic mountain extends from Gharesou of Ardabil in an east-west direction for roughly 60 km to Ghosheh-e dagh Mountain in the southern Ahar (Didon and Germain 1976; Dostal and Zerbi 1978).

Mishu and Moro Mountains

These are distributed along the Ghar-e Dagh Mountains in the south and southeast Marand. The highest peak (3155 m), is located at the Alamdar Mountain.

Sahand Mountain

This is a roughly circular mountain located 50 km south of Tabriz. Its highest peak, Jam-daghi is 3720 hight. It also contains 17 peaks with a height more than 3 km. The maximum length of Sahand in a west-east direction is 100 km (e.g. Scheffel and Wernert 1980).

Ghaflam Mountain

This mountain with a maximum hight of 1888 m is located at the southeastern Mianeh. Ghezel Avzan River divided it into two parts.

Azerbaijan is divided by Urmia Lake into eastern and western parts. Urmia Lake is the largest Iranian lake, having an extent of 4–6 km^3 and depth of up to 1375 m. After the Dead Sea, it is the 2nd saltiest lake of the world.

2.5.2 Geological Position

The geological position of Azerbaijan differs among general subdivisions presented by different researchers. However, this problem mostly depends on geologists' accepted concepts and beliefs concerning the geology of Iran.

Stöcklin (1968) included major parts of Azerbaijan as parts of Central Iran, the northeastern margin as the Alborz zone and the southwestern part as Sanandaj-Sirjan Zone. Nabavi (1976) in his structural-sedimentary division of Iran considered the main parts of Azerbaijan as the Alborz-Azerbaijan zone. He believed that this zone is bounded by the Alborz fault to the north, in northern Iran, and by the Tabriz-Urmia fault in the west. To the south it is bounded by the Semnan fault, and the eastern boundary is not still defined.

Innocenti et al. (1982) in a wider concept defined the structural units of western Iran, Azerbaijan, and Eastern to Central Turkey based on the two orogenic belts as follows:

1. Pantous- Lesser Caucasus and Alborz (Azarbaijan)
2. Taurus -Central Iran.

The northern part of Azerbaijan was considered as the Caucuses-Pantous Mountain in Turkey, the southern part as Central Iran and the western part as Iran to Taurus ranges of Turkey (Ghorbani 2013).

The Precambrian to Ordovician facies resembles those of central Iran. According to NIOC (2017), the lower Paleozoic movement created significant uplifts and local disconformity between Permian strata in some areas, e.g. Takab and Gharredagh. During the early Paleozoic, vertical movements caused abrupt facies change associated with a sedimentary gap, between the upper Cambrian Mila and lower Cambrian Lalun formations.

The Early Devonian event associated with major faulting has resulted in a well-defined facies division (Eftekharnezhad 1975). The Tabriz fault which starts with a northwest-southeast trend from the Abhar-Zanjan embayment to north Tabriz, Mishu, and Moro, and ends at northwest Azerbaijan and Caucuses was most probably created by this event.

The Early Devonian event has divided Azerbaijan into two blocks (like the structural subdivision of Innocenti et al. 1976). During Early Devonian time, the northeastern block was subsiding, whereas, the southwestern block remained uplifted until the late Carboniferous. The Permian succession in this zone starts with a red-colored clastic unit indicative of a continental environment and is followed by marine limestone. According to Rieben (1935), the Hercynian movements can be traced along Zenoz, Khoi, Moro, Mishu, Harzan-Darreh, so that, Permian or Permo-Triassic rocks overlie Devonian strata with an angular unconformity.

The Late Triassic orogeny has divided the Paleozoic platform of Azerbaijan into two distinct parts (Eftekharnezhad 1975). The dividing line, Zarrineh-Rud, is completely different to the dividing line of the Pre-Permian. These two lines presumably join together in northeastern, than continue in Caucuses. The western and southeastern part of this line forms a subsiding embayment in which pelagic sediments associated with submarine volcanics accumulated during the Late Cretaceous to early Eocene.

After the Late Triassic movements, geological processes formed the shale, coal-bearing Rhaetian-Lias sandstone of south Mianeh in a continental environment, and the Upper Jurassic rocks of the eastern and northeastern part of the area.

The main Alpine orogeny started with severe folding and faulting during the Late Cretaceous to Cenozoic. It was associated with Late Cretaceous volcanic activity but the main volcanic phase, mainly comprising submarine flows, occurred during the Eocene period (Eftekharnezhad 1975).

The onset of the Oligocene, "Pyrenean movements" resulted in magmatic rocks such as the Bozghush (Lotfi 1975) and Ahar syenite (Babakhani 1981) intruding Eocene volcanic rocks that subsequently resulted in the folding of western and southwestern Azerbaijan. Eftekharnezhad was of the opinion that Pliocene movements were significant in this area, citing the evidence of Plio-Quaternary volcanic activity and the reactivation of numerous faults cutting the pre-Pliocene rocks and even Quaternary alluvial deposits.

In reality, Azerbaijan cannot be defined as a distinct zone in the geology of Iran, and it should be mentioned that major parts are the Talaghi zone of other zones of Iran such as Alborz-Central Iran and Central Iran-Sanandaj-Sirjan. However, according to the present author, most of its areas are similar to central Iran.

2.6   Sanandaj-Sirjan

Based on its geological character, the Sanandaj-Sirjan Zone (Stöcklin 1968) is also named as Esfandagheh-Rezaieh (Takin 1971), Esfandagheh-Marivan (Nabavi 1976), Marivan-Manojan (Houshmandzadeh 1977). It extends as a NW-SE trending magmatic-metamorphic belt from Urmia in the northwest to Sirjan in the southeast and to neighboring Zagros and central Iran. The Sanandaj-Sirjan can be traced beyond the Iranian border into Turkey, Syria and the Caucuses. Its northern boundary is limited by Urmia-Dokhtar, Sirjan embayment, Marvadsht, Gavkhouni, the Mighan Desert, Kaboudarahang and Urmia Lake. The southern boundary is assumed to be the Zagros thrust.

Some researchers included the Sanandaj-Sirjan Zone as the Zagros Zone based on the same folding trend. However, it not only completely differs from Zagros in its magmatic, metamorphic, orogenic and lithofacies aspects, but it mostly resembles Central Iran (Fig. 2.9).

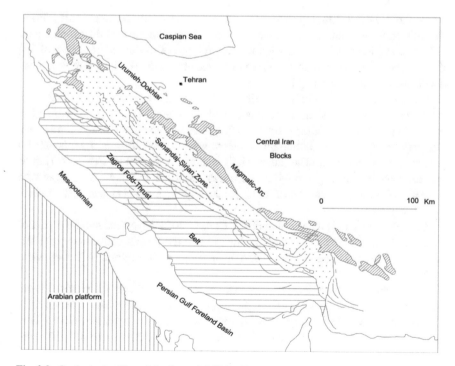

Fig. 2.9 Geological setting of the Sanandaj-Sirjan Zone

2.6.1   Geology

Stöcklin (1968) described and introduced this zone as follows.

The Sanandaj-Sirjan Zone, located on the northern side of Zagros thrust, is morphologically considered a part of Zagros Zone. However, based on its sedimentary regime, it is mostly similar to Central Iran, so that all of the Paleozoic to Mesozoic major unconformities detected in Central and northern Iran can be traced in the Sanandaj-Sirjan Zone but are never observed in the Zagros Zone. The Sanandaj-Sirjan Zone and Central Iran differ from Zagros by the presence of Cenozoic volcanism and lesser development of Cenozoic formations. Stöcklin (1968) believed that the Late Jurassic-Cenozoic intrusive rocks (diorite and granite) are of high significance in Sanandaj-Sirjan and Central Iran and cannot be observed in the Zagros. He additionally added that Mesozoic metamorphism affected most parts of Central Iran and Sanandaj-Sirjan. In fact, the Mesozoic intrusive rocks, especially those of Cretaceous-Paleogene, occur more extensively in Sanandaj-Sirjan than in other parts of Iran; whereas, the Mesozoic metamorphic rocks that were metamorphosed during the middle-late Cimmerian phase occur more widely than in Central Iran.

Eftekharnezhad (1980) named the northwestern part of the Sanandaj-Sirjan Zone, 'Urmia-Hamadan' and believed it to be distinguished from Zagros and Soltanieh-Mishu as it had been affected by the early Cimmerian orogeny during the late Triassic. Berberian and King (1981) introduced this zone as a narrow intracratonic basin during the Paleozoic time and an active continental margin during the Mesozoic. These authors considered it to be the southwestern margin of the Iranian microcontinent and believed that the Neo-Tethys Ocean was located to the south of this belt. They dated also the ocean opening to 240 Ma.

Further studies and sampling carried out by Eftekharnezhad (1996) resulted in subdivision of this zone into "northern" and "southern" subzones, of which the former is the so-called "Urmia-Hamadan" defined by the effects of the Cimmerian Orogeny. In this subzone, important intrusions such as Arak, Broujerd, Malayer, and Alvand can be observed. The latter, which extends from Golpayegan to Sirjan, is characterized by the orogenic evidence from Precambrian to Middle Triassic and the presence of intrusive masses such as Hajiabad, Eghlid and Sirjan, and Esfandagheh basic intrusions.

According to Houshmandzadeh (Pers. Comm. 2014), even the southern part of Sanandaj-Sirjan is different to Central Iran, in that the former contains upper Paleozoic volcanoclastic and epiclastic carbonate rocks with a minimum thickness of 4000.

In fact, the Paleozoic and Triassic rocks of the southern part are exposed more widely in comparison to the northern part. In the former area, evidence of the early Cimmerian orogeny can be seen. In comparison, the northern part shows more widely distributed and and thicker sequences of Jurassic to Cretaceous rocks which are intruded by a number of granites and diorites. In other words, the consequences of the middle to late Cimmerian and Laramide orogeny are more obvious.

The previously mentioned researchers regarded the emplacement of the middle Jurassic-late Cretaceous to Paleocene acidic intrusions of the Central Iran margin (i.e. Sanandaj-Sirjan) to be the results of the final stages of the subduction of the Neo-Tethys. The emplacement of upper Cretaceous ophiolites and radiolarites and their metamorphism to greenschist facies were attributed to the closure of the Neo-Tethys.

Berberian and Berberian (1981) studied the intrusions of Sanandaj-Sirjan and considered the subduction of the Neo-Tethys as an oblique type. According to this model, subduction started at the southeastern end of Sanandaj-Sirjan and the Middle to Upper Triassic igneous rocks of Eghlid area were products of the magmatic arc of this episode of subduction. With the northwestward development of subduction, the magmatic arc (Late Cretaceous-Paleocene) extended into this area. The present author is of the opinion that subduction in Sanandaj-Sirjan did not begin during the Triassic because the opening of the Neo-Tethys dates back to the Early Triassic and the subduction of the northern and southern parts cannot have such a big age discrepancy. On the one hand, as discussed below, there is no difference between the Paleozoic and Middle to Upper Triassic rocks of south and north Sanandaj-Sirjan, and, on the other hand, there is not any evidence of subduction-related magmatism and metamorphism (Ghorbani 2014). It seems that subduction was initiated during the Jurassic as evidence of intrusion can be observed in middle Jurassic rocks.

The final episode of subduction as well as the subsequent collision of the Iranian and Arabian plates is always disputed among researchers, with some (Mohajjel 1999; Stöcklin 1974b, 1977; Berberian and King 1981) assigning it to the Late Cretaceous, and others such as Sengör and Yilmaz (1981) and Dewey et al. (1973) considering it to be Miocene. In fact, according to present author, the geological evidence indicates that in most parts of Sanandaj-Sirjan belt, this was overprinted during the collision in Late Cretaceous-Paleocene. Subduction continued through the Miocene, though the southern end of the belt is an exception of this rule.

Alavi (1994) introduced the Sanandaj-Sirjan Zone for the middle part of the Zagros orogenic belt in the south, the Urmia-Dokhtar magmatic complex in the northeast, and the simple Zagros fold belt in the southwest. He believed the Sanandaj-Sirjan metamorphic belt to be the result of the closure of the Neo-Tethys opening during the Late Paleozoic and Mesozoic. According to the latter author, the belt is characterized by the following:

- Similarity of the Paleozoic to Mesozoic stratigraphical units in Sanandaj-Sirjan and Zagros. Therefore, Sanandaj-Sirjan, like Zagros, is believed to be part of the Gondwanan continent or Afro-Arabian plate.
- The ophiolites of the Sanandaj-Sirjan Zone can be classified into two groups having distinctive structural characters, especially that of northern part in Baft area where ophiolites are emplaced on Sanandaj-Sirjan as the result of northeastward subduction of Neo-Tethys ocean beneath Sanandaj-Sirjan.
- The Paleozoic-Triassic volcanic rocks of this belt resulted from opening of the Neo-Tethys.

The Iranian and Afro-Arabian suture zone is located at the boundary between the Urmia-Dokhtar magmatic belt and the Sanandaj-Sirjan Zone, i.e. at the Main Zagros thrust, rather than between Sanandaj-Sirjan and Zagros, where it is characterized by NW-SE trending structural depressions.

Following the subduction and subsequent collision of the Arabian and Iranian plates, the thickness of the Urmia-Dokhtar succession increased by 5–10 km, caused by magmatic activity and thrusting. The thickness of the Sanandaj-Sirjan sequence was increased by 10–15 km owing to the thrusting and tectonic shortening.

The Sanandaj-Sirjan basement is composed of approximately 40 km of Afro-Arabian continental plate.

Based on geophysical data, some parts of the Neo-Tethys ocean remained in the lower part of the Urmia-Dokhtar zone (Kadinsky-Cade and Barazangi 1982).

In fact, in terms of tectonics (general trends of faults and folds), the Sanandaj-Sirjan Zone resembles the Zagros. However, from a sedimentary facies, metamorphic and magmatic point of view, it not only differs from Zagros but also shows many affinities with Central Iran.

Mohajjel (2000) carried out Petro fabric studies on the Malonitic rocks of Dorud-Azna and introduced four rock types in these areas, attributing their occurrence to the dextral transpression tectonics of the Arabian-Iranian plates. He also described a thick crust tectonic model for Sanandaj-Sirjan. It is of note that Alavi (1994) and Hooper et al. (1994) also proposed a thick crust tectonic model for this belt. The former believed that it consists of numerous stacked nappes that have transported various rock sequences from the suture zone.

Mohajjel and Sahandi (1999) subdivided this zone into five NW-SE trending subzones, based on the tectonic and stratigraphic features.

1. **Radiolarite Subzone**

This is distributed throughout southwest Sanandaj-Sirjan in an anastomosing manner and its extent can be traced from western Mediterranean areas such as Cyprus, Greece and south Italy to areas such as the Havasina complex to the southeast. It begins at its northwest end of the Sanandaj-Sirjan Zone with an outcrop 35 km width and 250 km length and terminates in southeast Broujerd. Other exposures of this zone can be observed around southeast Kermanshah and in the Neyriz area. The Radiolarite sub-zone consists of Triassic–Cretaceous shallow marine limestone and dominant deep-marine radiolarite. Some olistoliths of Campanian-Maastrichtian age occur in this subzone.

2. **Bisotun Subzone**

Located in the northwest part of the radiolarite subzone, around Kermanshah, the Bisotun Subzone is composed of thick to massive bedded rocks of Upper-Lower Cretaceous age. Shallow and pelagic environments are recorded in the Upper Triassic and Lower Cretaceous, respectively. There are many thrust faults in this zone, some of which transect the radiolarite subzone and are thrust over Zagros Zone rocks. This thrusting event was active until Neogene time in southwest Sanandaj-Sirjan.

3. Ophiolite Subzone

Ophiolite and colored mélange are exposed along the southwestern edge of the Sanandaj-Sirjan Zone, Kermanshah, and Neyriz. This assemblage is composed of pillow lava, spilite, red and purple shale, and pelagic limestone. Ophiolites are thrust over Bisotun- Radiolarite subzones and a succession of Paleocene-Eocene.

From the temporal and compositional point of view, the ophiolites of the Kermanshah and Neyriz areas show many affinities with those of Oman, indicating a simultaneous age of subduction (Sengör 1990).

4. Marginal Subzone

This is located in the southwest of the Sanandaj-Sirjan Zone, with complex deformation. It is composed of Upper Jurassic-Lower Cretaceous volcanic rocks, subsequently metamorphosed to low-grade greenschist facies. The marginal subzone extends to all parts of the zone. The rocks of this zone were thrust over the ophiolite, and rocks of the the Radiolarite and Bisotun subzones, as well as of the Zagros Zone during Late Cretaceous to Pliocene time.

5. Complexly deformed Subzone

This is located to the northeastern side of the Marginal Subzone and is composed of metamorphic and highly-deformed rocks. It can be distinguished from the other subzones by the abundance of schist, phyllite, and amphibolite. It is also characterized by several deformation events, some of which are associated with metamorphic events as well as numerous igneous intrusions. With strong stratigraphical affinities with their equivalents in Central Iran and Alborz, the Neoproterozoic to Cambrian rocks are exposed in this area without any signs of metamorphism.

Middle and Upper Paleozoic rocks can also be observed in the Aligodarz, Dorud-Azna and Nahavand areas.

The Zhan Complex units and their equivalent are exposed in this subzone. The succession starts with quartzitic sandstone followed by dolomite and chert bearing limestone. This succession is overlain by mafic volcanic rocks, basalt and andesibasalt, and acidic volcanics with some intercalations of shale and limestone. This complex and its equivalents, the Totak, Kolikesh, and Surian in the Eghlid area and the Abbarik of Aligodarz, are overlain by the Hamedan Phyllite and its Upper Triassic-Jurassic equivalent rocks, capped by Cretaceous rocks with an angular unconformity. These rocks only crop out in erosional windows. Some researchers assign them to the Precambrian, others to the Mesozoic. These rocks, though being metamorphosed, are traceable everywhere in Sanandaj-Sirjan.

Most of the complexes cropping out in this zone, such as the Kolikesh, Surian, Totak in the southern Sanandaj-Sirjan, and metamorphic rocks of Northern Saman, Ab-barik, Aligodarz, northern Divandarreh, Saqqez and Kabudarahang, and Songhor series, belong to this group. These rocks are mostly of volcano-sedimentary protolith affinity, associated with small intrusions. These features imply an extensional basin formed by extensional episodes during opening of the Neo-Tethys. All of the complexes and associated rocks listed above have been metamorphosed to greenschist to amphibolite facies. The overlying Upper Triassic-Lower Jurassic Hamedan Phyllite was only metamorphosed to slate to phyllite grades, indicating that metamorphism mostly pre-dated their deposition.

2.6.2   Geodynamic Evolution

Sheikholeslami (2002) presented the following stages for the geodynamic evolution of Sanandaj-Sirjan.

1. Paleozoic intracontinental opening on the northern margin of Gondwana (Fig. 2.10).
2. Following the Neo-Tethys opening during the Permian time, Iran separated from the northern margin of Gondwanan.
3. At the beginning of the Late Triassic, because of subduction, the oceanic lithosphere of the Neo-Tethys begins to disappear (Fig. 2.10c). According to the present author, the subduction of the Neo-Tethys did not begin in the Triassic, because there is no adequate evidence to support such a claim. On the other hand, it is logically less possible to occur immediately after its opening, during the Triassic. As already stated, it began during the Early Jurassic, an interpretation confirmed by many lines of evidence described above.
4. The Neo-Tethys closure during the late Mesozoic. At this time, the old marginal edge of Iran, Sanandaj-Sirjan, with the associated metamorphic complexes and Tethyan ophiolite were thrust over the old Arabian-Gondwanean plate margin (Fig. 2.10d).
5. Considering the paleogeography, it is concluded that the Sanandaj-Sirjan Zone is a structural zone including Neoproterozoic to Recent strata, that was metamorphosed during the Early Cimmerian orogeny. It also encompasses the turbiditic Mesozoic basins that formed during the Late Triassic and closed during middle Cimmerian or Laramide event.

2.6.3   Lithostratigraphy

All of the rocks of this zone can be classified in three tectonostratigraphic units of Neoproterozoic- Middle Triassic; Late Triassic-Late Cretaceous and Cenozoic. Sabzehei (1996) was of the opinion that the oldest rocks of this zone are Precambrian metamorphosed ultramafic gabbro. Considering the Neoproterozoic-Paleozoic rocks of this zone, it can be stated that there is no apparent difference to Central Iran (Fig. 2.11).

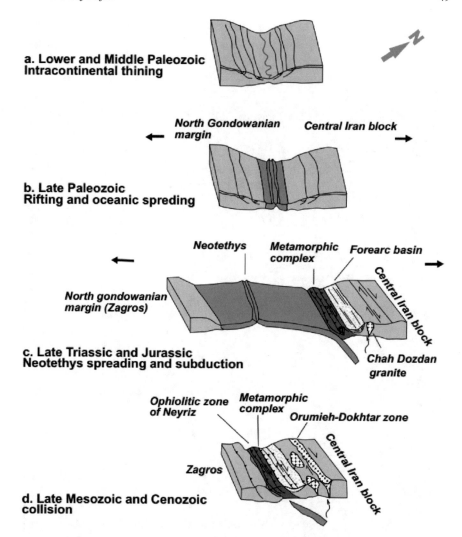

a. Lower and Middle Paleozoic Intracontinental thining

North Gondowanian margin

Central Iran block

b. Late Paleozoic Rifting and oceanic spreding

Neotethys   *Metamorphic complex*   *Forearc basin*

North gondowanian margin (Zagros)

c. Late Triassic and Jurassic Neotethys spreading and subduction

Central Iran block

Chah Dozdan granite

Ophiolitic zone of Neyriz   *Metamorphic complex*   *Orumieh-Dokhtar zone*

Central Iran block

Zagros

d. Late Mesozoic and Cenozoic collision

Fig. 2.10 Geodynamic reconstruction of the SW margins of central Iran and the north Gondowanan margin from Paleozoic to Cenozoic. **a** Lithospheric thinning during the middle Paleozoic and development of mantle-derived ultrabasic rocks. **b** Development of a rift type basin from Permian. **c** Development of a Neo-Tethys oceanic basin and beginning of subduction in the Late Triassic (Early Cimmerian phase) and generation of an accretionary prism, forearc basin and magmatic arc (granite/granodiorite of Chah Dozdan) over the subduction zone. **d** Collision and generation of an orogenic prism and development of the volcanic arc of the Urumieh-Dokhtar zone (After Sheikholeslami et al. 2003)

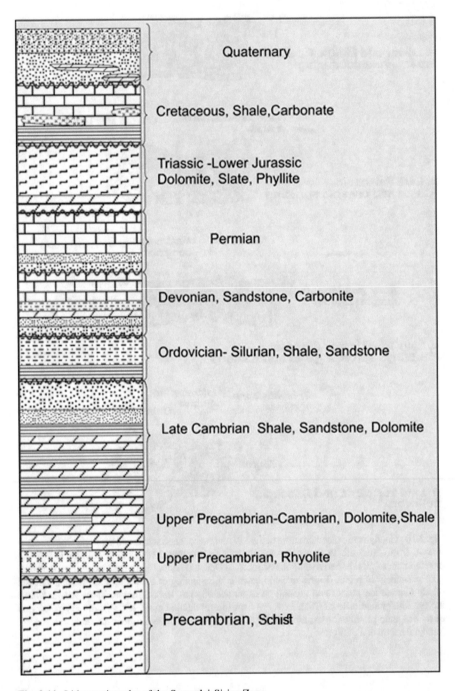

Fig. 2.11 Lithostratigraphy of the Sanandaj-Sirjan Zone

2.6.3.1 Precambrian

The Sanandaj-Sirjan Zone was a most active sedimentary-structural zone in Iran in which most rocks are metamorphosed. Previous studies assigned most of the metamorphic rocks of Sirjan, Golpayegan, Muteh, Mahabad, Marivan, Shahr-e Kord, and Urmia to the Precambrian, though; it now seems that there is not enough evidence to support this hypothesis. Apparently, most of the Precambrian clastic rocks associated with younger rocks were metamorphosed as the result of the Late Cimmerian orogeny during the Late Triassic.

Based on field observations, the present author is of the opinion that all of the rocks except those of Marivan, Mahabad, and Urmia covered by the Cambrian Lalun Formation, are actually Neoproterozoic-Early Cambrian in age.

2.6.3.2 Paleozoic

Towards the late early Paleozoic, Sanandaj-Sirjan was a subsiding basin in which clastic deposits accumulated. The middle-late Paleozoic succession is 4000 m thick and consists of volcanoclastic, carbonate and epiclastic rocks. Extensional phase associated with subsidence has caused the extrusion basaltic magma of continental alkaline type, and associated pyroclastics.

2.6.3.3 Mesozoic

The conformable Lower Triassic sequence consists of a succession of volcanoclastic, lava and pyroclastic rocks, mostly metamorphosed, observable in some of the erosional and tectonic windows of northern and southern Sanandaj-Sirjan.

The Upper Triassic-Lower to Middle Jurassic succession is very thick in this zone, especially in the northern part (Hamedan phylite). The Cretaceous rocks are mostly of carbonate to slightly clastic composition and are sometimes mixed with volcanic rocks, especially in the northern areas of Sanandaj -Saqqez-Piranshahr.

2.6.3.4 Cenozoic

Cenozoic rocks are sporadically exposed in this zone and are of localised distribution.

The general stratigraphical column of Sanandaj-Sirjan according to Alavi (1991) is as follows:

A: An alternating succession of siliciclastic and very shallow marine carbonates indicating multiple marine regressions and transgressions.
B: A succession of continental well bedded quartz sandstone capped by well bedded fossiliferous limestone.

C: A continuous succession of continental shallow siliciclastics, sporadically exposed fusulinid-bearing carbonate including lava lenses in the lower part.

D: A succession of andesitic and basaltic lava flow, and shallow marine carbonate with shale subordinates which overlie the older rocks with angular unconformity.

E: A succession of sandstone and well-bedded shale-bearing enclaves of older rocks. The lower part of this succession indicates a continental environment which changes upward to shallow water and continental crust.

F: A succession of shallow algal and oolitic limestone with shale, siltstone and sandstone intercalations.

G: The *Orbitolina* bearing limestone, marlstone and shallow water dolomite, locally accompanied by conglomerate and intraformational breccia and covered by coarse grained conglomerate and sandstone.

Locally, some grey to red coarse grain volcanoclastic rocks can be observed, indicating the time of plate thrusting during either the Cretaceous or most probably Cenozoic.

H: Syn-orogenic conglomerate with ophiolitic pebbles, sandstone, shale, limestone and marlstone.

2.6.4 Magmatism

Numerous studies by various researchers show that extensive magmatic episodes have occurred in the Sanandaj-Sirjan Zone, the most important being Middle-Late Triassic, Late Jurassic-Early Cretaceous and Late Cretaceous- Paleocene (Berberian and Berberian 1981).

Previously, the age of some intrusions, e.g. the Golpayegan and Muteh granites and the Ghushchi gabbro and diorite were assigned to the Precambrian-Cambrian. However, recent studies have rejected this age and have assigned a late Carboniferous age (Moghadam et al. 2015).

Triassic intrusions crop out locally in the southern Sanandaj-Sirjan, for example, the Chah-Bozorgan batholith, the southeastern Baft intrusions, and the Dehbid granite. The Chah-Bozorgan batholith is formed by two magmatic phases of gabbro, leucogranite-granodiorite (main and largest intrusion) and biotitic granite whose age of the formation is Late Triassic-Early Jurassic. In this part of the Sanandaj-Sirjan zone, the gabbroic rocks are older than granitic rocks, but not far apart.

According to the studies of Sabzehei (1996), the granites of these areas were the products of partial melting of the crustal lithosphere caused by thermal convection of basaltic heat.

The Jurassic-Cretaceous to early Cenozoic intrusions are exposed along the northern Sanandaj-Sirjan, Golpayegan-Urmia road. Their important characteristics are summarized in Ghorbani (2014).

It seems that the oldest intrusions of this group are Kolahghazi and Almogholan whose emplacement age is Middle-Late Jurassic (Valizadeh and Zarian 1976;

Shahbazi et al. 2010). The composition of the Kolahghazi body is granodiorite-monzogranite and sinogranite. The granitoid composition of this body is of calco-alkaline to per-alomine nature. The S type provenance was interpreted owing to the characteristic mineralogy, as well as geochemical composition, and enclaves with high mica content. Consequently, of the tectonic environment point of view, it was assigned to the syn- to post-orogenic grantooids (Noghreyan and Tabatabaei 1995).

The Almogholan pluton, located 12 km north of Asadabad, western Hamedan, intruded metamorphic Jurassic-Cretaceous rocks. It is composed of quartz-syenite, granite and diorite. The diorite was emplaced during the late Jurassic, 144±17 Ma: Rb-Sr (Valizadeh and Zarian 1976), whereas, the quartz syenite and granite were emplaced later (Amiri 1995). This pluton is of alkaline nature (Moein-Vaziri 1996) and its formation was linked to the subduction of the Neo-Tethys beneath the Iranian microcontinent (Amiri 1995).

The Astaneh pluton with a total areal extent of 30 km2, is one of the significant plutons of the Golpayegan-Urmia belt (Masoudi 1997). It is characterized by the presence of Andalusite crystals (Radfar 1987), indicating the palingenetic provenance of the pluton (Darvishzadeh 1991). According to Masoudi (1997), its Sr isotopic ratio is approximately 0.709, being comparable with S-type granites.

A large NW-SE trending batholith that crops out on the Boroujerd-Malayer road is considered to have a multistage emplacement history (Masoudi 1997). This pluton encompasses different rocks of Jurassic-Paleocene age. In addition to the age difference, its composition is also diverse. As a result, it consists of quartz diorite, granodiorite, granite, aplite, pegmatite and rarely norite (Berthier et al. 1974).

The associated I-type granite resulted from the partial melting of pre-existing metamorphic rocks (Berthier et al. 1974), caused by the subduction of the Neo-Tethys beneath the Iranian microcontinent (Shamanian-Esfahani 1974).

Being compositionally diverse, the Cretaceous-pre-Eocene granitoid plutons of the Boein to Miandasht area consist of diorite, granodiorite, granite and alkali granite.

Valizadeh and Ghasemi (1993) classified the intrusive rocks of these areas into three groups "dolerite, diorite, and granite" and were of the opinion that first the diorite-gabbro plutons and doloritic dikes, then granitic plutons, syenogranite and alkali-feldspar granite were emplaced. They considered the mother magma to be of basic and andesite types. The dolerite and diorite magmas resulted from the partial melting of the subducted oceanic crust or its overlying mantle wedge, and the acidic one from the partial melting of lower crust formed during post-orogenic conditions.

One of the largest plutons of the Golpayegan-Urmia area is the Alvand batholith which crops out between Asadabad-Hamedan and Toyserkan and is more than 500 km wide. The dating results indicate a discrepancy in the data, with the most recent results (Shahbazi et al. 2010) suggesting an age of 175 Ma, though younger ages had already been reported that suggested the emplacement of the pluton in Late Cretaceous-Early Paleocene time.

The main portion of Alvand Batholith is of granitic composition, whereas the basic types such as gabbro and diorite are only rarely exposed in the northeast near Ghassaban and Chapan villages, and on the southwest side of the pluton, on the Hamedan-Kermanshah road. Valizadeh and Sadeghian (1996) were of the opinion that it has resulted from the subduction of the Neo-Tethys Ocean crust beneath the Central Iran microcontinent they also classified the Alvand granite rocks as s-type granites, based on the following characters:

Mineralogy: the high normative corundum content, the presence of biotite, muscovite, tourmaline, graphite and some the refractory minerals such as garnet, andalusite, and k-feldspar megacrysts

Geochemistry: the high SiO_2 content, low sodium, high initial Sr87/Sr86 ratios, the peraluminous and calc-alkaline nature of pluton.

Sepahi-Garo (1999) considered the Alvand pluton to represent at least five magmatic phases, including gabbrodiorite, tonalite, fine grain porphiroid granites, coarse-grained porphiroid granite, hololcrate granitoid. This author postulated a first stage, during which a toloetic magma resulted in the formation of gabbro, diorite, and tonalite. Then, a wide spectrum of mafic to medium to tonalite was created. In the next stage, the mother magma, resulting from the melting of quartz diorite in the lower to middle crust, of holocrate granite was intruded into two stages.

The filed evidence, perographical and geochemical characters of the Alvand show many affinities with intrusive rocks of magmatic arcs, especially arcs of continental margins, so this pluton was probably a part of the "Sanandaj-Sirjan" magmatic arc, formed during Middle Jurassic to Early Cretaceous time, or more probably during the Paleocene.

In the northern part of Sanandaj-Sirjan, there are plutons such as Aghdarreh, Zidkandi, Mahmudabad, Blouz-Janbolagh, major Khazaei and Pichaghchi of the Shahindezh around as well as those of the Naghadeh-Piranshahr belt Most of these plutons are of Cretaceous to Paleocene age and are compositionally diverse.

Based on their affinities, they are included in the Golpayegan-Urmia intrusive belt whose main characters are given in Table 5.1.

The Ghushchi pluton is located in the northwestern part of the Golpayegan-Urmia intrusive belt and is exposed in northeastern Urmia. Like the other plutons of this belt, it is composed of acidic and basic components whose emplacement ages date back to the Paleocene to Late Oligocene. Although the granitic part intersects the basic one, both are of similar age (Behnia 1995; Asadpour 2000).

Recently, Asadpour (2000) carried out a study of the Ghushchi pluton and assigned very different ages to the two units, with the acidic unit dated to 500 Ma and basic unit to 300 Ma.

The granite rocks of this pluton were classified into five facies based on miner-alogical and textural features. The main Ghushchi granite was classified by Behnia (1995) as an A-type granite caused by high metasomatism of gabbro rocks. He was also of the opinion that biotitic granite and two-mica granites are the products of changes in biotitic schist and gneiss (S-type granite).

Behnia (1995) also determined the composition of basic to medium rocks in this pluton to include pyroxenite, olivine gabbrodiorite, gabbronorite, quartz diorite, quartz monzonite, monzonite-syenite, syenite, alkali-syenite and acidic ones to tonalite, alkali-granite.

Asadpour (2000) assigned the provenance of the basic rocks of the area to mantle deming and believed that this phenomenon and its associated fractures resulted in the rift development in this area. The partial melting of mantle and basaltic magma as well as its intrusion in lower crust have caused the crustal melting and formation of granite magma.

As seen above, most of the plutons of the Golpayegan-Urmia intrusive belt can be classified in composite pluton type, encompassing compositionally and temporally various rocks. Most of the plutons are of Cretaceous-Paleocene age but the Kolah-Ghazi and Alvand plutons have been assigned to the Jurassic. However, some plutons such as Boroujerd contain Jurassic rocks. In addition to the age variation in this belt, various rock types can be observed; so that most plutons contains both basic to acidic parts, for example, the Borujerd, Alvand, and Ghushchi batholiths. Some of them include rock types of alkaline syenite to alkaline granite or lococrate granite. The formation of the basic rocks precedes the acidic ones. The major compositional differences within this belt indicate the various provenances. This is well documented, with most researchers noting the presence of I, S and H type granites and rarely H type. In these studies, the provenance of the basic to intermediate rocks and I-type granites were linked to partial melting of the mantle, primary or evolved, subducted oceanic crust, whereas the acidic, S and I type granites, were derived from crustal partial melting.

Form the point of view of tectonic provenance, most researchers believe that the Golpayegan-Urmia belt is a result of subduction and its intrusive rocks show affinities to active continental margins. In contrast, some studies assign the intrusive rocks to an extensional environment.

Berberian and King (1981) and Moein-Vaziri (1996) considered the age of some plutons such as Golpayegan, Muteh and the gabbrodiorite of Ghushchi to the Precambrian, though this is rejected by most recent studies. This will be explained in more detail in chapter five.

2.6.5 Metamorphism

Most Neoproterozoic to Mesozoic rocks were metamorphosed at low to the medium grade of greenschist. The youngest rocks affected by metamorphic events are the fossiliferous Upper Cretaceous rocks and this fact indicates that at least the youngest metamorphic event occurred during Late Cretaceous time.

However, there is some evidence documenting a hydrothermal and metamorphic event of Middle Triassic age that altered continental crust into marmarite, gneiss schist and quartzite.

Metamorphic and magmatic activities in the Middle to Late Triassic were all linked to an extensional episode, corresponding to the formation of the Neo-Tethys,—Tethys III.

Low-grade metamorphism of the late Cimmerian event in Late Jurassic occurred in the northern part of Sanandaj-Sirjan with some rocks showing recrystallization and others reaching low-grade greenschist facies. It can be demonstrated conclusively that a metamorphic event occurred during the Late Cretaceous-Paleocene, the evidence for which is as follows:

Plutons such as Alvand, Malayer, and Arak have the same trend as Zagros and cut the Jurassic rocks, causing contact metamorphism.

In general, four metamorphic phases can be distinguished in Sanandaj-Sirjan as follows:

1. Neoproterozoic to early Cambrian observed in the Soltanieh Formation and the underlying rocks.
2. A simultaneous metamorphic event caused by the Early Cimmerian Orogeny whose evidence, as already mentioned, can be traced in all Sanandaj-Sirjan.
3. A Middle Cimmerian metamorphic event with very low-metamorphosed schist that affected Hamedan phyllitic rocks.
4. A Late Cretaceous-Paleocene event simultaneous with the Laramide event.

2.7 Zagros Zone

Constituting the northeastern edge of the Arabian plate, this zone extends from the south of Iran, Bandar-Abbas, continuing to the northeast in Kermanshah and terminating in Iraq. The Zagros basin is distinguished from other Iranian basins by its continuous subsidence history though multiple unconformities have been detected.

The main characteristics of this zone are as follows:

*   The absence of Permian to Recent magmatic and metamorphic activity
*   Low distribution or near absence of Devonian-Carboniferous rocks
*   The presence of numerous large scale anticlines and small scale synclines
*   From the beginning of the late Permian to late Miocene, no large sedimentary breaks have been detected, though there are some unconformities and disconformities.

The Zagros Zone encompasses the most important oil and gas reservoirs of Iran. It is a sedimentary basin that extends to the north and northeast to the Main Zagros Thrust, to the southeast to the Oman Mountains and to the southwest to the Arabian shield. It is confined by the Arabian shield and Sanandaj-Sirjan Zone. The lower Cambrian to Pliocene succession is continuously exposed in this area (Table 2.1 and Fig. 2.12). The lower Cambrian to Triassic rocks indicates shallow platform conditions, sometimes affected by epeirogenic movements. From the, More than 10 km of Late Triassic to Miocene sediments were deposited in this consistently subsiding basin. These sediments are mostly composed of carbonates and rarely shales, sandstones, and marls. The subsidence rates decrease from south to north, where the thickness of Paleozoic, Mesozoic and Cenozoic rocks is less than 5 km.

The Zagros, which is located within the Alpine-Himalayan orogenic belt, is one of the young, active orogens formed by the north-south collision of the Arabian and Eurasian plates (Vernant et al. 2004). Interpretations of the spatial and temporal

Table 2.1 **a** Paleozoic Formations of the Zagros zone (Setudehnia 1975; Motiei 1993; Ghavidel-Syooki 1995; Aghanabati 2004); **b** Mesozoic formations of the Zagros zone (Motiei 1993; Aghanabati 2004); **c** Cenozoic formations of the Zagros zone (Setudehnia 1975; Motiei 1993; Aghanabati 2004)

Eon	Period	Formation	Lithology
(a)			
Paleozoic	Permian	Dalan	Fossiliferous limestone, dolomite, dolomitic limestone, thick bedded anhydrite interbedding with lateritic dolomite
		Faraghan	Sandstone, dolomite and limestone with shale intercalations
	Devonian	Zakin	Sandstone with cement
	Silurian	Sarchahan	Black shale
	Ordovician	Siahoo	Conglomerate sandstone, alternation of shale, siltstone, sandstone and limey sandstone
		Zardkuh	Shale and sandstone
		Ilbeyk	Alternation of shale, thin bedded grey thin bedded limestone, shale and sandstone
	Cambrian	Mila	Alternation of dolomitic shale to carouse grained massive dolomite with thin to thick bedded shale intercalations
		Lalun	Red to white and green sandstone
		Zaigun	Colored shale
		Barut	Thin bedded dolomite and red shale
Proterozoic	Neoproterozoic	Hormoz	Salt, anhydrite, gypsum, thin bedded limestone, cherty dolomite, shale, claystone and oxide rocks
(b)			
Mesozoic	Cretaceous	Tarbour	Massive to cliff-forming limestone with rarely anhydrite
		Gurpi	Blueish grey shale and marlstone with thin-bedded clayey limestone intercalations
		Ilam	Light to dark grey, fined grained limestone and well bedded limestone with thin bedded shale intercalations
		Surgah	Light to dark grey pyritic shale alternating with fine grained limestone
		Sarvak	Dark clayey limestone, massive- to thick-bedded limestone
		Kazhdomi	Dark grey shale locally black and containing
		Garou	Alternation of dark grey to black bitumen bearing pyritic and cherty shale
		Darian	Grey to brown thick bedded to cliff-forming limestone
		Gadovan	Alternation of dark grey limestone and grey to brownish grey marlstone
		Fahlian	Massive- to thick -bedded

(continued)

Table 2.1 (continued)

	Jurassic	Surmeh	Massive to thick-bedded limestone, clayey limestone, marl and marly limestone
	Triassic	Neyriz	Thin bedded dolomite, sandy and silty dolomite, clayey to shaley thin-bedded limestone
		Khanehkat	Dark grey, very fine grained
		Dashtak	Evaporative rocks,
		Kangan	Clean carbonates, claystone and shale, evaporative carbonates

(c)

Cenozoic	Pliocene	Bakhtiari	Conglomerate
		Aghajari	Alternation of sandstone, brown limestone and red marlstone with gypsiferous veins and siltstone
		Mishan	Grey marlstone and clayey limestone
	Miocene	Razak	Silty marlstone associated with rare silty limestone
	Oligocene	Gachsaran	Anhydrite, colored marlstone and limestone with shale intercalations
		Asmari	Resistant crème to brown, occasionally dolomitized limestone with shale intercalations
	Eocene	Jahrom	Dolomite and dolomitic limestone
		Shahbazan	Dolomite and dolomitic limestone
		Pabdeh	Clayey sediments divided into five shaleyl and clayey units
		Kashkan	Red clastic sediments, siltstone, sandstone and conglomerate
		Talezang	Grey to brown medium to massive bedded limestone
	Paleocene	Sachoon	Evaporate sediments, gypsum, marl, dolomite and limestone
		Amiran	Siltstone, sandstone, some conglomerate beds with chert pebbles and some limestone beds

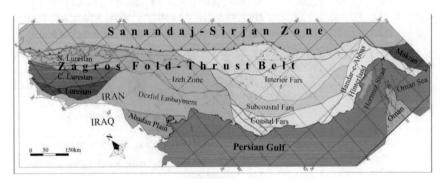

Fig. 2.12 Main structural subdivisions of the Zagros zone (modified after Jahani et al. 2009)

distribution of active deformation, as well as the seismological behaviors of all parts of the Zagros belt, confirm the weak structure of this belt.

Based on the possible occurrence of earthquakes in an area (Berberian 1995), recent data (Tatar et al. 2002; Walpersdorf et al. 2006) and convergence rates, it is a simple folded zone including foreland fold and thrust belts. Taking into account GPS data, a shortening rate of 8–10 mm/year has been determined, which is comparable with 40–45% of total convergence rate, 21 mm/year, of the Eurasian with Arabian plates. However, according to Vernant et al. (2004) and Sella et al. (2002), there is a N-S active convergence with a rate of 25–30 mm/yr on the eastern edge of the Arabian plate.

Despite the geological and geophysical data, there are ambiguities concerning the onset of deformation in the Zagros crust. Asymmetrical folds which can be observed in the Zagros Mountains generally imply the N-S compression of the upper 10 km of crust, covered by Phanerozoic sediments (Berberian 1995). Based on recent studies, the unmetamorphosed sediments of the Zagros crust are considered to be 10 to 16 km thick (Jahani et al. 2009). The overlying folds are of a detachment basal surface and are located in the Neoproterozoic-early Cambrian crystalline basin covered by folded Phanerozoic sedimentary rocks (Berberian 1995). Although the folded Zagros crust was measured by Berberian (1995) and Oveisi et al. (2009) to be more than 10 km thick, based on the reports and subsurface data of NIOC for every lithostratigraphic unit, the depth of the base of the deformed zone is estimated to be at least 14 km, roughly confirming the data of Jahani et al. (2009; i.e. 16 km).

The asymmetrical topographic surface indicates the presence of large basement faults (Berberian 1995). However, structural cross-sections of all portions of Zagros show that the fold geometry varies significantly, both horizontally and vertically, and is closely related to the changes in the mechanical behavior of the lithostratigraphic horizons and, in particular, to the presence of intermediate decollements within the sedimentary pile (Sherkati et al. 2006). Nevertheless, other structural models to a large degree indicate structural steps having similar results compared to the cross-sections (McQuarrie 2004).

2.7.1 Stratigraphy

The strata of the Zagros can be divided into two groups "Precambrian metamorphic basement" and "overlying sedimentary cover". Stöcklin (1968) introduced following stages in the development of the Zagros basin:

- Continental shelf, Neoproterozoic to Triassic
- Geosyncline, Middle Triassic-Pliocene
- Post orogeny, Pliocene to Recent

Alavi (1994) based on the lithostratigraphic facies and consequences of geological events, subdivided the Zagros strata into several tectonostratigraphic units as follows:

1. Neoproterozoic to Middle Triassic Gondwanan platform
2. Jurassic to Upper Cretaceous Neo-Tethyan Continental Shelf
3. Latest Turonian to Recent Pro-foreland Basin, marine and non-marine sediments accumulated in NW- to SE-trending, forward and backward migrating, late Cretaceous to Recent proforeland basin.

The stratigraphical studies of this zone indicate that this zone in Iran was a part of the Gondwana supercontinent during Precambrian to Middle Triassic time. After the Middle Triassic, with the development of the Neo-Tethys Ocean, it was dominated by special marine conditions (Alavi 1994).

During the Cenozoic, after the closure of Neo-Tethys and collision of the Central Iran and Zagros plates, depositional environments change to Syn-orogenic type. Though the orogenic evidence is not recorded to be younger than Pliocene, GPS and geological data indicate its continuation till Recent (Talebian and Jackson 2002; Vernant et al. 2004; Masson et al. 2005).

A thick evaporitic unit, the Hormoz salt, overlies the Precambrian-early Cambrian basement of Zagros in the eastern Zagros and the Persian Gulf, especially in the coastal Fars (Harrison 1931; Kent 1958; Player 1969; Edgell 1996). The Lower Paleozoic Hormoz salt is overlain by more than 10–15 km of younger sediments (Fig. 2.13). The exposures of salt domes in folded Zagros and some Persian Gulf Islands can be seen in Fig. 2.14 (Jahani et al. 2009).

The surface and subsurface stratigraphical evidence indicate the high distribution and thickness of Paleozoic, especially the Mesozoic and Cenozoic. A summary of the lithostratigraphic units and their main features is presented in Fig. 2.13.

2.7.2 Tectonic Subdivisions

The structural pattern of Zagros is not uniform everywhere, so the lithostratigraphy of the areas, though being similar, differed in terms of thickness of sedimentary cover and facies.

Structural studies show that the sedimentary cover overlying basement displayed various responses to compressional stresses, so that following subzone can be recognized (Fig. 2.12).

2.7.2.1 High Zagros (Northern Zagros)

This subzone is a belt 10–65 km wide, forming the highest mountains of Zagros, the "High Zagros". Its northeastern boundary is the Zagros Major Thrust and its southwestern boundary by an important thrust intersecting the northern Kino and southern Dehangan and Sabzu mountains (Motiei 1995).

There is some evidence that two folding phases affected these subzones, occurring during the Late Cretaceous and Late Miocene-Recent, respectively (Falcon 1974).

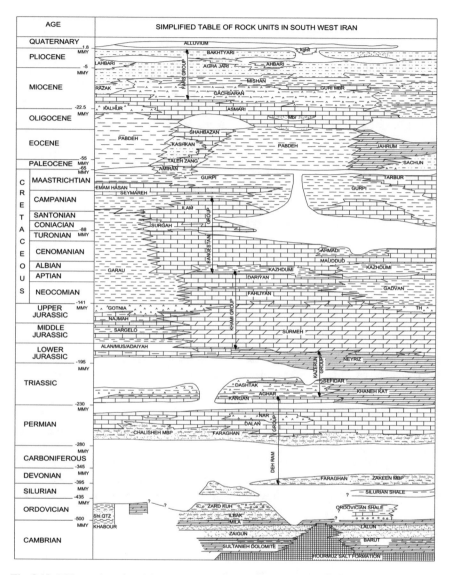

Fig. 2.13 Lithostratigraphy of the Zagros zone (modified after Motiei 1993)

The maximum intensity of the second phase occurred during Pliocene time. These two folding phases created many folds with an amplitude of more than 5 km and a wavelength of more than 8 km. The NW-SE trending folds are of oblique axial surface and NE-SE plain dip. All of the folds classified as closed folds are related to posterior and anterior folds by some SW trending thrusts.

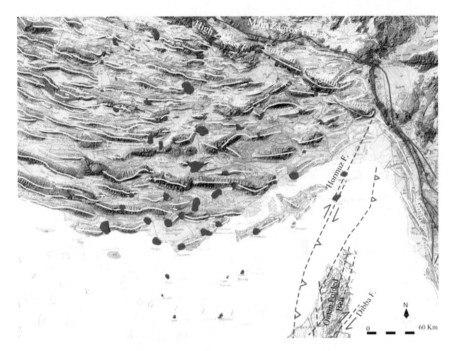

Fig. 2.14 Geological map of the eastern Fars, Iran (NIOC 1969) showing distribution of the Hormoz Salt diapirs (red areas), buried salt structure (light grey areas) and main faults (red lines)

The faults in this zone are mostly NE dipping thrusts and rarely normal faults. The eminent morphology and increasing of continental crust thickness were linked to thrusting (Motiei 1995).

The High Zagros is characterized by the presence of abundant NE dipping thrust faults whose displacements cannot be easily established and are only estimated by measuring the displacements caused by thrusting of Cambrian over Pliocene strata.

2.7.2.2   Folded Zagros

This includes areas located between a thrust intersecting Oshtorankuh, Zardkuh, Dehangan and Sabzou, and the southern margin of Susangerd, Abtiomour and Mansouri anticlines. The Folded Zagros (Stöcklin 1968) is also named 'Simply Folded Zone' (Falcon 1974) and 'Mountainous Folded Belt' (Favre 1975). The existing data indicate that the geological structure of this area is unclear, due to the effects of numerous basement faults, the presence of salt diapers, blind thrusts, embayments, and folds, so that the name "Faulted and Folded Zagros" is more appropriate.

The sedimentary cover overlying the basement is present in elongated NW-SE trending synclines with meandering and oscillating axial planes, giving the folds a sigmoidal geometry. The general structural trend follows the trend of Zagros but in the ductile Miocene sediments, basement faults activate the movement vector change of the Arabian plate in relation to Iran and finally the salt diapers' modified the general trend and system of folds.

The most local deformation can be observed in and around the salt diapers, especially along the Kazerun and Minab faults whose right-lateral displacements resulted in the intensification of orbital movements and structural inclining of fold axes linked by some researchers to the two distinct phases of deformation as follows:

- compression and creation of NW-SE trends
- shear right lateral deformation related to strike slip faults created the E-W trends.

However, these could also have resulted from continuous deformation.

The timing and mechanism of deformation is still disputed. Stöcklin (1968) and Haynes and MaQuillan (1974) believe that the main movements which resulted in Zagros folding occurred during the late Miocene or early Pliocene, i.e. long after the unification of Zagros and Central Iran, whereas, the structural evidence shows that Zagros deformation began during Late Cretaceous time, reaching its maximum during the Pliocene, resulting in size shortening by 20% (4% in Dezful embayment and 16% or a little more in the Folded Zagros (Jamali 1991). It is of note that, due to the northward compression by the Arabian Plate, the Zagros folding continued. The horizontal and vertical movements of this area were measured to be approximately 2 mm and 3.5–4.8 cm, respectively.

The Folded Zagros can be divided into 6 subzones as follows:

1.   **Izeh Zone**

This is limited to the north by the southern boundary of a thrust zone, to the south by the northern boundary of the Dezful embayment, to the east by the Kazerun Fault and to the west by the hypothetical continuation of the Balarud fault or flexure. The width of the Izeh Zone varies between 40 km in the west, 115 km in the northern Behbahan and 70 km in the east.

It is characterized by the presence of the Izeh fault which, like the Kazerun fault, is a right-lateral strike-slip transverse fault whose activities were resulted in subdividing the Izeh zone into northwestern and southeastern parts. In the former, the anticline core is composed of the Cretaceous Bangestan Group, without any oil reservoirs, whereas the latter contains Olio-Miocene rocks in the anticline cores showing less erosion and uplift. The presence of numerous oil and gas fields is a conspicuous feature of the southeastern part.

2.   **Dezful embayment**

The Dezful embayment is a structural graben in the southwestern Zagros thrust. In the western part, it is separated from Lorestan by the left-lateral Balarud fault and in the east, it is separated from the Fars area by the right-lateral Qatar-Kazerun fault. The Dezful embayment is confined by the following structural elements:

Balarud flexure

This E-W trending flexure separates the Fars Group from the Lorestan area where the Asmari formation and older rocks are exposed. Most importantly, this subzone contains a major structural depression in the south (3000-5000 m) resulting from faulting or continuation of the steep flanks of anticlines.

Mountain front flexure

The northeastern limit of the Dezful embayment is called "Mountain front flexure" (Falcon 1967). It is a NW-SE lineation beyond which the elevated Zagros anticlines disappear and are underlain by Fars Group sediments.

Qatar-Kazerun fault zone

The roughly N-S trending Qatar-Kazerun fault zone with its sinusoidal style and roughly 219 km length is a right-lateral fault belt that is associated with the left-lateral Balarud fault that created a vertical strike-slip movement which resulted in the development of the Dezful embayment. The Dezful embayment is characterized by the thick Cenozoic sequence compared to Lorestan and Fars platform. Another feature of this zone is the surface exposure of the Asmari Formation.

Based on the studies of Adams and Bourgeois (1967), the Dezful embayment formed after the Aquitanian. However, Motiei (1995) disputed that opinion and assigned it to the Burdigalian or younger. In contrast, Ghalvand (1996) based on studies of the upper Barremian to Upper Albian rocks of the Zagros basin, is of the opinion that during Early Cretaceous, late Aptian, right-lateral lineation of Qatar-Kazerun and the left-lateral Balarud fault activities resulted in a large subsiding basin developed in the Dezful embayment that is located between Lorestan and Fars. The Dezful embayment is distinguished from the surrounding areas on the basis of geological and topographic mapping and areal imagery. There are many palaeo-highs in the folded Zagros, such as Mish, Khami, Bangestan, and Haftgel. The presence of palaeo-highs, including Hendijan and Bushehr, in the Khuzestan plain, are now confirmed, although, they have geological histories; with some having been affected by Aptian erosion and others having been eroded in Cenomanian time. According to some researchers, these palaeo-highs resulted from some trends of the Arabian plate (NIOC 2012).

On moving from the mountain front area towards the Khuzestan plain, compressional impacts are gradually reduced and the style of folding changes to a more gentle type. Sometimes, axial trend changes are observed, so that the general trends of Khorramshahr and Darkhovian is north-south, like Arabian structures. The Dezful embayment is characterized by following features:

The Dezful embayment is a part of the folded Zagros and is located at the southwestern end of the thrust zones. It includes most of the young oil fields of Iran.

It is confined between three important structural elements, the left-lateral Balarud flexure zone, the Mountain front flexure zone and the right-lateral Kazerun flexure/fault (Berberian 1995).

It is characterized by approximately 3-6 km of subsidence. However, compared to surrounding areas, it was more stable and is less folded.

The activities of the right-lateral Qatar-Kazerun lineation and the left-lateral Balarud are responsible for the formation of the Dezful embayment in Aquitanian-Burdigalian and Early Cretaceous time. Confirmation of this dating requires further investigation.

3. **Fars Subzone**

Geographically, this subzone is divided into the interior and exterior Fars, with the latter being further subdivided into coastal and subcoastal Fars parts (NIOC 1977). Although most researchers consider the subzone as the area located between the Kazerun and Minab zones, Motiei (1993) believes that the western part of the Fars subzone is very different and refers it to as the "Bandar-Abbas Hinterland". It is limited to the west by the Kazerun fault zone, to the east by a hypothetical line from Bandar-e Nakhilu, across the northern Bandar-Abbas to the thrust zone. The northern and southern boundaries of the Fars subzone are defined by the thrust zone and the coast of the Persian Gulf, respectively. The most important feature of the subzone is its platformal nature, formed by the continuation of one of the Arabian basement's highs, the "Qatar-Fars high" that extends from Qatar to Fras.

4. **Bandar-Abbas Hinterland**

This is actually a part of the Fars subzone that shows considerable differences to the rest of the Fars subzone. The Bandar-Abbas Hinterland is the southeastern limit of Zagros. Its eastern and southern boundaries are believed to be Minab fault and the Zagros Fold front, respectively (NIOC 1999). The most important geological features of this subzone are as follows:

- Most anticlines contain Fars Group deposits.
- The thickness of sedimentary cover is assumed to be more than that recorded in the Fars area, actually it is similar to that in the Dezful embayment.

5. **Abadan Plain subzone**

Located at the southwestern end of the Zagros Zone, the Abadan Plain is a subzone that is limited to the north and northeast by Zagros fold-front, the southern edge of Sousangerd, Abtimour and Mansour anticlines, and to the south by the Persian Gulf and Saudi Arabia. It is a part of the Mesopotamia plain that constitutes the northern termination of the Arabian platform. It is characterized by the following:

Non-seismicity; the absence of anticlines' surface traces; a general N-S trend that contrasts with the general SE-NW trends of Zagros but compares with structures in southern Iraq, Kuwait, the northern Persian Gulf, and Northeastern Saudi Arabia; the presence of anticlines created by basement faults rather than by compressional forces.

6.   Lorestan arch subzone

The Lorestan arch is delimited to the north and southwest by the political boundary of Iran and Iraq, to the north and northeast by thrust faults, and to the south by the Balarud flexure. Its general trend is parallel to the thrust zone and its northern and geologically northwestern boundaries are compared to the Zagros anticlines (Motiei 1995).

2.8   Kopet-Dagh Basin or Zone

Located in the northeastern part of Iran, the Kopet-Dagh basin was dominated by an epicontinental sea from the Middle Jurassic onwards (Berberian and King 1981). The water depth of the western part was greater than in other areas, owing to subsidence. There is a thick succession of approximately 10 km, of marine and non-marine sediments in this zone. No significant sedimentary gaps or volcanic activity has been recorded during the Jurassic-Oligocene time interval. The present shape of this basin resulted from uplift that occurred during the late Miocene. It is also of note that this basin is very suitable condition for oil accumulation and production.

The Kopet-Dagh basin is a NW-SE trending belt that is located between Turan shield to the north and Central Iran to the south. It is up to 3000 m high, some 2000 m higher than the adjacent Turkmen plain. The minimum and width of the belt and and its highest areas are both observed in the eastern and central parts, with increasing width and decreasing height to the west. It finally terminates in the low area of the southwestern Caspian margin. It is located between N. 54°61′ and latitude E. 36° 30′.

2.8.1   *Structural Division of the Mountainous Part of the Kopeh-Dagh Orogenic Belt*

In general, the Kopeh-Dagh basin can be structurally subdivided into eastern, central and western parts as follows:

2.8.1.1   Eastern Kopeh-Dagh

Being a fold and thrust belt, the eastern Kopeh-Dagh is confined by the Sarakhs plain to the east and Quchan to the the west and evidently contains thrust faults. This subzone was affected by Paleo-Tethyan and Neo-Tethyan-related after sedimentation ceased. The final tectonic phase that affected eastern Kopeh-Dagh and produced its modern morphology is assumed to have occurred during the latest Miocene. It is the highest part of Kopeh-Dagh.

2.8.1.2 Central Kopeh-Dagh

The Central Kopeh-Dagh extends from Quchan to Bojnourd and is characterized by a high crush zone that resulted from the abundant parallel NW-SE trending, right-lateral faults (Afshar-Harb 1979). Due to the joint movements of blocks that resulted from the activities these faults, extension resulted from the opening of the Caspian and these blocks were rotated in an anti-clockwise manner, controlling the structural features of this area (Afshar-Harb 1994).

2.8.1.3 Western Kopeh-Dagh

The E-W trending western Kopeh-Dagh is located between Bojnourd and Gonbad-e Kavous. The maximum width of Kopeh-Dagh is observed in this area. Its southern part can be traced in Iran, whilst its northern part extends into Turkmenistan. The Western Kopeh-Dagh is affected by Paleo-Tethyan- and Neo-Tethyan-related events and the south Caspian opening. As a result, Jurassic sediments have been affected by the latter (Afshar-Harb 1979).

2.8.2 Tectono-Sedimentary History

The Kopeh-Dagh basin, some 600 km long and 200 km wide, is an intra-continental basin formed by the collision of Iran with the Turan microcontinent during the Middle Triassic Eo-Cimmerian Orogeny (Berberian and King 1981). During the Liassic, the Kopeh-Dagh basement located between Iran and Turan was covered by fluvial and deltaic to shallow marine coal-bearing sediments of the Kashafrud Formation (Madani 1977). The unconformity between the Triassic and Jurassic rocks (the Kashafrud Formation), can be observed in the Aghdarband.

area, assigned by researchers to the Eo-Cimmerian or even late Hercynian (Huber, 1977). Adabi and Ager (1997) considered the Kopeh-Dagh to be a part of the European plate during the Late Jurassic, a hypothesis that had been previously proposed by Ager (1988) based on studies of the Mesozoic brachiopods of Turkey. He further proposed that this basin is similar to those of the Russian platform, especially the Urals.

The Kopeh-Dagh basin contains a Jurassic to Cenozoic succession deposited in various settings from marine to continental. A geological map of Kopet Dah basin can be availed of in Fig. 2.15.

The relatively continuous and thick sedimentary succession of this basin has been divided into 15 formations, based on the series of marine transgressions and regressions observed (Huber 1977; Kalantari 1987).

The first main marine transgression from the Northwest occurred during the Middle Jurassic in the Sarakhs area deposited the thick deltaic-marine siliciclastic

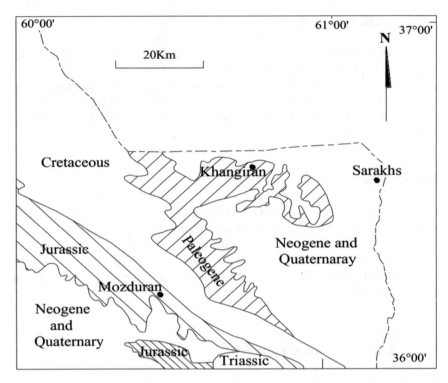

Fig. 2.15 Simplified geological map of the eastern Kopeh-Dagh (modified after Adabi and Mosavi-Harami 1989)

Kashafrud formation. This phase of deposition continued until the end of the Tithonian when the Chamanbid and Mozduran sediments were deposited (Fig. 2.16).

Towards the end of the Jurassic to Early Cretaceous (Berriasian), the Kopeh-Dagh Sea retreated to the northwest owing to the epeirogenic movements of the late Cimmerian orogeny (Huber 1977). This process subsequently caused the deposition of the thick siliciclastic fluvial sediments of the Shurijeh Formation in the eastern Kopet Dah (Moussavi-Harami and Brenner 1992). During Late Jurassic-Early Cretaceous time, the coastal areas of Mesozoic sea were located in the eastern Sarakhs. The second transgression occurred during the Early Cretaceous Barremian stage, during of which it grades into Upper Cretaceous, From the Early Cretaceous to the Upper Cretaceous, all parts of Kopeh-Dagh were dominated by a uniform sea in which thick sediments of Upper Cretaceous Tirgan-Sanganeh, Abderaz and Abtalkh Formations accumulated, as the basin steadily subsided (Kalantari 1987).

An angular disconformity between the Turonian Atamir and the Cenomanian Abderaz formations was caused by epeirogenic processes. A third regression started in the Kopeh-Dagh during middle Maastrichtian time (Fig. 2.16). The high clastic content was probably due to the shallowing cycles of the Kopeh-Dagh basin.

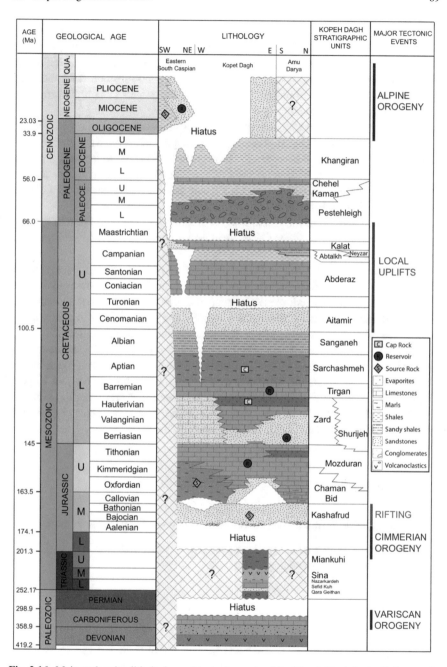

Fig. 2.16 Main rock units, lithologies and tectonic events of the Kopeh-Dagh basin (Robert et al. 2014)

Following the third transgression, late Maastrichtian movements which produced a regional unconformity at the base of the Cenozoic deposits in Iran, only caused a short marine regression (the Upper Maastrichtian Neyzar-Kalat detrital materials and the Paleocene Pesteligh Red Beds) associated with a minor disconformable contact between the Upper Maastrichtian Kalat and the Paleocene Pesteligh formations (Berberian and King 1981).

During the early Paleocene, a marine transgression occurred in the northern and eastern part of the Kopeh-Dagh basin, resulting in deposition of the Khangiran Formation (Huber 1978). After the last marine transgression occurred in the area, a regression affected the basin, with deposition of the siliciclastic Khangiran Formation, the youngest shallow marine or brackish water deposit in the eastern and central Kopeh-Dagh (Afshar-Harb 1979).

After the deposition of the Khangiran Formation, the final regression in the area resulted the sedimentation of the Cenozoic non-marine rocks, due to the uplift caused by epeirogenic movement (Berberian and King 1981).

During the Jurassic, the Kopeh-Dagh basin began to subside along the major faults recognized by Afshar-Harb (1979) as four basement faults. This subsidence caused the deposition of a thick sedimentary sequence in this basin. The thermal history of the eastern part indicates that the main reason for the very thick sedimentary cover in this area is sediment loading rather than tectonic subsidence (Fig. 2.17).

Moussavi-Harami and Brenner (1992) calculated a relatively high subsidence rate in the Upper Jurassic (98.2 m/Ma). This high rate resulted in more sedimentation of carbonates than siliciclastics. The same rate was reported from the western Kopeh-Dagh (Afshar-Harb 1979). In addition to the proposed interpretation, the high sea level recorded during the Late Jurassic and Early Cretaceous was caused by the high accommodation space (Haq et al. 1987; Hallam 1988).

2.9 Eastern Iran

Geologically, eastern Iran can be subdivided into two major parts as follows:

- The Lut block which actually encompasses most of eastern Iran
- The Flysch or colored mélange zone

2.9.1 Lut Block

The Lut block is located at the western part of Zabol-Baluch zone. In terms of geometry, it is NW trending, elongate, and approximately 900 km in length, limited to the north and south by the Darouneh and Jazmurian basins, respectively (Stocklin and Nabavi 1973). The western boundary is assumed to be the Nayband fault and the Shotori Mountains separate it from Central Iran, Tabas block. Its eastern boundary is the Nehbandan fault, being its limit with flysch zone. The oldest rocks in this

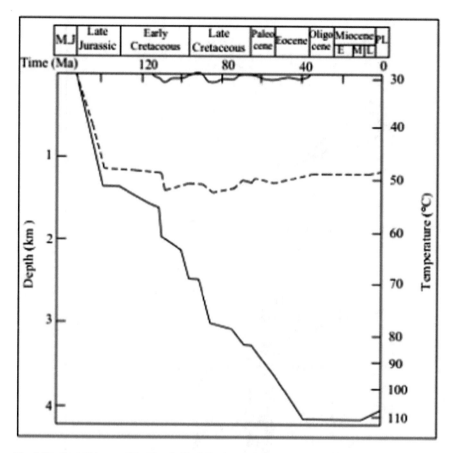

Fig. 2.17 Burial history of the Kopeh-Dagh Basin, total subsidence = continuousness line, tectonic subsidence and uplifting = dashed line (Moussavi-Harami and Brenner 1992)

block are believed to be the Neoproterozoic-lower Cambrian metamorphic, mainly schistose rocks of the southern Bazman, which is covered by either Permian or other Paleozoic units. As shown in Fig. 2.18, the Lut block is included as part of the Central Iran microcontinent.

2.9.2 Flysch or Zabol-Baluch Zone

This is confined to the west and east by the Lut and Helmand blocks, Afghanistan, respectively and is composed of thick deep sea sediments, such as siliceous to clayey shale, radiolarite, pelagic limestone; igneous rocks of basalts, spilitic basalts and diabase, and other volcanic rocks including andesite, dacite, rhyolite, acidic tuff associated with the enclaves of mostly serpentinized ultramafic rocks. In contrast to

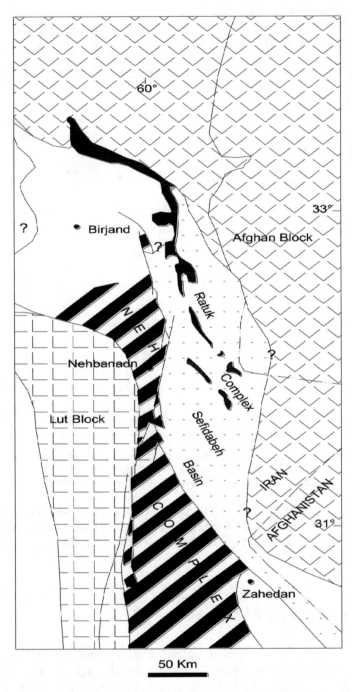

Fig. 2.18 A general view showing the structure of the eastern Iranian Flysch Zone (Tirrul et al. 1983)

the Lut Block, the Flysch zone is severely tectonized. The rocks of this zone can be classified as follows:

- Ophiolite series rocks
- Flysch sediments
- Volcanic, volcano-sedimentary and intrusive rocks.

2.9.3 Development and Evolution

Until the early Cretaceous, a uniform continent existed in this part of Iran and western Afghanistan. Towards the mid-Cretaceous, this was divided into two parts, the eastern Lut block and the western Afghan block, by the development of a N-S trending rift. These gradually diverged, resulting in the formation of a small, narrow ocean, with new oceanic crust. Then, with the Late Cretaceous movements, great masses of ophiolite-mélange were emplaced along the Makran active continental margin of Central Iran in the southern Lut, apparently due to arc-trench collision and tectonic accretion at the base of the active trench slope of the eastern High-Zagros Ocean. At the same time, the Zabol-Baluch ophiolite-mélange was emplaced along the eastern Lut margin of Central Iran. The Eocene flysch that unconformably covers the ophiolite-mélange sequences resulted from the subsequent uplift caused by collision of Afghan and Lut blocks. This collision is also resulted in development of some fractures, folding and uplift.

The important calco-alkaline plutons such as the Zahedan grantoids, were emplaced during or possibly following collision, probably resulting from the partial melting of sediments deposited between two blocks.

Finally, new magmatic activity occurred during the Neogene, concentrated along the margins of major faults. The ages of important deformation phases and magmatism in the eastern Iranian flysch zone is shown in Fig. 2.18. Based on the figure, a part of flysch zone which was deposited between the Lut and Afghan blocks can be subdivided into three subdivisions of Retok, Hozeh-Sefidabeh and Noh (Tirrul et al. 1983).

2.10 Makran Zone

Located at the southern end of the Jazmurian basin, the Makran zone is limited to the west by the Minab fault, to the south by the Oman Sea and to the east by the western border of Pakistan, though it actually extends beyond this border. At the northern limit of this zone, some E-W trending faults and thrusts can be observed, among which the most important is the Bashagerd fault. Besides these fractures, well-distributed colored mélanges are exposed, which are the oldest rocks in this zone and are Late Cretaceous-Paleocene in age. These are followed by a thick (ca. 5000 m) succession of flysch-like Eocene-Oligocene rocks consisting of sandstone, shale, and marlstone.

These rocks experienced a folding mechanism during the Oligocene. The Neogene succession is also very thick (Nabavi 1976).

According to Berberian and King (1981), Neogene sedimentation changed in the northern Makran with contemporaneous folding and thrusting, molasses and deltaic sediments sourced from the erosion of ophiolitic highs and the overlying flysch were deposited. The Miocene movements resulted in a southward shift of the basin axis and subsequent regression. The axial shift of the basin, associated with the development of younger molasses and flysch increased the northern highs and subsequently the Jazmurian embayment was formed.

Below and between the delta fans in coastal Makran, a considerable thickness of gypsiferous mudstones and marls of Middle Miocene age were deposited in a shallow but subsiding basin (Berberian and King 1981). The absence of carbonates and the presence of argillaceous sandy detritus suggest a considerable relief and subaerial erosion of northern mountains (Huber 1978).

These studies also showed that:

- The oceanic crust of the Oman Sea is subducting beneath the Lut block at a rate of 5 cm/year (Le Pichon1968).
- The Makran area is considered as a gradually subsiding area, so that the thickness of clayey and sandy sediments may exceed 1000 m (Stöcklin 1975), evidently indicated by gravitational tectonic effects (Reyer and Mohafez 1972).
- The ocean crust flexure beneath the Makran is probably located at the beginning of Jazmurian embayment where it is subducting with steep dip (Brosses and Moein-Vaziri 1980).
- Young volcanoes of northern Jazmurian were created by ocean crust melting at depths exceeding 100 km (Brosses and Moein-Vaziri 1980).
- An abyssal plain probably existed during Late Cretaceous o Early Miocene time whose continental margin is the present Jazmurian. The evidence for this is first because this zone contains rare occurrences of oceanic crust, colored mélange, and second because ophiolitic enclaves can be observed between conglomeratic rocks. Therefore, the basin floor of Makran is probably of oceanic type and the flysch rocks were probably derived from the erosion of older oceanic crust.

References

Adabi MH, Ager DV (1997) Late Jurassic brachiopods from north-east Iran. Palaeontology 40:355–362

Adabi MH, Moussavi-Harami R (1989) Geomorphology of eastern Kopet-Dagh Basin. In: Popoli MH (ed), International Geographical Seminar, Mashhad, Iran 1:87–104

Adams C, Bourgeois E (1967) Asmari biostratigraphy: Geological and Exploration Div., Iranian Oil Offshore Company: Report 1074

Afshar-Harb A (1979) The stratigraphy, tectonics and petroleum geology of Kopet-Dagh Region. Unpublished Ph.D. thesis, Imperial College of Sciences and Technology, University of London, p 316

Afshar-Harb A (1994) Geology of Kopet Dagh. In: Treatise on the geology of Iran, Hushmandzadeh A (ed). Tehran: Geological Survey of Iran, p 275

Ager DV (1988) Mesozoic Turkey as part of Europe. In: Audley-Charles MG, Hallam A (eds) Gondwana and Tetheys, Special publication no. 37. Geological Society, London, pp 241–245

Aghanabati A (1977) Etude geologigue de la region de Kalmard (W. Tabas). Geological Survey of Iran, Report No.35, p 230

Aghanabati A (2000) The main sedimentary-structural basins of Iran. Geological Survey of Iran, Tehran (in Persian)

Aghanabati A (2004) Geology of Iran. Geological Survey of Iran, Tehran

Alavi M (1991) Sedimentary and structural characteristics of the Paleo-Tethys remnants in northeastern Iran. Geol Soc Am Bull 103(8):983–992

Alavi M (1994) Tectonics of the Zagros orogenic belt of Iran: new data and interpretations. Tectonophysics 229(3–4):211–238

Amidi SM, Emami MH, Michel R (1984) Alkaline character of Eocene volcanism in the middle part of Central Iran and its geodynamic situation. Geol Rundsch 73(3):917–932

Amiri M (1995) Petrography of the Almouglagh. University of Tarbiat Moalem, Tehran, p 231

Asadpour M (2000) Petrology and geochemistry of ultramafic to intermediate rocks of Ghoshchi area. M.Sc. Thesis, Shahid Beheshti University, Tehran, Iran

Assereto R (1963) The Paleozoic Formations in Central Elburz(Iran) (Preliminary Note). Riv Ital Paleontol Stratigr 69:503–543

Babakhani A (1981) Study of petrography and geochemistry of Azerbaijan Nepheline-syenite and Phenolites. MSc thesis, Tehran University, p 135

Behnia P (1995) Petrogenesis of granitoids of Qooshchi area: A process of alkaline metasomatism. Master's Thesis, Tehran University, p 210

Berberian F, Berberian MJ (1981) Tectono-plutonic episodes in Iran. Zagros Hindu Kush Himalaya Geodyn Evol 1;3:5–32

Berberian F, Muir ID, Pankhurst RJ, Berberian M (1982) Late Cretaceous and early Miocene Andean-type plutonic activity in northern Makran and Central Iran. J Geol Soc 139(5):605–614

Berberian M, King GCP (1981) Towards a paleogeography and tectonic evolution of Iran. Can J Earth Sci 18(2):210–265

Berberian M (1995) Master blind thrust faults hidden under the Zagros folds: Active basement tectonics and surface morphotectonics. Tectonophysics, 241:193–224

Berthier F, Billiault JP, Halbronn B, Maurizot P (1974) Etude stratigraphique, pe´trologique et struc-turale de la re´gion de Khorramabad (Zagros cen-tral). Ph.D. thesis, Univ. Joseph Fourier,Grenoble, France, p 181

Boulin J (1991) Structures in Southwest Asia and evolution of the eastern Tethys. Tectonophysics 196:211–268

Brosses R, Moein-Vaziri H (1980) Le volcanisme du Kouh-e-Tchah-e-Shah, au nord du Makran (Iran), 1:69

Clark GC, Davies RG, Hamzehpour B, Jones CR (1975) Explanatory text of the Bandar-e-Pahlavi quadrangle map, 1:25O,OQO: Geological Survey of Iran, Tehran, Iran, p 198

Crawford AR (1972) Iran, continental drift and plate tectonics. In 24th International Geological Congress, vol 3, pp 106–112

Darvishzadeh A (1991) Geology of Iran. Neda Publication, Tehran, p 901

Dedual E (1967) Zur geologie des mittleren und unteren Karaj-Tales, zentral-Elburz (Iran). Mitt Geol Inst ETH Univ Zurich 79:45–75

Dewey JF, Pitman Iii WC, Ryan WB, Bonnin J (1973) Plate tectonics and the evolution of the Alpine system. Geol Soc Am Bull 84(10):3137–3180

Didon J, Germain YM (1976) Le Sabalan, Volcan Plio-Quaternaire de l Azerbaidjan oriental (Iran): E tude geologique et petrographique de le difice et de son environment regional. Grenoble: Université de Grenoble. Dimitrijevic, M D (1973) Geology of Kerman Region. Gel Surv Iran, Report Number Yu/52, p 334

Dostal J, Zerbi M (1978) Geochemistry of the Savalan volcano (northwestern Iran). Chem Geol 22(1978):31–42

Edgell HS (1996) Salt tectonism in the Persian Gulf basin. Geol Soc, Lond, Spec Publ 100(1):129–151

Eftekharnezhad J (1975) Brief description of tectonic history and structural development of Azerbaijan, Field excursion guide, No 2 Note A Sym, Geodynamic of southeast Asia, Tehran, pp 469–478

Eftekharnezhad J (1980) Subdivision of Iran into different structural realms with relation to sedimentary basins (in Farsi). Bull Iran Pet Inst 82:19–28

Eftekharnezhad J (1996) Geology of Iran and neighboring countries. Shahid Beheshti University curriculum

Eftekharnezhad J, Behroozi A (1991) Geodynamic significance of recent discoveries of ophiolites and Late Paleozoic rocks in NE-Iran (including Kopet Dagh). Abh Geol B-A 38:89–100

Emami MH (1981) Geologie de la region de Qom-Aran (Iran);Contribution a petude dynamique et geochimique du Volcanisme tertiaire de l, Iran central. Theses. Sciences naturelles Univ Sc Et Medicale de Grenoble, p 489

Falcon NL (1967) The geology of the north-east margin of the Arabian basement shield. Adv Sci 24:31–42

Falcon NL (1974) Southern Iran: Zagros Mountains. Geol Soc, Lond, Spec Publ 4(1):199–211

Favre G (1975) Structures in the Zagros Orogenic Belt OSCO (No. 1233E52). Report

Forster H, Fesefeldt K, Kiirsten M (1972) Magmatic and orogenic evolution of the Central Iranian volcanic belt. Proc 24th Int Geol Congr 2:198–210

Gaetani M, Angiolini L, Ueno K, Nicora A, Stephenson MH, Sciunnach D, Rettori R, Price GD, Sabouri J (2009) Pennsylvanian-Early Triassic stratigraphy in the Alborz Mountains (Iran). Geol Soc, Lond, Spec Publ 312(1):79–128

Ghalvand H (1996) Lithostratigraphy and biostratigraphy of the Darian and Kazhdomi in the southwestern Iran, Fars and Dezful Embayment. MSc thesis, Faculty of Earth Sciences, Shahid Beheshti University

Ghasemi A, Talbot CJ (2006) A new scenario for the Sanandaj-Sirjan zone (Iran). J Asian Earth Sci 26:683–693

Ghavidel-Syooki M (1995) The Sarchahan Formation. Quat Geosci J 4(15–16):74–89.

Ghorbani G (2005) Petrology and Petrogenesis of igneous rocks in south of Damghan. PhD thesis, Faculty of Earth Sciences, Shahid Beheshti University

Ghorbani M (1999) Petrological investigations of Tertiary-Quaternary magmatic rocks and their metallogeny in Takab area. PhD thesis, Shahid Beheshti University

Ghorbani M (2002) Petrological study of Tertiary-Quaternary magmatic rocks and metallogeny of Takab area. PhD. Thesis, Shahid Beheshti University

Ghorbani, M (2013) Economic geology of Iran, vol. 581. Mineral deposits and natural resources: Springer

Ghorbani M (2014) Geology of Iran. The magmatism and metamorphism vol., Arian zamin publication

Ghorbani, M. (2019) Lithostratigraphy of Iran. Springer

Haghipour A, Pelissier G (1968) Geology of the Posht-e Badam-Saghand area, (East- central Iran), Geol Surv Iran, Note No 48, 144, 51 Figs, 3 Pls, Map

Haghipour A, Farooqui M, Amidi SM, Aghanabati A (1979) The destructive Tabas earthquake of September 16/1978, East Iran, a preliminary report. Proc Annu Eng Geol Soil Eng Symp. 17:1–12

Hallam A (1988) A re-evaluation of the Jurassic eustasy in the light of new data and the revised Exxon curve. In: Wilgus CK, Hastings BS, Kendall GSC, Posamentier HW, Ross CA, Andvan Wagoner JC (eds) Sea- level changes: an integrated approach: Tulsa, Society of Economic Paleontologists and Mineralogists, Special Publication 42, pp 71–108

Haq BU, Hardenbol J, Vail P (1987) Chronology of the fluctuating sea levels since the Triassic. Science, 235:1156–1167

Harrison H (1931) Salt domes in Persia. J Inst Petrol Technol 17:300–320

Hassanzadeh J (1993) Metallogenic and tectonomagmatic events in the SE sector of the Cenozoic active continental margin of central Iran (Shahr-e Babak area, Kerman Province). Ph.D. thesis, Los Angeles, University of California, 204 p

Haynes SJ, McQuillan H (1974) Evolution of the Zagros suture zone, southern Iran. Geol Soc Am Bull 85(5):739–744

Hooper RJ, Baron I, Hatcher Jr RD, Agah S (1994) The development of the southern Tethyan margin in Iran after the break up of Gondwana: implications of the Zagros hydrocarbon province. Geosciences 4, 72–85

Houshmandzadeh A (1977) Ophiolites of Southeast Iran and their genetic problems. Geological Survey of Iran, Tehran, Internal Report (in Persian)

Huber H (1977) Geological cross sections, north-central Iran: Tehran, National Iranian Oil Company, scale 1:500 000

Huber H (1978) Geological map of Iran: Tehran, National Iranian Oil Co., 6 sheets, scale 1:1,000,000, digital version in vector format

Huckriede R, Kürsten M, Venzlaff H (1962) Geotektonische Kartenskizze des Gebietes zwischen Kerman und Sagand, (Iran), Bundesanstalt für Bodenforschung

Innocenti F, Manetti P, Mazzuoli R, Pasquare G, Villari L (1982) Anatolia and north–western Iran. In: Thorpe RS (ed) Andesites. Orogenic Andesites and Related Rocks. Wiley, Chichester, pp 327–349

Innocenti F, Mazzuoli R, Pasquare G, Di Brozolo FR, Villari L (1976) Evolution of the volcanism in the area of interaction between the Arabian, Anatolian and Iranian plates (Lake Van, Eastern Turkey). J Volcanol Geotherm Res 1(2):103–112

Jahani S, Callot J, Letouzey J, Frizon de Lamotte D (2009) The eastern termination of the Zagros Fold-and-Thrust Belt, Iran: Structures, evolution, and relationships between salt plugs, folding, and faulting. J TectonS 28

Jahani S, Frizon de Lamotte D, Letouzey J (2009) Salt activity and Halokinesis in the Zagros Fold-thrust Belt and Persian Gulf (Iran), Publisher European Association of Geoscientists & Engineers, Pages cp-125-00012, Shiraz 2009-1st EAGE International Petroleum Conference and Exhibition

Jamali F (1991). How and level of the crust shortening in southwestern Iran. J Earth Sci, Geological Survey of Iran, First Year (1991 (No. 2)

Jenny JG (1977) Géologie et stratigraphie de l'Elbourz oriental entre Aliabad et Sharud, Iran: these, Imprimerie nationale

Jung D, Kursten M, Tarkian M (1976) Post-Mesozoic volcanism in Iran and its relation to the subduction of the Afro-Arabian under the Eurasian plate. In: Pilger A, Rosler A (eds) A far between continental and oceanic rifting. Schweizerbartsche Verlagbuchhandlung, Stuttgart, pp 175–181

Kadinsky-Cade K, Barazangi M (1982) Seismotectonics of southern Iran: the Oman line. Tectonics 1(5):389–412

Kalantari A (1987) Biofacies map of Kopet Dagh region. Exploration and Production, National Iranian Oil Company (NIOC), Tehran

Kent PE (1958) Recent studies of south Persian salt diapirs. AAPG Bull 42:2951–2972

Kheirkhah M, Neill I, Allen MB (2015) Petrogenesis of OIB-like basaltic volcanic rocks in a continental collision zone: Late Cenozoic magmatism of Eastern Iran. J Asian Earth Sci 106:19–33

Le Pichon X (1968) Sea-floor spreading and continental drift. J Geophys Res 73:3661–3697

Lotfi M (1975) Geological and petrological studies of the northeast of Mianeh region (East Azerbaijan). MSc thesis, Tehran University, p 217

Madani M (1977) A study of sedimentology, stratigraphy and regional geology of the Jurassic rocks of Eastern Kopet-Dagh, NE Iran. Ph.D. thesis, University of London, 246 pp. (Unpublished)

Masoudi F (1997) Contact metamorphism and pegmatite development in the region SW Anark, Iran. PhD Thesis. University of Leeds

Masson F, Chéry J, Hatzfeld D, Martinod J, Vernant P, Tavakoli F, Ghafory-Ashtiani M (2005) Seismic versus aseismic deformation in Iran inferred from earthquakes and geodetic data. Geophys J Int 160(1):217–226

McQuarrie N (2004) Crustal scale geometry of the Zagros fold–thrust belt, Iran. J Struct Geol 26(3):519–535

Moein-Vaziri H (1996) An introduction of geology of magmatism in Iran. Tehran University of Teacher Education, Tehran, p 440

Moghadam HS, Khedr MZ, Arai S, Stern RJ, Ghorbani G, Tamura A, Ottley CJ (2015) Arc-related harzburgite–dunite–chromitite complexes in the mantle section of the Sabzevar ophiolite, Iran: a model for formation of podiform chromitites. Gondwana Res 27(2):575–593

Mohajjel M (1999) Analysis of the relationship between metamorphism and deformation in magmatic rocks of Dorud Azna region: a guide to the interpretation of tectonic events in Sanandaj-Sirjan, Third Conference of the Geological Society of Iran, Shiraz University, pp 582 586

Mohajjel M (2000) Fabric of Azna syn-tectonic granite in Sanandaj-Sirjan zone, western Iran. 31st International Geological Congress, Abstract, Rio de Janeiro, Brazil

Mohajjel M, Sahandi MR (1999) Tectonic evolution of the Sanandaj-Sirjan zone in northwest Iran, introducing the sub-zones. Sci Q J Geosci 31–32:28–49

Motiei H (1993) Geology of Iran: stratigraphy of Zagros. Geological Survey of Iran, Tehran

Motiei H (1995) Petroleum geology of Zagros. Geological Survey of Iran (in Farsi), p 589

Moussavi-Harami R, Brenner RL (1992) Geohistory analysis and petroleum reservoir characteristics of Lower Cretaceous (Neocomian) sandstones, eastern Kopet-Dagh Basin, northeastern Iran. AAPG bulletin 76(8):1200–1208

Nabavi M (1976) An introduction to geology of Iran. Geological Survev of Iran, Tehran, p 109

NIOC (1969) Geological map of South-West Iran, 1:1000000. Geological and Exploration Division, Sheet No. 5

NIOC (1977) Tectonic map of southwest Iran, 1: 2500000 Scale. NIOC, Tehran, Iran

NIOC (1999) Geological Map of Iran quadrangle No. I-13 Bandar Abbas,1:250000. Unpublished

NIOC (2012) Fault and lineament demarcation and maping in Coastal Fars Province (Between Kazeroon and Razak Faults), Pars Geological research center (Client: Exploration Directorate of National Iranian Oil Company)

NIOC (2017) Paleontology and biostratigraphy of Paleozoic rocks of Zagros and Central Iran Basins, Pars Geological research center (Client: Exploration Directorate of National Iranian Oil Company)

Noghreyan M, Tabatabaei SM (1995) Study of Kolah Ghazi Ganitoitoid with emphasis on its contact metamorphism. Sci Q J Geosci, Geological Survey of Iran

Nogol-e-Sadat A (1993) Seismotectonic Map of Iran, scale, 1: 1,000,000; Trea-tise on the geology of Iran, 128. Geol Surv Iran, Tehran (in Persian)

Oveisi B, Lavé J, Van Der Beek P, Carcaillet J, Benedetti L, Aubourg C (2009) Thick-and thin-skinned deformation rates in the central Zagros simple folded zone (Iran) indicated by displacement of geomorphic surfaces. Geophys J Int 176(2):627–654

Pearce JA, Harris NBW, Tindle AG (1984) Trace element discrimination diagrams for the tectonic interpretation of granitic rocks. J Petrol 25:956–983

Player RA (1969) Salt plug study, N. I. O. C., Report 1146

Radfar J (1987) Geology and petrology study of Granitoid rocks of Astaneh-Gusheh area (Shazand region). MSc thesis, Science Faculty, Tehran University

Reyer D, Mohafez SA (1972) First contribution of the NIOC-ERAP agreements to the knowledge of Iranian geology. Edition Techniqs, Paris, p 58

Rieben, H., 1935. Contribution à la géologie de l'Azerbeidjan Persan (Doctoral dissertation, Neuchatel.)

Robert AM, Letouzey J, Kavoosi MA, Sherkati S, Müller C, Vergés J, Aghababaei A (2014) Structural evolution of the Kopeh Dagh fold-and-thrust belt (NE Iran) and interactions with the South Caspian Sea Basin and Amu Darya Basin. Mar Pet Geol 57:68–87

Ruttner A, Nabavi M, Hajian J (1968) Geology of the Shirgesht area (Tabas area, East Iran). Tehran: Geological Surveyof Iran. Report 4, 133 pp

Sabzehei M (1996) An introduction to general geological features of metamorphic complexes in southern Sanandaj-Sirjan zone. Geol Surv Iran (Unpublished)

Scheffel RL, Wernert SJ (1980) Reader's digest natural wonders of the world. readers digest

Sella GF, Dixon TH, Mao A (2002). A model for recent plate velocities from space geodesy. J Geophys Res 107(B4):11-1–11-30

Şengör AC, Yilmaz Y (1981) Tethyan evolution of Turkey: a plate tectonic approach. Tectonophysics 75(3–4):181–241

Sengor AMC (1990) A new model for the late Palaeozoic-Mesozoic tectonic evolution of Iran and implications for Oman. In: Robertson AH, Searle MP, Ries AC (eds) The geology and tectonics of the Oman region. Geological Society Special Publication 49:797–831

Sepahigaro AA (1999) Petrology of Alvand plutonic rocks with special putlook on granitoids. Ph.D thesis, Tarbiat moallem university

Setudehnia A (1975) The Paleozoic sequence at Zard Kuh and Kuh-e-Dinar. Bull Iran Pet Inst 60:16–33

Shahabpour J (2007) Island-arc affinity of the central Iranian volcanic belt. J Asian Earth Sci 30(5–6):652–665

Shahabpour J, Doorandish M (2008) Mine drainage water from the Sar Cheshmeh porphyry copper mine, Kerman, IR Iran. Environ Monit Assess 141:105–120

Shahbazi H, Siebel W, Pourmoafee M, Ghorbani M, Sepahi AA, Shang CK, Vousoughi-Abedini M (2010) Geochemistry and U-Pb zircon geochronology of the Alvand plutonic complex in Sanandaj-Sirjan Zone (Iran): New evidence for Jurassic magmatism. J Asian Earth Sci 39(6):668–683. https://doi.org/10.1016/j.jseaes.2010.04.014

Shamanian-Esfahani GH (1974) Geochemistry, Mineralogy and Fluid Inclusionsin Tangestan Mining, Nezam Abad, Markazi province, Iran. MSc thesis, Shiraz University

Sheikholeslami MR (2002) Évolution structurale et métamorphique de la marge sud de la microplaque de l'Iran central: les complexes métamorphiques de la région de Neyriz (zone de Sanandaj-Sirjan), Thèse de doctorat en Géosciences marines

Sheikholeslami R, Bellon H, Emami H, Sabzehei M, Pique A (2003) New structural and K-40-Ar-40 data for the metamorphic rocks in Neyriz area (Sanandaj-Sirjan zone, southern Iran). Their interest for an overview of the Neo-Tethyan domain in the Middle East. Comptes Rendus Geosci 335(13):981–991

Sherkati S Letouzey J, Frizon de Lamotte D (2006) Central Zagros fold-thrust belt (Iran): New insights from seismic data, field observation, and sandbox modeling. Tectonics, 25(4)

Stampfli GM (1978) Etude géologique generale de l'Elbourz oriental au sud de Gonbad-e-Qabus (Iran NE). Doctoral dissertation, Université de Genève

Stöcklin J (1968) Structural history and tectonics of Iran: a review. Am Assoc Pet Geol Bull 52(7):1229–1258

Stöcklin J (1974a) Northern Iran: Alborz Mountains. Geol Soc, Lond, Spec Publ 4(1):213–234

Stöcklin J (1974b) Possible ancient continental margins in Iran. In The geology of continental margins. Springer, Berlin, Heidelberg, pp 873–887

Stöcklin J (1975) On the origin of ophiolite complexes in the southern Tethys region. Tectonophysics 25:303–322

Stöcklin J (1977) Structural correlation of the Alpine ranges between Iran and Central Asia

Stöcklin J, Nabavi MH (1973) Tectonic map of Iran. Geol Surv Iran 1:5

Takin M (1971) Geological history and tectonics of Iran, a discussion of continental drift in the Middle East since the Early Mesozoic. Geological Survey of Iran, internal report

Talebian M, Jackson J (2002) Offset on the Main Recent Fault of NW Iran and implications for the late Cenozoic tectonics of the Arabia-Eurasia collision zone. Geophys J Int 150(2):422–439

Tatar, M., Hatzfeld, D., Martinod, J., Walpersdorf, A., Ghafori-Ashtiany, M. and Chery, J., 2002. The present-day deformation of the Central Zagros from GPS measurements. Geophysical Research Letters, Geophys Res Lett. 29. 10.1029/2002GL015427

Tirrul R, Bell IR, Griffis RJ, Camp VE (1983) The Sistan suture zone of eastern Iran. Geol Soc Am Bull 94:134–150

Valizadeh MV, Zarian S (1976). A petrological study of the Almogholagh, Asadabadand Hamedan plutons. J Sci, Islam Repub Iran 8, 49–59 (in Persian with

Valizadeh M, Ghasemi H (1993) Petrogenesis of intrusive rocks of Boiin-Miandasht, Southeast of Alighoodarz. Sci Q J Geosci 7:74–83

Valizadeh MV, Sadeghian M (1996) Petrogenesis of Alvand granitoid complex. Geosci Iran 5(19):14–31

Vernant P, Nilforoushan F, Hatzfeld D, Abbassi MR, Vigny C, Masson F, Nankali H, Martinod J, Ashtiani A, Bayer R, Tavakoli F (2004) Present-day crustal deformation and plate kinematics in the Middle East constrained by GPS measurements in Iran and northern Oman. Geophys J Int 157(1):381–398

Walpersdorf A, Hatzfeld D, Nankali H, Tavakoli F, Nilforoushan F, Tatar M, Vernant P, Chéry J, Masson F (2006) Difference in the GPS deformation pattern of North and Central Zagros (Iran). Geophys J Int 167(3):1077–1088

Zanchi, A., Zanchetta, S., Berra, F., Mattei, M., Garzanti, E., Molyneux, S., Nawab, A. and Sabouri, J., 2009. The Eo-Cimmerian (Late? Triassic) orogeny in North Iran. Geological Society, London, Special Publications, 312(1), pp 31–55

Zandkarimi K, Najafian B, Bahrammanesh M, Vachard D (2014) Permian foraminiferal biozonation in the Alborz Mountains at Valiabad section (Iran). Permophiles 60:10–16

Zandkarimi K, Najafian B, Vachard D, Bahrammanesh M, Vaziri SH (2016) Latest Tournaisian–late Viséan foraminiferal biozonation (MFZ8–MFZ14) of the Valiabad area, northwestern Alborz (Iran): geological implications. Geol J 51(1):125–142

Zandkarimi K, Vachard D, Cózar P, Najafian B, Hamdi B, Mosaddegh H (2017a) New data on the Late Viséan-Late Serpukhovian foraminifers of northern Alborz, Iran (biostratigraphic implications). Rev Micropaléontol 60(2):257–278

Zandkarimi K, Vachard D, Najafian B, Hamdi B, Mosaddegh H (2017b) Viséan-Serpukhovian (Mississippian) archaediscoid foraminifers of the northern Alborz, Iran. Neues Jahrbuch für Geologie und Paläontologie-Abhandlungen 286(1):105–123

Zandkarimi K, Vachard D, Najafian B, Mosaddegh H, Ehteshami-Moinabadi M (2019) Mississippian lithofacies and foraminiferal biozonation of the Alborz Mountains, Iran: Implications for regional geology. Geol J 54(3):1480–1504

Chapter 3
Faults and Tectonic Phases of Iran

Abstract In this chapter, the faults of Iran are classified and described within the respective structural zones, and then the orogenic phases of Iran are defined. The behavior and characteristics of each fault are also designated. The tectonic and orogenic phases are described in this chapter, and their role in formation and alteration of various structures are analyzed. The seismotectonics of Iran are also discussed and finally, all earthquakes are listed, together with their date of occurrence and magnitude; moreover, a distribution map of earthquakes in Iran is given separately.

Keywords Orogenic phases · Faults · Seismotectonics · Earthquakes · Iran

3.1 Faults of Iran

The faults of Iran are a group of linear geological structures, some of which significantly affected the geology of the country, bounding most of the structural zones, metallogenic provinces and districts. Most facies variations, and all aspects of basin evolution have resulted from fault movements. The intrusion of magmatic masses and volcanic activity have also been directly or indirectly related to the faults, specially their junctions.

The faults of Iran can be classified with reference to various characters but here they are classified here based on their trend and geographical position (Table 3.1).

The Iranian faults are classified based on the trends as follows:

1. **N-S trending faults**

These faults resulted from the Neoproterozoic Pan-African or Katangian Orogeny (Nabavi 1976), and are of right-lateral strike-slip mechanism, similar to some of geological structures of Iran and the Middle East, e.g. longitudinal extension of the Lut desert, southern Islands of Iran, Qatar, United Arab Emirates and Oman

© The Author(s), under exclusive license to Springer Nature Switzerland AG 2021
M. Ghorbani, *The Geology of Iran: Tectonic, Magmatism and Metamorphism*,
Earth and Environmental Sciences Library,
https://doi.org/10.1007/978-3-030-71109-2_3

Table 3.1 Classification of Iranian faults based on their structural position (based on data of Berberian 1976a, 1981, 1995; NIOC 1977; Afshar-Harb 1979, 1994; Sepehr 2001, Sepehr and Cosgrove 2004, 2007; Aghanabati 2004; Jahani et al. 2009)

Alborz	Central Alborz	Caspian (Khazer); North Tehran; Mosha-Fasham, North Alborz, Semnan-Garmsar; Kandovan; Baijan; Jirandeh; Takistan; Taleghan; Firuz-Kuh; Ray; Kahrizak; Aivanaki; Pishva; Alamout-Rud; Northern Qazvin; Abyek; Galandroud; Kojur; Hezar; Leleh-Band; Orim; Zarin-Kuh; Astaneh; Kahar
	Eastern Alborz	Atari; Badeleh; Damghan; Jajarm; Khosh-Yeilagh; Chahar-Deh
	Western Alborz and Azerbaijan	Soltanieh; Urmia-Zarrinroud; Astara; Tarom; Tabriz; Zanjan; Mianeh; Roudbar; Neor; Lahijan; Manjil
Kopeh-Dagh		Kor-Khud; Nabia; Takal-Kuh; Maraveh-Tapeh; Ghelli; Jajarm; Kashaf-Roud; Baghan-Robat; Sherkan-Lou; Quchan-Chapanlou; Khaboushan; Gharreh-Cheh; Rahvard; Palkano; Lojeli; Ghulanlou
Zagros		Main Zagros Thrust; Balaroud; Dena (Dinar); Minab; Ardal; Zard-Kuh; Aghajari; Maroun; High Zagros; Zagros mountain front; Kazerun; Izeh; Mengharak; Sabzpoushan; Sarvestan; Shiraz; Nezam-Abad; Razak; Main recent Zagros; Dorud; Nahavand and Garoum; Sahneh; Bakhtiari-Kuhrang
Central Iran and Sanandaj-Sirjan		Mayami (Shahroud); Doruneh; Binaloud; Anjilou; Toroud; Kalmard; Posht-e-Badam; Qom-Zafreh; Aindes; Dehshir; Sarvestan; Shahdad; Kuhbanan; Jorjafak; Golbaft; Nayband; Tabas; Avaj; Eshtehard; Ipak; Sabzevar; Neyshabur
Eastern and Southeastern Iran		Nehbandan; Bashagard; Harirud; Khuzestan; Gowk

(Aghanabati 2004). These include the Harirud, Nehbandan, Nayband and Qatar-Kazerun faults.

2. **NW-SE trending faults**

These faults which are attributed to consequences of the Pan-African orogeny have the same trend as the main Zagros Fault. Like group 1, the faults are right-lateral strike-slip faults. It can be assumed that the latter group is the same as the former group, the trends of which were modified by subsequent orogenic phases. A lot of old and new structures of Iran such as faults and fractures of Paleozoic, Mesozoic and Cenozoic age have the same trend as this group, confirming the trend changes of these faults.

3. NE-SW trending faults

Having the same trend as the northern Alborz, these faults are mostly seen in northern and northwestern Iran and are not observed in the southern parts. They are the products of a shearing left-lateral mechanism. Many researchers (e.g. Nabavi 1976; Aghanabati 2004) believe that they were created by the Caledonian orogeny, while the present author attributes them to the Early Hercynian orogeny, an interpretation supported by their main concentration in northern Iran. In fact, Iran, being far from the Caledonian collision, suffered only epeirogenic movements and regional regression (Berberian and King 1981) without any major faulting. On the other hand, if they are assigned to the Caledonian orogeny, no other geological activities accompanying the orogeny are observed. The faults of Iran can be classified considering their geographical position as Fig. 3.1. The major active Iranian faults are illustrated in Fig. 3.2.

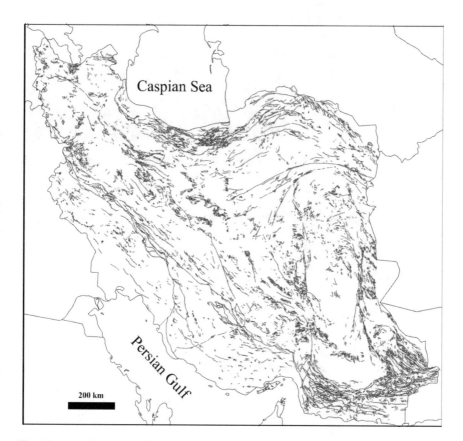

Fig. 3.1 General scheme of the Iranian faults (based on the tectonic data of GSI)

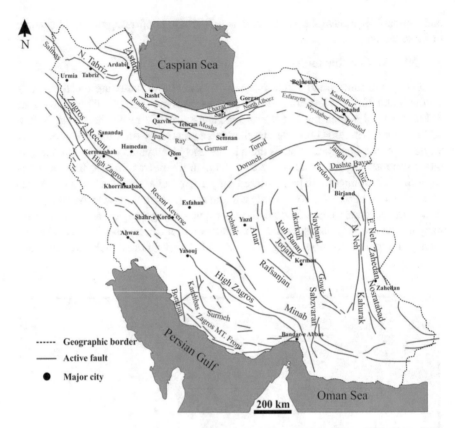

Fig. 3.2 Main active faults of Iran (adopted after Berberian 1976a, c 1981, 1995, 1997, 2005; Berberian and Yeats 1999, 2001)

In the following, the faults of each structural zone of Iran are described:

3.2   Alborz

3.2.1   Central Alborz

Caspian (Khazar) fault
This fault forms the border between the northern Alborz and the south Caspian plains (Berberian et al. 1992). Presumably indicating the Paleotethyan suture, it is approximately 550–600 km long, has an E-W trend and northward dip, and separates Alborz from the southern Tertiary Caspian basin. This fault was presumably active from the Cambrian (Nabavi 1976). The first relative downward movement of the northern segment is believed to have occurred in the Eocene (Aghanabati 2004),

while the beginning of subsidence dates back to Jurassic time. However, it seems that the Caspian fault is now highly active, causing most of the earthquakes affecting northern Iran. For example, Mazandaran and Guilan resulted from the displacements that occurred along with it. Having southward dip, the fault resembles a southwardly-inclined arc owing to the bending of its central part. In the western Lahijan, a left-lateral, NE-SW trending fault "Sefidroud" has displaced the Caspian Fault. The Caspian fault which consists of several segments is believed to play a significant role in the morphotectonics of Alborz (Nazari and Shahidi 2011).

North Tehran fault

Located on the southern slopes of Central Alborz, northern Tehran, this is considered to be one of the most hazardous faults of Tehran. The North Tehran fault is characterized by a 110 km length, a NE-SW trend, and a 10–80°N dip. It seems that the fault is inactive in the western areas as a solitary structure and attached to the Mosha-Fahsham Fault in the eastern areas. The Northern Tehran fault brings Eocene volcanics into contact with Pliocene-Quaternary alluvial deposits. It shows a mostly compressional mechanism, changing eastward to an E-W trending, left-lateral strike-slip compressional fault (Nazari and Shahidi 2011).

It seems that a roughly 3 m-long fracture affected the Holocene-Upper Pliocene sediments in the central part. This fracture with a N115 E trend, 30 N dip of the fault plane and reverse displacement resulted from alternating compressional phases.

Mosha-Fasham fault

The Mosha-Fasham Fault (Dellenbach 1964) which is located on the southern slopes of the Alborz is characterized by being approximately 180 km length, forming a sigmoidal pattern and a depth of *ca.* 10 km. It apparently affected the drainage pattern of the area. Its strike changes along its length, so that it shows an E–W trend in the eastern segment and WNW–ESE in the southwestern one, while the central segment is characterized by a double-bend towards a northwest strike. The Mosha-Fasham Fault is primarily considered as a north-dipping thrust (35°–70°) which changes to a left-lateral strike-slip fault with a normal component owing to the transformation of stress vector in Late Pliocene- Early Quaternary time (Ritz et al. 2006; Nazari and Ritz 2008; Nazari 2006). A left-lateral displacement occurred along the fault, which continues nowadays at a rate of 3 mm/year. However, the exact timing is not well defined. It seems that this fault is one of the most active faults in Iran, having caused earthquakes more than 7 (ms) in magnitude in 1830, 1958 and 1986.

North Alborz fault

The North Alborz fault is almost 400 km length (Berberian 1976a) and extends from the Aliabad, Golestan Province, to the surroundings of Tonekabon. It is parallel to the Caspian fault and is located 10 km to the south of it. These faults meet each other in the Chalus and the resulting fault extends to near the Lahijan fault of western Alborz. The North Tehran fault is also considered to be a south-dipping thrust fault, bringing Mesozoic rocks into contact with the Cenozoic. It forms the southernmost border of the Neogene deposits of the Caspian basin (Stöcklin 1974a). Considering the geographical position of their epicenters, several earthquakes have been attributed to the activity of the North Alborz Fault.

Semnan-Garmsar fault

Located in northern Semnan, eastern Tehran, the Semnan-Garmsar fault is a seismically-active, NE-SW trending, SE dipping left-lateral fault (with a 20-km depth depocenter). Its first activity presumably dates back to the Paleozoic, owing to the differently metamorphosed Bahram Formation in the two adjacent fault walls. According to Nabavi (1976), the lithostratigraphic features of Paleozoic strata (especially those of Devonian rocks) of two adjacent fault walls are completely different. This fault is considered to be the boundary between central Iran and Alborz.

Kandovan fault

The Kandovan fault is a north and northeast dipping thrust fault, 151 km long that extends from the surroundings of Baijan in the east to the Alamkuh area in the west. Several strikes can be observed along the same fault, varying from NW-SE in the east; through E-W in the center, to NW-SE in the west. Along the Kandovan Fault, Neoproterozoic rocks, Kahar Formation, and Paleozoic rocks are thrust over the Eocene volcanics. The central shows strike slip features from the Cenozoic (Ghasemi and Ghorashi 2004; Nazari 2006).

Baijan fault

The Baijan fault is a mainly E-W trending fault whose western part at Damavand Mountain trends NW-SE (Fig. 3.3). In the northern Damavand Mountain, near Nandel Village, it is covered by volcanic rocks.

Some researchers are of the opinion that the Kandovan fault in the northwestern part of Damavand Mountain is the continuation of the Baijan Fault. The 25th June 1963 Kurph earthquake was regarded as the activity of this fault. The landslides that resulted from this earthquake destroyed the hot springs located in this area. Along the old part of the fault, Jurassic calcareous units are thrust over volcanic rocks. All of this evidence indicates a currently active fault.

Jirandeh (Khashachal, Zard-Gol or Kepteh) fault

Located on the northern part of the Shahroud River, the south-dipping Jirandeh fault has a length of 178 km, extends from the western Alamkuh Mountains in the east to its junction with the Manjil fault in the west. It trends NW in its eastern part and W-NW in the west, with a dip direction towards the south.

On 20th June 1990, the Rudbar earthquake was associated with the rupture of the western part of the Jiraneh Fault and the Zard-Gol and Kepteh micro-faults with a left-lateral strike-slip mechanism (Ms. 7.3) (Ghasemi and Ghorashi 2004; Berberian et al. 1992).

Takestan fault

This is a N-NE trending, west dipping fault which extends from Kondar Village in the north to southwestern Takestan. The Takestan fault has Eocene volcanic rocks

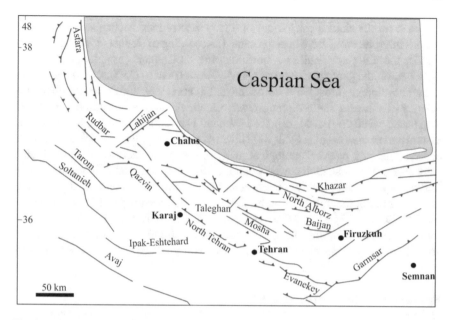

Fig. 3.3 The Quaternary faults of Alborz (modified after Berberian 1976a, 1976c, 1981; Nazari and Shahidi 2011)

in the hanging wall and Quaternary deposits in the footwall (Ghasemi and Ghorashi 2004).

Kahar fault

The Kahar fault is an E-W trending fault with a length of 68 km which extends from the Kahar Mountains to the northwestern Abyek. The main trend is the E-W. However, it inclines in the eastern termination changing to the NW-SE. the Fault trace is curved at the eastern end, and has the NW-SE trend. The Kahar Fault, which joins the Mosha Fault in the eastern termination, juxtaposes the Neoproterozoic (Kahar Formation), Paleozoic and Mesozoic rocks with the Eocene and Neogene. It forms the southern boundary of one of the Neogene intermontane basins, of northern Alborz. Ghasemi and Ghorashi (2004) considered the 8th November 1966 Taleghan (Zamghabad), earthquake to be related to the activity of one of the Kahar, Abyek and Taleghan faults.

Taleghan fault

The Taleghan fault is 70 km long and is located near the southernmost border of central Alborz *ca.* 50 km north-west of Tehran (Fig. 3.3). To the south, it borders the Taleghan and Azadbar Valleys, juxtaposing the Lower Cenozoic pyroclastic rocks (Eocene) with Paleozoic and Precambrian deposits. The Taleghan Fault is regarded as an E-W trending, north-dipping thrust fault by Assereto (1966). It joins the Qazvin fault in the West and the central Mosha fault in the East.

Based on the studies carried out on the geometry and mechanism of the fault, various interpretations have been presented. According to Dedual (1967) and Meyer (1967), the Taleghan Fault is a north-dipping thrust fault (Annells et al. 1977). However, on the geological map of Qazvin-Rasht (scale: 1/250000), it is described as a south-dipping thrust fault. Berberian and Yeats (1999, 2001) also described the Taleghan Fault as a south-dipping reverse fault. Finally, on the recent geological map (scale 1:100,000) of Marzanabad (Vahdati Daneshmand and Nadim 2001), the eastern part of the Taleghan Fault was shown as a north-dipping thrust fault. The field observations and paleoseismological analysis obviously indicate a south-dipping compressional reverse fault corresponding with a compressional fault mechanism of Eocene deposits (footwall) in the north and Jurassic rocks and older (hanging wall) of the south. So, the Taleghan fault might have been the origin of the earthquakes of 958 AD (ms. 7.7) 1428, 1608, 1808 and 1966. A lot of recent movements on a south-dipping fault surface have been recorded along the same fault, indicating a normal component associated with a predominantly left-lateral component (Nazari et al. 2009).

Firuzkuh fault (eastern central Alborz)
The Firuzkuh fault is approximately 55 km long, extending from the eastern end of the Mosha Fault to the west of "Aminabad" village to the Gaduk neck (Fig. 3.3; Nazari 2006; Nazari et al. 2010). The Kumes earthquake of 865 AD (ms. 7.9), as well as those of 1969, 1973, 1975, 1979, 1989, 1990 and 2008 are attributed to movements of this fault (Ambraseys and Melville 1982). The 1990 Gaduk earthquake (ms.5.8) is also considered to be the largest instrumentally recorded earthquake of the fault (Nazari and Shahidi 2011).

Ray faults (North and south Ray faults)
To the south of Tehran, near the Ray City, two parallel segments with 20 km length and WNW-ESE strikes were firstly mentioned by Pedrami (1981) as the "Valiabad" and "Salehabad" faults. However, Berberian et al. (1985) changed their names to the North and South Ray faults and described it as a northward dipping fault. According to Berberian and Yeats (1999), many earthquakes such as those of 853, 855, 856, 958, 1177 and 1364 resulted from the movement of this fault.

Nazari et al. (2011) believed that the North Ray fault is a continuation of the Parchin fault and could be its westward extension. In spite of urban development in the area, one still observes the escarpments on SRTM images. The latter authors also mentioned that, according to recent studies, no obvious evidence confirms the occurrence of the south Ray Fault. These faults can be attributed to the central faults.

Kahrizak fault
The Kahrizak fault forms the southernmost border of the alluvial piedmont of Tehran, with an approximately E-W trend. It is located 20 km south of Tehran, 10 km south of the city of Ray.

Eyvanaki (Parchin) fault

The Parchin fault marks the boundary between the southernmost piedmonts and the Anti-Alborz Mountains to the southeast of Tehran. This fault has a total length of 70 km and extends from the southern Eyvanaki in the east to the Bibi-Shahrbanou Mountain in the west. It is a NW-SE trending and NE dipping reverse fault whose northeastern part, known as the Parchin Fault, cuts the Pliocene and Quaternary deposits. Its eastern part is accompanied by faulting related folding. The eastern segment joins to the Garmsar fault (Nazari et al. 2011).

The destructive historical earthquake of 743 AD in the eastern Ray possibly occurred due to the movement of this fault. The southern part of Central Alborz and northern Central Iran is limited by these faults.

Varamin (Pishva) fault

The Varamin fault, also known as Pishva Fault, with 42.5 km length is a NW-SE trending, NE dipping reverse fault which extends from Kuh-e-Sorkh in the southeast to the Varamin County in the northwest. Its southern part borders Central Iran and the Alborz. Another NW trending fault can be observed along the middle inclined part of Varamin fault, geometrically and mechanically resembling to the western part of the latter fault.

Nazari and Shahidi (2011) considered it as a seismically active fault based on the paleoseismic studies and field observations.

Alamut Rud (Shahroud) fault

Juxtaposing Eocene volcanic rocks and Neogene rocks, the 110 km long Alamut Rud fault has two main strikes; E-W and NW-SE.

According to Nazari and Shahidi (2011), the large magnitude earthquake of 20th April 1608 AD (ms = 7.4) in the Taleghan area was presumably related to the Alamut Rud Fault. The last movement dates back to the late Quaternary and Holocene, based on geomorphological features such as stream arrangement, morphological features and fault movement traces (Nazari et al. 2011).

North Qazvin fault

This is a 78 km long, W-NW trending, north-dipping fault which extends from Shekarnab village in the western Ziaran to Kuhin Village. Considering the structural characters, the main component indicates thrust type fault (Fig. 3.3; Nazari et al. 2011).

Abyek fault

Bordering the plain and mountain in northern Qazvin, the NW-SE trending, NE dipping Abyek fault is *ca.* 91 km long, extending from Khur village in the east to Aghababa village in the west. The fault has juxtaposed the Paleozoic to Cenozoic rocks with the Quaternary. The Abyek fault joins the northwestern part of the North Tehran Fault, as well as the Mosha and Kahar Fault terminations in the western part (Ghasemi and Ghorashi 2004).

Galand Rud fault

The Galand Rud fault is 73 km long, with a W-NW trend that dips north. It extends from the southern Chamestan County in the east to Pol-e-Kalat on the Tehran-Chalus Road in the west. Along the same fault, Mesozoic rocks are thrust over Cretaceous and Neogene rocks (Ghasemi and Ghorashi 2004). It also joins the North Alborz Fault in the west.

Kojour fault

This is a south-dipping, W-NW trending fault whose western part is a fault zone along which Neoproterozoic rocks and the Kahar Formation are thrust over the Jurassic Shemshak Formation and the whole package over the Cretaceous and the volcanic assemblages of Alamkuh area (Nazari 2006). The Kojour Fault is *ca.* 95 km in length and extends from the southeastern Varazan in the Kojour area to the Alamkuh intrusion in the west. The fault trace is inclined, with the eastern part showing a W-NW trend. The western termination is believed to joins the Mosha fault.

Hazar fault

This is a W-NW trending, north-northeast dipping fault which extends from the southern Sheikh-Musa in the east to the southern Chak-Bozeh, western Chalus valley in the west. The Hazar fault is 135 km long and has resulted in thrusting of the Kahar Formation over Paleozoic and Mesozoic rocks. The Hazar fault terminates with the Lelehband fault in the east and the Kojour fault in the west (Ghasemi and Ghorashi 2004).

Lelehband fault

This is an E-W trending, south-dipping thrust fault which extends from the eastern Pol-e-Sefid, Firuzkuh road, in the east to the eastern Kahroud-e Paeen, Haraz road, in the west. It mainly cuts the Shemshak Formation. Nevertheless, in some areas also has also affected Triassic, Jurassic and Cretaceous rocks.

The destructive 2nd June 1957 Bandpey earthquake, at Sangchal, in Mazandaran province is assumed to be related to the activity of the Lelehband fault and the Northern Alborz Fault.

Orim fault

The Orim Fault is a NE trending, southeast-dipping reverse strike-slip fault juxtaposing Paleozoic and Mesozoic rocks with the Cenozoic. This fault with 124 km in length extends from Astaneh in the northeast to Gaduk in the southwest. Nabavi (1976) reported the fault to have a left-lateral mechanism in some areas.

The 5th March 1935 Talarroud earthquake presumably resulted from the activity of either the Mosha or the Orim fault.

Zarrin-Kuh fault

Being 78 km in length, the Zarrin-Kuh Fault is a W-NW trending fault which extends from the Firuzkuh Fault in the East to the Haraz road, eastern Plour in the west. Located in the hanging wall of Mosha fault, the Zarrin-Kuh fault is parallel and close to former fault, joining it in the east. The fault dip was interpreted as being towards the northeast on the geological map, scale: 1/100000. However, considering the geological and morphological features of the area, it is most probably a southwest-dipping fault that caused the Jurassic and Cretaceous rocks to be thrusted over the Eocene and presumably Quaternary (Ghasemi and Ghorashi 2004).

Astaneh fault

The Astaneh fault is 139 km in length and is a NE trending, southeast-dipping thrust fault which extends from the northeastern Astaneh in the east to the Basham, north-western Semnan, in the west. Paleozoic, Mesozoic and Eocene rocks are juxtaposed with the Eocene, Neogene and Quaternary (Hollingsworth et al. 2010). It also thrust Barut Formation and other Paleozoic and Mesozoic rocks over Eocene volcanic and volcanoclastic rocks, and Neogene sediments.

The 3rd October 1966 Sangsar and Shahmirzad earthquakes (Ms. 4.5, Mb. 4.9) are considered to have been caused by the activity of the either the Astaneh Fault or the compressional Basham microfault in the west. The 12th December 1968 Sangsar and Shahmirzad earthquakes can also be related to the movement of the latter fault.

Paleoseismological studies carried out on the Astaneh Fault show that its reactivation caused the occurrence of the 856 AD large historical earthquake (M = 7.9, x) of Kumes (Hollingsworth et al. 2010). This was the most disastrous historically recorded earthquake of Iran, which destroyed both urban and rural regions from the Kumes (Ghumes or Gumesh) to Neyshabur, and killed approximately 2 million people.

3.2.2 Eastern Alborz

Attari fault

This is a NE-SW trending, southeast dipping fault (Alavi-Naeini 1972) that extends from Mohammad-Abad, western Maiami in the northeast to the east Damghan county in the southwest (212 km in length; Nazari and Shahidi 2011). Crossing the Jam-Abkhori zone, the basement Attari fault has affected both ends of the basin since Cambrian to Late Cretaceous time (Alavi-Naeini 1972). Berberian (1995) regarded it as a thrust fault that has thrust the Eocene Karaj Formation in the south over Miocene clastic and evaporite and Pliocene-Pleistocene clastic rocks in the north. The Paleoseismological history of Semnan revealed that the right-lateral, strike slip Attari fault with NW-SE and NE-SW trends is an active fault that has caused very hazardous earthquakes, including Kumes in 856 AD and Damghan in 1967.

Damghan fault

Located 10 km north of Damghan, the Damghan fault is a south-southeast dipping fault crossing the Quaternary deposits. It is composed of two segments; the eastern and western segments, of which the former crosses Upper Neogene conglomerate and the younger and older alluvial fans or the Quaternary clayey siltstone (Omidi et al. 2002). No instrumentally-recorded seismic data related to the activity of the Damghan Fault is available, and the epicenter of the 1900 and 1976 earthquakes was not located along the Quaternary Damghan Fault. However, the Damghan city was severely destroyed, owing to the activities of this fault during the April 1830 and December 1856 earthquakes. It appears that the 22nd December 1856, December 22 and 9th June 1982 earthquakes were related to the activity of the Damghan fault.

Badeleh (Shahkuh) fault

This is an E-NE trending, north-northwest dipping fault, *ca*. 180 km in length, which extends from the western Khosh-Yeiylagh, northeastern Shahroud in the east to the Aznai, western Kiyasar, in the west. The Upper Paleozoic rocks, especially those of Devonian Khosh-Yeiylagh Formation, are juxtaposed with the Jurassic Shemshak Formation, Eocene and Neogene rocks. The Badeleh fault is located in the hanging wall of the North Alborz Fault and is parallel to it.

Jajarm fault

This is an ENE-ESW trending, left-lateral strike-slip fault, which is considered to be one of the faults of eastern Shahroud fault system. With a length of 130 km, it located 20 km north of Jajarm County.

Satellite images show Quaternary deposits displaced by the Jajarm Fault with left-lateral movement.

Khosh-Yeilagh (Shavar) fault

The NE-SW trending, north-dipping Khosh-Yeilagh fault (227 km in length) extends from Asghar-Abad, southern Chamanbid in the east to Tash-e Paeen, northwestern Shahroud, in the west, where it terminates at the junction with the Badeleh Fault. The most displacement is recorded in the western segment, where Paleozoic rocks are thrust over the Mesozoic and Cenozoic rocks.

The 11th June 1890, Tash-Shahroud, the 9th October 1981, Abar-Ghatari and the 11th May 1984, Abar earthquakes can be related to the activity of one of the Khosh-Yeilagh, Badeleh or North Alborz faults (Ghasemi and Ghorashi 2004).

Chahardeh fault

The NE trending, north-dipping Chahardeh fault (24 km in length) which extends from the Chahardeh in the northeast to the Astaneh in the southwest, plays an important role in the geology of southwestern Alborz, though its mechanism cannot be easily defined (Nazari et al. 2011). The junction of this fault corresponds to the trend variation of main faults of the area; so that the Basham, Shahroud and Tazareh faults, are of an E-W trend in the eastern side of the fault and a NE trend in the west.

3.2.3 Azerbaijan and Western Alborz

Soltanieh fault

The NW-SE trending, southwest dipping, right-lateral Soltanieh fault which is located 8 km south-southwestern of Soltanieh County is presumably related to the formation of the Abhar-Zanjan embayment. This fault separates the higher mountains of the Soltanieh from Precambrian to Mesozoic from the Eocene volcanic belt in the northwest (Berberian 1976c). The 1803 Soltanieh earthquake is presumably related to the movements of this fault, which apparently continues to the Takestan Fault and joins the Boein Zahra fault.

Urmia-Zarrineh Rud fault

This is a NW-SE trending fault (400 km in length) which extends from the Maku area and after crossing the western flank of Urmia Lake, to the Zarrineh Rud River. The depression of Urmia Lake is assumed to be related to the activity of this fault (Eftekharnezhad 1980). Its trend corresponds to the basement faults but its effects are not recognized in the Precambrian. Activities of this fault led to the formation of ancient grabens (Precambrian in age, reaching Permian in some localities) in the western part of the Recent Urmia Lake. Due to the effects of the Eo-Cimmerian orogeny, the western Urmia Lake experienced uplift, resulting in the absence of Triassic and Jurassic sediments. The fault movement is recorded to be mainly vertical, and horizontal movement is not recognized. The formation of Urmia Lake, 6500–8500 years ago, was the result of the most recent activity of this fault.

Astara fault

The reverse Astara fault, which juxtaposes Paleozoic strata with younger deposits, extends from the western Guilan plain, where it forms the boundary of plain and mountain, to Astara and the Caucuses (Jackson et al. 2002). The fault length was measured 200 km from its eastern end to the border of Iran with Azerbaijan and 400 km to its northern termination. It has played a significant role in the morphological evolution of the area, with the depression of the eastern Caspian Sea having formed due to its activity. This fault has earthquake potential, with the 1953 and 1987 Caucuses earthquakes are resulting from its compressional. It is also assumed to have been the cause of historical earthquakes in Rasht City in 1709 and 1713 AD (Berberian and Yeats 1999, 2001; Barzegari et al. 2017).

Tarom fault

Being the north border of the Tarom embayment in the Manjil-Ardabil area, the NW-SE trending Tarom fault has juxtaposed Paleozoic and Mesozoic rocks with the Tertiary, and has formed the Tarom embayment. The thrusting is directed southward, and continues to the western Khalkhal. It is a seismically-active fault that has played a significant role in the earthquake history of the area.

Tabriz fault

The N115°E trending, vertical-dipping Tabriz Strike-Slip fault (150 km in length) extends from Mishu Mountain in the west to Bostan-Abad in the east and divides

Azerbaijan into two blocks, the "northeastern" and the "southwestern", of which the former acted as a horst and latter a graben. The first activity of this fault presumably dates back to the Devonian, though an older age is also possible.

There is no historical evidence indicating the activity of Tabriz fault during past century. However, according to Berberian and Arshadi (1976) it will be possibly be reactivated in the near future, causing a destructive earthquake.

The Tabriz fault can be divided into two segments (eastern and western) having a right-lateral strike-slip mechanism (Berberian and Arshadi 1976; Hessami et al. 2003; Masson et al. 2006; Ritz et al. 2011). The horizontal movement rate of the Tabriz fault is considered to be 3.1–6.6 and 8 mm/year, depending on the method of calculation used (Rizza et al. 2013).

According to geomorphological indicators as well as the photo- luminescence method of dating, compared to GPS network data, the movement rate was regarded to be 7 mm/year (Nazari and Shahidi 2011).

Zanjan fault

The northeast-dipping, right-lateral Zanjan fault (Nazari and Shahidi 2011), juxta-poses Eocene volcanic rocks with Neogene deposits and forms the border between the plains and mountains from Ghazvin, in the east, to Mianeh in the west. It is divided into "eastern" and "western" segments of which the former trends NW and the latter N-NW. The southeast termination joins the Takestan fault and the northwestern one the Mianeh fault (Ghasemi and Ghorashi 2004).

Mianeh fault

The N-S trending, east-dipping Mianeh fault (145 km in length) which extends from the Khan-Yordi in the south to the eastern Laroud in the north juxtaposes metamorphic rocks (Paleozoic?), Eocene volcanic and Neogene deposits with Neogene-Quaternary rocks.

The 13th May 1844, Mianeh-Garmrud and 22nd March 1879 Bozghush-Garmroud earthquakes are considered to be related to this fault (Ghasemi and Ghorashi 2004; Nazari and Shahidi 2011). According to the present author, the Tabriz, Mianeh, Zanjan and Takestan faults can be considered together as a single fault, showing some variation along its extension.

Rudbar fault

The W-NW trending, north-dipping Rudbar fault (57 km in length), extends from Nusha, southern Tonekabon in the southeast to Rudbar in the west, bringing Permian rocks into contact with the Jurassic and Eocene (Nazari and Salamati 1998). Considering the regional geology, it seems that this fault is a left-lateral strike-slip reverse structure. The August 1485, Mazandaran-Guilan earthquake was related to activity on the eastern segment (Ghasemi and Ghorashi 2004).

Neor fault

The N-S trending, strike-slip Neor fault (55 km in length) extends from the eastern Ardabil to the border of Iran and Azerbaijan in the north and crosses Eocene volcanic rocks. The Neor Lake of the southeastern Kargan is assumed to be related to the activity of this fault (Ghasemi and Ghorashi 2004).

Lahijan fault

The N60°E trending Lahijan fault (70 km in length) forms the boundary between Cretaceous and Jurassic rocks in the southeast and Quaternary deposits in the northwest. It is only recognized based on geological and physiographical characters.

The 3rd February 1967, Lahijan earthquake was probably related to either the Lahijan fault or the western termination of the north Alborz Fault (Ghasemi and Ghorashi 2004).

Manjil (Sangavar) fault

This is a northeast-east dipping thrust fault which extends from southwestern Jirandeh in the east to the vicinity of Namin in the northwest. It strikes E-NW in its eastern main part.

The destructive 30th December 1863, Hir-Ardabil and 4th January 1896, Khalkhal-Sangabad earthquakes were possibly related to the activity of Manjil fault (Ghasemi and Ghorashi 2004).

3.3 Kopeh-Dagh

Afshar-Harb (1979) divided the Kopeh-Dagh faults into two main groups as follows:

1. Faults created by the vertical movements of basement that have played an important role in controlling sedimentation, vertical movement, facies and lithostratigraphic variations (basement fault).
2. Faults that appeared as thrust faults created by the orogenic phases.

Group 1 can be further subdivided into two subgroups based on their movements and consequences.

Subgroup 1 faults were active during the sedimentation and caused facies variations and sedimentary gaps, whereas those of subgroup 2 were created by basement movements during severe Late Alpine folding; most of these are still active at the present day.

Subgroup 1 faults are detected solely based on detailed sedimentary studies of strata on both sides of faults but the prerequisite for these investigations is the existence of sufficient outcrop. In the central and eastern Kopeh-Dagh, this condition is met only in sequences younger than the latest Jurassic; there are no adequate exposures of older strata (Fig. 3.4).

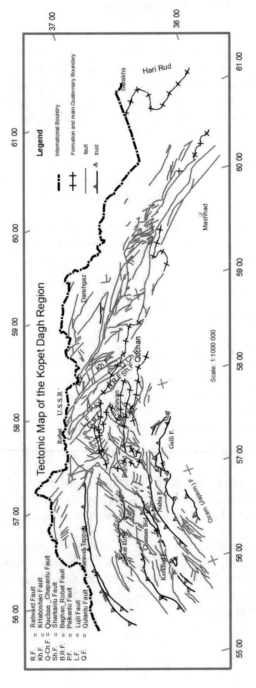

Fig. 3.4 Main faults of the Kopeh-Dagh area (modified after Afshar-Harb 1994)

3.3.1 Active Syn-Depositional Basement Faults

These faults are represented by approximately W-SW and E-NW trends that change eastwards to E-W. The oldest fault of this subgroup was active during the Late Jurassic and the youngest during the Late Albian.

The relative movements of these faults are assumed to play a critical role in the facies variation and sedimentary gaps of the fault segments. They probably were primarily E-W trending faults inclined by rotation tectonic, syn-folding and post-folding events (Afshar-Harb 1994).

Korkhud fault

This fault was activated in the late Kimmeridgian when the northern block started to subside (Afashar Harb 1979). Around this time, a marine regression occurred in all areas except for the northwestern part. Fault movement ceased in the late Neocomian and two blocks subsided relative to the other blocks. Subsequently, the Tirgan limestone was deposited over these two blocks. The Korkhud fault played a significant role in controlling sedimentation in the northern and southern areas. At the beginning of the late-Aptian, activity of the fault resumed, and subsidence of the northern block continued until the late Albian. On the northern block over 1200 metres of Aptian and Albian marls and shales accumulated, whilst on the southern block, these sediments are only a few tens of metres thick (Afshar-Harb 1979).

Nabia fault

Being parallel to the Korkhud fault, the Nabia fault was active during the middle Aptian, while the northern block was subsiding. This fault was reactivated during the Coniacian, when the Abderaz Formation was deposited on the northern block, though the southern block continued to be exposed. The southern block, south of the Nabia fault, is limited to the south by the Jajarm Fault. The present width of the Nabia-Jajarm block is about 22 kilometres. Very rapid uplifting of the block south of the Jajarm fault provided the source of the thick conglomerates deposited during Paleocene time on the Nabia-Jajarm block.

In the late- Paleocene, the Nabia-Jajarm block sank to become level with the Khorkhud-Nabfa block and both were covered by the shallow Eocene sea (Afshar-Harb 1979).

Takal Kuh fault

This NE-SW trending fault was activated in the Middle Aptian. The northern block was subsiding relative to the southern block, until the beginning of the middle Cenomanian. The thickness of Sarcheshmeh and Sanganeh formations deposited on the former block is more than the latter, measured 2500 m in thickness. Comparing the northern and soutthern blocks, the thickness and facies of the Sarcheshmeh Formation are quite different; the northern sequence consists of marly limestone with more carbonate composition wheras relatively thin shale accumulated form the southern sequence (Afshar-Harb 1979, 1994).

It seems that during the late Turonian, the fault movement was reversed and consequently the southern block was subsiding, leading to the deposition of the Abderaz Formation, followed by the exposure of the two blocks in the middle Coniacian. The primary movement of the late Turonian was repeated in the middle Maastrichtian and the northern bock started to subside. Finally, during the Late Cretaceous and Paleocene, the Kalat Limestone, the clastic-continental Pesteligh Formation, and the Chehel-Kaman limestone were deposited on the northern block while the southern block was affected by erosion.

Maraveh-Tapeh fault

This E-NE trending fault clearly affected Albian sedimentation. As a result, on the northern block, the total thickness of the Albian Aitamir Formation (including 600 m thickness of sandstone intercalations) is *ca.* 1000 m, while on the southern block, it is less than a few tens of meters. In the Campanian to middle Maastrichtian, all parts of the southern block experienced erosion, while marine deposition continued on the northern block (Afshar-Harb 1994).

At the beginning of Tertiary, all blocks south of the fault emerged from the sea. In the early Paleocene, continental red beds of the Pesteligh Formation were deposited on the low land areas of the southern blocks and deposition of carbonates continued on the northern block (Afshar-Harb 1979).

Ghelli fault

This fault was activated from the beginning of the Paleocene when the northern block was rapidly uplifted. Its rapid erosion provided the source of the pebbles of Pesteliq Formation on both the southern and northern blocks. This severe erosional process is indicated by the presence of pebbles derived from the Cretaceous Tirgan and Jurassic Mozduran Formation. This conglomerate is 1700 m and 150 m thick on the Ghelli-Jajarm and Nabia-Korkhud blocks, respectively (Afshar-Harb 1994).

Jajarm fault

This fault forms the southern boundary of the Nabia-Jajarm and Ghelli-Jajarm in Cretaceous and Tertiary times, respectively. Geological data show that, in the southern part of this fault, there are no sedimentary rocks having the same facies as the Upper Jurassic, Cretaceous and Tertiary formations (Afshar-Harb 1994).

Kashafrud fault

Being parallel to Kopeh-Dagh Mountains, and clearly crossing the Quaternary deposits of the Mashad plain, the sinusoidal, NW-SE trending reverse Kashafrud fault (120 km in length) is a basement fault which extends along the northern Mashhad Plain.

The dolomite and limestone beds of the Jurassic Mozduran Formation and red and brown gypsiferous shale as well as the sandstone of the Lower Cretaceous Shurijeh Formation in the northeast were juxtaposed with the alluvial fans and Quaternary silty plain of the southwest, by this fault. The presence of the triangular facet morphology along the fault, breccia in the recent alluvial deposits, abundant springs and the fault trace all hint at the recent activity of the compressional Kashafrud fault (Berberian and Ghorashi 1989). The 30th June 1673, (Ms. 6.6) and 23rd April 1687, (Ms. 5.5) historical Mashhad earthquakes are assumed to be related to the reactivation of this fault (Berberian 1981).

The location of the epicenter of 28th September 1988 earthquake near the Kashafrud fault presumably indicates the activity of this fault in the twentieth century (Berberian and Ghorashi 1989).

Fault zones

Afshar-Harb (1979) defined some fault zones, each include several faults in the Kopeh-Dagh area as follows: the Quchan-Robat, northeastern Bojnourd, Ghatlish, and northwestern Ashkhaneh fault zones.

A.   Quchan-Robat fault zone

This is the largest of the Kopeh-Dagh area, including abundant faults as follows:

Baghan-Robat (Baghan-Garmab) fault

This is a right-lateral strike-slip fault (80 km in length) creating roughly 8.5 km horizontal displacement without any trace of vertical movements. The fault can be traced for *ca.* 3 km in the alluvial deposits of the Quchan Plain from satellite imagery.

Considered to be the most active fault in this zone, the Baghan-Robat fault is assumed to be related to the 1st May 1929, Baghan-Garmab earthquake. It extends from the old Quchan city (Quchan-Shirvan plain) in the south to Turkmenistan in the north (Tchalenko 1975).

Sherkanlu fault

Being approximately parallel to the Baghan-Robat fault, the Sherkanlu fault is located some 7–8 km from it. On satellite images, it can be traced about 5 km in the alluvial deposits of the Quchan Plain. It is also a right-lateral strike-slip fault, 50 km in length, with a maximum displacement of 2.5 km (Afshar-Harb 1994).

Quchan-Chupanlu fault

This is a right-lateral strike-slip fault which extends from Turkmenistan in the northwest to the Quchan-Shirvan Plain in the southeast. Located 5 km east of the Sherkanlu fault, the Quchan-Chapanlu fault crosses the Quchan fault, resulting in 1.5 km displacements (Afshar-Harb 1994).

Khabushan-Gharecheh fault

Located 7.5 km east of the latter fault, this is a right-lateral strike-slip fault joining the Quchan-Chupalu in the northwest. The Khabushan-Gharecheh Fault (60 km in length) is characterized by a 1.5 km vertical displacement.

Rahvard fault and associated eastern faults

The Rahvard fault (40 km length) is a strike-slip fault located 5–7.5 km from the eastern Khabushan-Gharecheh fault. Its maximum displacement was measured to be 1.5 km. Two right-lateral strike-slip faults associated with this fault created a total displacement of 2 km (Afshar-Harb 1994).

Palkano fault

This is located west of the Baghan-Robat Fault which it joins in the northwest. The Palkano Fault is a right-lateral strike-slip fault (42 km in length) with a 0.5 km displacement. The maximum distance between Palkano and Baghan-Robat fault is about 5 km (Afshar-Harb 1994).

Lujeli fault

The length of Lojeh fault is 50 km, it is a right-lateral strike-slip fault with a displacement of 1.5 km. More or less parallel to the Palkano Fault and to its west. Its distance from the Palkano Fault varies between 5 and 7 km.Ghulano fault

This is a right-lateral strike-slip fault (22 km in length), located in 3–5 km west of the Lujeli fault. The maximum lateral displacement is measured to be 3.2 km.

B.   **Northeastern Bojnourd fault Zone**

This is located in the northeastern part of Bojnourd city and northeastern Sheykh Syncline, and is characterized by an approximate length of 10 km and width of 18. It is composed of four active right-lateral strike-slip faults creating about 4 km displacement.

The most geologically-important fault of the zone is assumed to be the western fault that created *ca.* 3 km displacement and possibly considerable vertical displacement.

C.   **Ghatlish fault Zone**

This fault zone includes three faults; the Aiiob, Ghatlish and Pirboz faults, whose effects can be traced in the northern areas of Ghatlish village, 37 km north of Bojnourd. The Ghatlish fault zone is *ca.* 20 km long and 4 km wide. The Aiiob Fault is a right-lateral strike-slip, N-S trending fault and the most easterly fault of this zone. Its maximum displacement is 0.5 km in the Ghazi Syncline (Afshar-Harb 1994).

The Ghatlish Fault is also a right-lateral strike-slip fault, roughly parallel to the Aiiob fault. Its maximum displacement was measured to be 2 km in Arnaveh Anticline. This fault and its subordinates have created the N-S trending streams of Darband-Mianzou and Ghazi-Ghatlish (Afshar-Harb 1994).

The NNW- SSE trending Pirboz fault and its western branch are right-lateral faults. Its western branch was assumed to have vertical displacement.

D.   **Northwestern Ashkhaneh fault Zone**

This zone, which is located west to northwestern of Ashkhaneh city on the Danghuz Dagh anticline, is characterized by being *ca.* 30 km long and 25 km wide. Contrary to the mentioned faults in Kopeh-Dagh, they are left-lateral strike-slip faults that have caused 5 km total displacement. The largest fault which has created approximately 2.5 km displacement crosses the Danghuz Dagh anticline. Geologically, this fault zone compares closely to the northeastern Bojnourd fault zone (Afshar-Harb 1994).

Left-lateral strike-slip faults with a maximum displacement of 2.7 km have been reported from western Turkmenistan (Tchalenko 1975).

3.3.2 Thrust Faults

The "thrust faults" or "faults of the Mesozoic Era" are known to be the most important faults of the western Kopeh-Dagh. These thrust faults are form concentric arcs with the inner arcs having the maximum displacement. Afshar-Harb (1994) believed that displacement along the basement fault caused deformation, folding and subsequent flexure, accompanied by subordinate fractures with a reverse mechanism.

Jajarm thrust
The NW dipping Jajarm overthrust fault (120 km in length) juxtaposes Devonian (Padeha Fm.) and Neogene rocks.

Ghelli overthrust
The north-dipping Ghelli thrust, with a length of *ca.* 100 km, is located to the north of the Jajarm overthrust, and brings Upper Cambrian rocks into contact with the Neogene rocks of the Jajarm Plain.

Nabia thrust
The Nabia fault (130 km length) juxtaposes the Shemshak Formation and Neogene rocks. The 17th September 1922 earthquake (Ms. 6.5–7) is assumed to be related to activities of this fault.

Korkhud-Robat-e Gharebil thrust
This is a thrust zone consisting of severalclosely-spaced, northwest dipping faults. The two most important faults are the Robat-e Gharehbil and Chamanbid faults. The Robat-e Gharebil fault juxtaposes Silurian and Neogene rocks. The Chamanbid thrust located to the west of Chamanbid village, bringing Middle Jurassic rocks into contact with Devonian and Carboniferous strata. To the east of the same village and north of Juzak village, it has juxtaposed the Middle Jurassic and Eocene rocks. The total length of this zone is 170 km.

Sugha thrust
The Sugha thrust (70 km long) brings Jurassic rocks into contact with Albian rocks.

Takalkuh-Galudagh-Pishkamar thrust
This fault has a total length of 170 km. It has thrust the Jurassic Mozduran Formation over the Cretaceous Sarcheshmeh Formation. In recent centuries, several earthquakes occurred along this thrust (Ms. 5–5.5). The 30th June 1970 earthquake, whose

epicenter was located in the southern Maraveh-Tappeh is assumed to have been related to the activity of this fault.

Maraveh-Tappeh thrust

This fault located in the northern Maraveh-Tappeh with a total length of 170 km crosses the northern part of Atrak River. This thrust with low displacement has thrusted the Aitamir Formation on the Sanganeh Formation in the northern Maraveh-Tappeh and Aitamir Formation on the Sarcheshmeh Formation in the Northern Boz-Dagh.

Tchalenko (1975) illustrated left-lateral strike-slip faults of Ghara-Ghaleh area of Turkmenistan and placed the Maraveh-Tappeh fault in the same group; although the geological understanding of the present author is not in agreement with this classification.

3.4 Zagros

Notwithstanding the time of fault formation, the major fault zones of the Zagros were divided into two groups based on their mechanism at the present time (Fig. 3.5) as follows:

Group A is characterized by a thrust or reverse mechanism and a trend parallel to that of the Zagros belt.

Group B is characterized by a strike-slip mechanism and oblique trend relative to that of the Zagros belt.

The present day morphotectonics of the Zagros area are the consequence of the activity of these fault zones which were reactivated during the shortening caused by the continental collision of Central Iran and the Arabia microcontinents in Tertiary time.

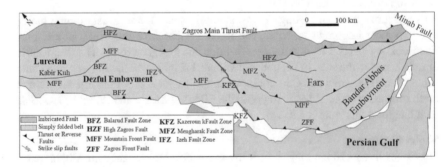

Fig. 3.5 A generalized map showing main faults and structures of Zagros Basin (modified after Sepehr 2001: compiler)

3.4.1 Main Zagros Thrust

This is a right-lateral strike-slip fault which extends from Minab in the east to southern Marivan in the west. It enters Iraq in the west and after crossing it, trends towards Iran in the Sardasht area and finally enters Turkey. The fault strikes N130°E in its western segment, from the Turkey border through eastern Hajiabad, Bandar Abbas, and then strikes N170°E in its southern segment, which is also named the Oman line, the Zendan fault or the Minab fault. This fault formed during the Precambrian Pan-African orogeny and afterward was the main basin control in the area. The right-lateral strike-slip MZT is the border between the high Zagros and the Sanandaj-Sirjan Zone. The exposures of oceanic crust along the MZT in the Neyriz and Kermanshah of Iran as well as the Taurus of Turkey indicate the continental collision of Central Iran in the northeast and the Afro-Arabian plate in the southwest.

The MZT has played a significant role in seismicity of Iran, especially its north-western part where many historical and instrumental earthquakes have been recorded. The young faults corresponding to the MZT have been named, "Recent Faults". These include the Dorud, Nahavand, Garun (Gharun), Sahneh, Morvarid and Piranshahr faults, all of which are derived from the MZT.

Balarud fault
The NW-SE trending Balarud fault is a right-lateral strike-slip fault which is considered as the northern border of Dezful embayment. The formation of Dezful embayment was related to the activity of the same fault. It also divides the Zagros basin into the various stratigraphical successions and geological profiles. This division is indicated by different structural patterns which extend along Zagros belt and caused by subsequent collision. In the Latest Cretaceous, the NW-SE trending segment of Balarud fault divided the existed fold belt into two basins including main foreland in the southeast and piggy back in the northeast (Sepehr 2001).

Dena (Dinar) fault
The Dena fault is located on the margin of the Sisakht summits, northern Lordegan. The NW-SE trending and E-NE dipping Dena fault is one of the main fractures of the Zagros Precambrian basement, dividing it into two parts with different geological, seismical and morphological characteristics. The right-lateral strike-slip displacement of this fault has resulted in the folding of the northern portion of Dena Mountain, and the Hezar Darreh and Charo mountains were possibly affected by the right-lateral displacement of Dena Mountain. In the western Dena Mountain, the Cambrian Zaigun and Lalun Formations are thrust over Cretaceous rocks (Setudehnia 1975). Being seismically active at the present time, this fault has played an important role in the sedimentation of Phanerozoic rocks (Setudehnia 1975). However, no study has reported continued movement on this old and important fault. Cretaceous rocks have been thrust over the plain deposits or Pliocene Bakhtiari Formation by the main Zagros thrust (Berberian and Ghorashi 1986). It is worth mentioning that the thrust faults of the southern Dena Mountain form the border between the high Zagros and the folded Zagros. These faults are probably separate from the Dena fault.

Minab fault

The NW-SE trending, strike-slip Minab fault (*ca.* 300 km long) forms the border between the Zagros and Makran structural zones. Its eastern segment has moved southward. The right-lateral horizontal movement presumably occurred during the Late Cretaceous-Early Tertiary (Falcon 1967). However, the presence of salt diapers in the Zagros and the Persian Gulf and their absence in the Makran area has caused some geologists to attribute this fault to the Pan-African Orogeny. Nabavi (1976) and Aghanabati (2004) interpreted a right-lateral and over-thrusting mechanism for this fault, respectively. This right-lateral thrust fault connects the main Zagros fault to the Makran Fault and the Sabzevaran-Kahnouj-Jiroft fault system, the latter of which limits the Jazmurian embayment to the west. InSar (Interferometric Synthetic Aperture Radar) and block models have been used to produce some data about this fault. These models suggest that relative inertia during past 200 years has yielded a fractional slip of about 2 meters height.

Ardal fault

Located in the northern Ardal-Naghan area, the Ardal fault, length 80 km, is a W-SE trending, north dipping fault, the dip of which increases towards the northeast. The Paleozoic succession of the Bangestan group in the northeast is thrust over Cretaceous rocks and plains in the southwest. In the northwestern Ardal, some salt diapirs are exposed.

The Ardal fault is overlain by Quaternary deposits. There is no evidence of seismic activity movement, such as low-angle thrust, or active subsidence during this fault position.

Zardkuh fault

Located along the Bazoft River in Chaharmahal-e Bakhtiari, the NW-SE trending, northeast dipping Zardkuh fault (130 km in length) brings Cambro-Ordovician rocks into contact with the Bakhtiari Formation (Setudehnia 1975). Being a thrust fault, the Zardkuh fault forms part of the boundary between the High Zagros and the Folded Zagros.

Aghajari fault

The NW-SE trending, northeast dipping Aghajari thrust fault (150 km in length) is located in the northern Aghajari, and has thrust the Aghajari and Pazenan over the alluvial plain of Aghajari. The Gachsaran, Mishan and Aghajari formations associated with the Quaternary sediments of Aghajari plain are located at a fault plane.

Maroun fault

Located to the northwest of the Aghajari fault, Folded Za gros, the Maroun NW-SE trending fault has thruste the Maroun anticline over the adjacent plain.

There is no instrumentally recorded data from this fault.

High Zagros fault

The High Zagros fault is characterized by a thrust which separates the High Zagros Zone from the Folded Zagros (Sepehr and Cosgrove 2007). It defines the front part of a highly-deformed zone (flake-shaped belt) including thrusted anticlines. It is the southwest border of the highest parts of Zagros; especially northward towards the Dezful embayment where Upper Paleozoic rocks are exposed in anticline cores.

The eastern termination of the Zagros fault in a NW-SE direction includes three segments; namely Gahkum, Faraghan and Khosh-Kuh which are 46, 45 and 50 km in length, respectively. The core of these structures is composed of Paleozoic rocks exposed along the right-lateral strike-slip faults, that are juxtaposed with the foreland basin deposits.

The High Zagros Fault has up to 6 km of vertical displacement (Berberian 1995). The main character of this fault is assumed to be its exposure of Paleozoic rocks. The 14th November 2017 Sarpol-e Zahab earthquake occurred very close to this fault (Berberian 2017).

Zagros Mountain Front fault

The Zagros mountain front fault which forms the southern border of the Folded Zagros, is a NW-SE trending blind thrust fault (with several segments 15–115 km in length) characterized by special structural, morphological and seismic features (Jackson and McKenzie 1984). Its total length is measured to be 1350 km in Iran and seismic activity has resulted from the 10–20 km-depth faulting.

Being as an active border since Miocene time, the period during which the Zagros foreland basin underwent extension to southwest, the Zagros Mountain Front Fault is one of the most important structures in the structural evolution of Zagros and has made a significant contribution to sedimentation in this foreland basin. This fault zone subdivides the Simply Folded Belt and has influenced sedimentation in the Zagros Foreland Basin since the early Paleocene (Sepehr and Cosgrove 2004; Sherkati and Letouzey 2004).

Zagros front fault

This fault, which is also named the "Zagros Foredeep fault" (Berberian 1995), separates the Zagros foreland from the folded Zagros (Sepehr and Cosgrove 2007). This fault partly controls the morphology of the northern Persian Gulf defining the Dezful and Fars embayments by long linear anticlines in the Lorestan area. The Zagros front fault is seismically less active than the Zagros mountain front fault (Talebian and Jackson 2004).

Kazerun fault

The N-S trending Kazerun fault is distributed along a valley with a 300 km length from High Zagros Fault of the north to the Persian Gulf of the south. According to Berberian (1995) this fault is parallel with a specific magnetic lineation well seen on the airborne magnetic map.

Strike-slip movement was firstly reported by Kent (1958). Falcon (1969, 1974) considered the Kazerun fault as the basement fault, with evidence of surface right-lateral movement. Pattinson and Takin (1971) divided the fault into two segments and interpreted the right-lateral transcurrent movement along the Kazerun fault system as the response to crustal shortening resulting from the Zagros orogeny. Favre (1975) was of the opinion that the N-S trending faults, in relation to NW-SE faults, existed before the Zagros orogeny occurred.

Being seismically active, the Kazerun fault is believed to be responsible for most of the medium-magnitude earthquakes that occurred along this fault, at the junction of the fault with Zagros Mountain Front fault. The structural characters and focal mechanism solution of earthquakes show a right-lateral strike-slip displacement along the fault (Berberian 1995; Sepehr 2001). Separating the Fars area from the Dezful embayment, the Kazerun fault is divided into the following four segments (Sepehr 2001).

1.   The Sisakht Segment

This segment is the most northerly segment of the Kazerun fault, also named the Dena fault (Berberian 1981). The Sisakht Segment with a length of 100 km is extends from the southwestern Brojen to Sisakht, and corresponds to the western slope of Dena Mountain, the highest point of the Zagros belt at 4409 meters in height. It is a part of the high Zagros fault system and is seems that the displacement along the Sisakht segment was initially controlled by the High Zagros Fault. The segment is where several earthquakes have occurred in the twentieth Century (Sepehr 2001; Fig. 3.6).

2.   The Yasouj Segment

The N-S trending Yasouj Segment, with a length of 90 km, extends from the high Zagros fault (Sisakht) to southern Nurabad. It can be observed along the main road of Yasouj-Noorabad. It seems that it was active during latest Cretaceous and early Tertiary time (Sepehr 2001). Movement had taken place along the Sisakht and Kamarij segments during the Miocene when the High Zagros and Mountain Front Faults, were the active deformation fronts and controlled the subsidence of the Zagros foreland basin (Sepehr 2001).

3.   The Kamarij Segment

This segment forms the southern boundary of the Dezful embayment and Fars area, from which the N-S trend of Kazerun fault changes to N 20°, and is roughly perpendicular to the Zagros structures. This segment is a continuation of the Yasouj fault and the Asmari Formation is downthrown about 6 km to the west (Pattinson and Takin 1971).

4.   The Borazjan Segment

The morphological features of this segment include large-scaled folded structures and pronounced topographic height in the eastern part of the segment (Fars area). Like the Komarij segment, the Borazjan segment is characterized by fold terminations at

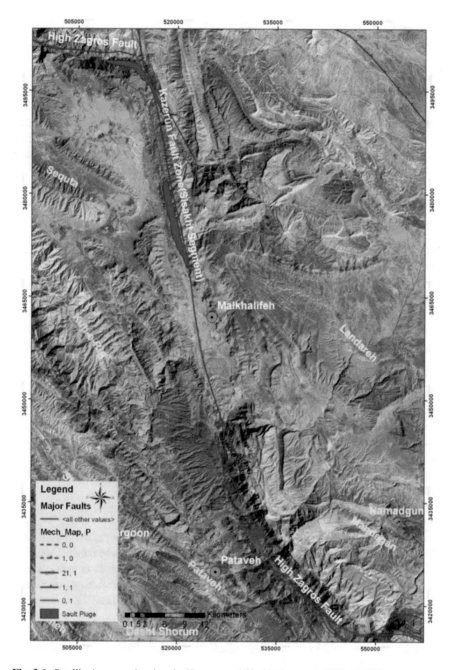

Fig. 3.6 Satellite imagery showing the Kazerun and Sisakht Segments (NIOC 2012)

the fault zone. To the west of the fault, a widespread embayment can be observed (Pattinson and Takin 1971).

Izeh Fault

The N-S trending Izeh fault is located at the western end of the Kazerun fault. The right-lateral displacement of the Zagros Mountain Front fault and the seismically highly active zone can be observed along the Izeh fault. This fault zone is located in the simple fold zone of Zagros, northeastern Dezful Embayment, is known as the Izeh Zone by Motiei (1995) and Sherkati and Letouzey (2004). The same authors divided the Izeh Zone into eastern and western parts, of which the latter shows more folding and thrusting than the former. Seemingly, the Izeh fault zone which is associated with the Kazerun Fault was activated during Cretaceous and separated the Fars basin from the Lorestan basin (Fig. 3.7).

Mengharak fault

The NNW-SSE trending Mengharak fault, which was also referred to as the "Kare-hbas fault" by Ricou (1974), is located in the Fars area, eastern part of the Kazerun fault. Although being oblique to the Zagros belt, the Mengharak fault is characterized by different features in comparison to those of the Kazerun and Izeh faults (Sepehr and Cosgrove 2007). It constituted the lateral ramp of the High Zagros fault and the Zagros Mountain Front fault. Large vertical displacement was not recorded along it.

Sabz Pushan fault

The active NNW-SSE trending right-lateral strike-slip Sabz Pushan fault with a length of 220 km extends from northwestern Shiraz (western continuation of the Ardakan fault) to the eastern and southeastern Ghir County. The mechanism of this fault was described by Safari et al. (1999) as right-lateral strike-slip and by Andalibi et al. (1998) as right-lateral strike-slip with a thrust component. In the Fars area, it is considered to be a fault zone rather a simple fault, consisting of several N170°–180° trending, en-echelon faults.

Sarvestan fault

The NNW-SSE trending, southwest dipping Sarvestan fault is a right-lateral strike-slip fault (90 km in length) located in the eastern part of the the Mahralu-Sarvestan depression. According to Berberian (1995) this fault displaced an anticline about 20 km right-laterally. In the southwestern-western Fasa and Eastern-southeastern Sarvestan, it is roughly parallel to the Kazerun-Borazjan and Karehbas transverse faults.

The Sarvestan fault, with a length of *ca.* 35 km is directly or indirectly traceable from the Gheshghar (Kolah-Ghazi) salt diapir in southwest to near the Sarvestan salt diapir in the northwest (Samadian 1984).

No direct seismic evidence associated with motion along the Sarvestan fault has been recorded. The 25th March 1890 earthquake (Ms: 6.4) of Fasa may be associated with movement on the Sarvestan fault. It is classified as a possibly active fault (Berberian 1995).

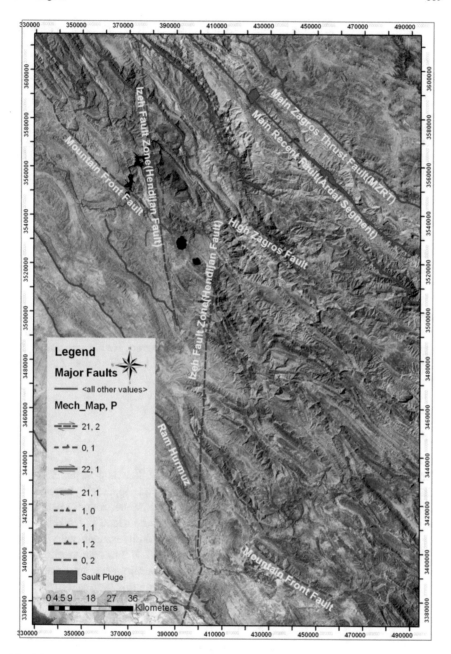

Fig. 3.7 Satellite image showing the Izeh, Kazerun, Sabz Pushan. Fault segments and a part of the Mountain front fault (NIOC 2012)

In the eastern part, Paleocene rocks are thrust over Quaternary alluvial sediments by the action of the Sarvestan Fault, which cuts the Quaternary sediments and salty and clayey zones in the middle part. It seems that this fault is a continuation of the Gowk fault zone, and created the active seismic zone in this part of Iran (Berberian 1976a).

Nezamabad fault
The N060° trending, left lateral strike slip Nezamabad fault (length 256 km) is one of the transverse-shear structures of the Zagros fold and thrust belt (Motiei 1995). This fault extends with a NE-SW trend from the northern Zagros Zone to the central parts of Fars province. It intersects and displaces the axial trace of the Kaftar anticline and Hormoz formation outcrops.

Razak fault
Barzegar (1981) described this NNE-SSW trending fault (230 km in length), which extends from the vicinity of Nakhilu Port to the adjacent fault zone on the eastern flank of Todaj Mountain, to the Razak fault adjacency to Razak Village.

Iranpanah (1986) considered the lineation as an east dipping, normal fault and believed that this fault which is associated with the Kazerun fault is a dendritic-shaped structure, overlaying the central and southern Zagros basins, along of which significant variation of thickness and lithology has occurred. He also, regarded the Razak fault as a deep basement fault, that has been periodically active during the geological history of the area.

3.4.2   The Main Recent Faults

This fault and its segments are located in the Zagros and Sanandaj-Sirjan Zones. In addition to the Main Zagros Reverse fault (MZRF) which is apparently traceable on geological maps and satellite images, there is another structure, roughly parallel to the latter, named the "Main Recent Fault" (MRF; Braud and Ricou 1971; Tchalenko and Braud 1974). It is an assemblage of NW-SE trending fault segments with a length of 800 km that extends from the Iran-Turkey border (37 N°) to southeastern Iran. The MRF is younger than the MZRF and cuts it in several places. In the areas of Neyriz and Kermanshah, studies have been carried out on this fault and a series of consecutive fault with an almost vertical slope have been encountered, the average length of each piece is about 80 km or more. Their En-enchelon arrangement, the presence of conjugate faults, small folds and thrusts are characteristics of the right-lateral strike-slip movement along the MRF (Tchalenko and Braud 1974). These authors reported this fault to have 8 segments, namely the Piranshahr, Marivan, Sar Takht, Morvarid, Dinour, Sahneh, Nahavand and Dorud. Following these studies, two segments were added to the MRF "Bakhtiari-Kuhrang" and "Ardal", which extends from the Dorud segment to the northern termination of the Sisakht fault (Jahani and Nogol-e Sadat 1999).

The most complete study carried out on the MRF was by Tchalenko and Braud (1974), who analyzed it from N 33° to N 35° and carefully examined the seismicity and structure of its segments.

Jahani (1998) studied the southeastern termination of the Dorud fault in his thesis and introduced a new segment, the "Bakhtiari-Kuhrang".

Jahani and Nogol-e sadat (1999) also introduced a new segment already named as the Ardal fault. Their field evidence indicated a right-lateral strike-slip mechanism.

The MRF is attached to the Kazerun fault system, and displacement transfers to the Zagros thrust by activation of the Kazerun fault and its associated faults (Authemayou et al. 2005).

Segments of the MRF

The MRF consists of several segments forming a seismically-active zone. The Piranshahr, Marivan, Sartakht, Morvarid, Dinour, Sahneh, Nahavand and Dorud faults as well as two new faults, the "Bakhtiari-Kuhrang" and "Ardal" faults are recognized as the main segments of MRF.

The MRF is not a distinct, single structure; it is a complex of independent right-lateral strike-slip fault segments with T, P and R shear systems, located within a narrow zone. The MRF is roughly traceable through all of the northern Zagros Zone but is less well seen in the southeastern Zagros.

Dorud Fault

This is one of the seismically-active Quaternary faults. The Dorud fault (130 km in length) trends 315°N and dips 90 towards N 33° in the southeast. After crossing Dorud City it terminates in the vicinity of Borujerd City to the northwest, where joins the Ghaleh-Hatam fault. It is divided into two segments, the northwestern segment which continues to the Silakhor Valley and the southeastern segment which, after crossing the southern margin of Oshtorankuh and Gahar Lake ends in the Ghalikuh heights at the Absefeid River (Jahani 1998). The division of the Dorud fault into two segments is only based on the 1909 earthquake, the effects of which caused ca. 40 km fault displacement in the northeastern part which is not traceable in the southwest (Tchalenko and Braud 1974). Based on the recognisable beds that are displaced by the fault, 10–60 km right-lateral displacement is recognized along the MRF (Gidon et al. 1974).

The 1909 Silakhor earthquake was associated with the activity of this fault. A 1 m vertical displacement was measured in the northeastern part but the several-meters topographic step that can be observed presumably resulted from previous movements (Tchalenko and Braud 1974).

The surface trace of the fault extends as a straight line from the southeastern mountains to the northwestern alluvial valley and indicates the verticality of the fault plain (Jahani 1998).

To a large extent, the drainage pattern around the northwestern part of the Dorud fault results from recent tectonic deformation.

Nahavand fault

The Nahavand fault begins in the southeastern Venai, western Borujerd, and strikes N 320° strike to the Ghosheh area, northwestern Nahavand. The Nahavand fault (55 km in length) consists of several segments which are named based after the closest village, viz. Venai, Amiarbad-Chakavol, Peyzaman and Ghosheh. The northeastern part of the Venai segment is downthrown and right-lateral movement inferred by the presence of slip-scratch marks is also very evident. The azimuth of the Amirabad-Chakavol fault slightly differs from that of the other segments. The marble outcrop along the fault in the Chakavol area, along the northeastern part of the fault indicates an uplifting of at least several hundred meters compared to the upthrow of the Miocene flysch that presumably resulted from the initial orogenic phase (Tchalenko and Braud 1974).

Garun fault

The Garun fault (length 25 km) is approximately parallel to the Nahavand fault and 10 km to its southwest. It extends from the Taznab area in the southeast to its junction with the Sahneh fault and Gam-Asiab River in the northwest. In the latter area, topographic terraces, each several metres high, are evident in the alluvial valley (Tchalenko and Braud 1974). Recent activity of the Nehavand and Garun faults is shown by the displacement of Quaternary and modern alluvial formations, as well as by the occurrence of recent earthquakes (Tchalenko and Braud 1974).

Sahneh fault

The Sahneh fault (azimuth: N 115°–120°) is 100 km long and connects the Garun fault in the southeast to the Morvarid fault in the northwest. The Sahneh Fault may be divided into three sections of approximately equal length, referred to here as the Southeastern, Central and Northwestern sections. Its strike slightly differs from the other MRF segments (Tchalenko and Braud 1974). The Central section starts near the town of Sahneh, approximately in the continuation of (or possibly displaced by 1 or 2 km to the NE) the previous section. It separates a narrow rectilinear limestone ridge, the Kuh Gilbesar in the southwest, from the main mountain range (Kuh Dalakhan and Kuh Nakuchal) in the northeast. The Northwestern section, which coincides over much of its length with the Zaman Valley, consists entirely of Mesozoic and Tertiary formations. The northeast side, formed of basic volcanic rocks, contains several short N trending faults terminating at the Sahneh Fault. South of the fault, the same volcanic rocks are thrust over an E-W band of Tertiary flysch. The Southeastern section starts at the intersection with the Garun Fault and follows the Gamasiab Valley, separating the Kulil Khalineh mountains from the Kangavar Valley (Tchalenko and Braud 1974).

Bakhtiari-Kuhrang fault

This was first introduced by Jahani (1998) and named to as the "Bakhtiari fault" owing to its location in the Bakhtiari Mountains. The horizontal displacement measured along the fault is more than 20 km.

The MRF discontinuously extends along the total length of Zagros and it may be possible to find new MRF segments in the southeastern part of these mountains. Studies carried out on the MRF in southwestern Iran as well as shown on geological maps generally consider it to be a reverse fault, parallel to the MZRF. However, field evidence indicates a right-lateral strike slip mechanism. Clearly the mechanism in the whole Zagros area needs more comprehensive investigation (Jahani 1998).

3.5 Central Iran

The Central Iran faults are generally recognized basement faults, which have played a major role in the formation of the Central Iran sedimentary basins. They resulted in the formation of Paleozoic horsts and grabens as well as having controlled variations in formation thicknesses and lithology. Some of them are considered to be the active faults, that caused historical and instrumental earthquakes in Central Iran.

Maiami (Shahroud) fault
This fault with (length 370 km) extends from the Garmab, eastern Sabzevar, in the east to the northern Aligholi (Chah-Jam) desert in the west. The left-lateral (Wellman 1966) Miami fault brings the Cretaceous Sabzevar ophiolite complex and Eocene volcanics into contact with Neogene-Quaternary rocks. The western end of this fault has truncated the mud plains of the northern Aligholi (Chah-Jam) desert. There have been no large instrumental earthquakes on the Shahroud fault system (SFS) in the last 50 years, but destructive earthquakes are known from historical records, including an event directly north of Shahroud in 1890 (Ambraseys and Melville 1982). The evidence indicates that this fault acted as a right-lateral fault until Pliocene time but is now a left-lateral one (Mousavi et al. 2015).

Doruneh (Great Kavir) fault
The arch-shaped, NE-SW to E-W trending Doruneh fault, with an approximately length of 700 km, extends from Naein and southwestern Kashmar (Doruneh) to the Afghanistan-Iran border (Wellman 1966). There is evidence which indicates that this fault continues to the area around Isfahan but the surface of the fault plane is covered by the Cretaceous and younger deposits. Some geologists have considered the Naein-Baft fault to be the continuation of the Doruneh fault, and believing that the latter fault is actually a Pan-African N-S trending fault, the strike of which has subsequently was changed by the Caledonian orogeny. The eastern continuation of the right-lateral Doruneh fault is assumed to be the Harat fault, detached and displaced by the Harirud fault (Aghanabati 2004). The fault was also named the "Great Kavir fault" by Stöcklin and Nabavi (1973).

Binalud fault
The arch-shaped, NE dipping, NW-SE trending Binalud fault (length of 92 km) is located 15 km east of Neyshabur, on the southwestern slopes of the Binalud Mountains. The activities of this fault have caused the sharp morphological differentiation between the northern Neyshabur Mountains and adjacent plains. It is characterized by several duplex structures and south dipping reverse mechanism (Aghanabati 2004).

The Binalud fault thrusts the Jurassic metamorphic rocks over Neogene deposits (Hollingsworth et al. 2010).

Anjilu fault
The NE trending, NW dipping, right lateral strike slip Anjilou fault (107 km in length) which extends from the southern Chah-Jam in northeast to the Torud in the southwest, truncates the Paleozoic, Mesozoic and Cenozoic rocks. This fault was mainly active during the Paleozoic and Mesozoic; especially during Cretaceous time (Kazemi-Safa 2011). According to Houshmandzadeh et al. (1978), this fault is associated with the Torud fault which has affected the area since the Cambrian.

Torud fault
The active, E-NE trending, left-lateral Torud fault with a length of 150 km, extends from the southern Dochah Mountain in the east to southwestern Rashm in the west. Its main part is located in Neogene-Quaternary deposits. It is characterized by a high-angled dip, from its linear surface trace but stratigraphic variations on the fault flanks suggest that this fault exhibited dip-slip movement during some periods of activity (Ghasemi and Ghorashi 2004).

Kalmard fault
Resembling the other Precambrian faults with a N-S trend, the Kalmard fault, located in the Tabas area, has an arch-shaped pattern. It has a N-NE trend in its northern part in the Shirgesht area and a SE trend in its southern part (Aghanabati 1975). The Kalmard fault has a E 75–80° dip and formed due to the effects of Pan-African orogeny (Aghanabati 1975). The Kalmard fault juxtaposes the Shirgesht-Tabas embayment and the Kalmard graben. In the Shirgesht area, the Permian-Triassic rocks deposited on the both flanks are lithologically different. In this area, the Kalmard fault separates two facies in the Permian and Triassic: thin and degenerated in the east; thick and fully developed in the west (Berberian 1976c). The most northerly Quaternary deposits are transected by this fault, proving its very young movement. The Kalmard fault is one of the old and deep faults of Central Iran. The bending of its northern part was linked by Nabavi (1976) to the Caledonian orogeny, while Aghanabati (2004) disputed this and assigned it to the Eo-Cimmerian orogeny. The 1933, 1939, 1991 and 1994 earthquakes were associated with the fault movements.

Poshtbadam fault
The N-S trending, right-lateral strike slip (or reverse) Poshtbadam fault which is located in the eastern Lut desert, has caused the formation of several horsts and grabens as well as the facies differentiation in the Poshtbadam area. The east inclined Poshtbadam fault is believed to have result from the Pan-African orogeny (Nabavi 1976) and is considered to be parallel to the Chapdoni fault (Tirrul et al. 1983).

Qom-Zafreh fault
The NW-SE trending, east dipping right-lateral Qom-Zafreh fault extends from the southwestern Qom to the Gavkhuni Swamp. The Cretaceous rocks of the Natanz area

are displaced about 2 km along this fault (Nabavi 1976). According to the available geological map, the Qom-Zafreh fault is associated with the Kashan, Saveh and western Ardestan faults and has played an important role in the occurrence of the volcanic rocks of the Urmia-Bazman belt. The last activity of this fault was during the time of formation of the Sahand-Bazman volcano.

Indes fault
The reverse Indes fault, located 18 km southwest of Saveh, (length of 70 km) is arch-shaped and NW-SE trending. It has caused the formation and subsidence of Saveh plain. Consisting of several parallel faults, the Indes fault forms the boundary beween the southwestern mountains and plains of the Saveh area.

The 19th December 1980 and 22nd Decembe 1980 earthquakes in the Salafchegan area were presumably associated with movements of this fault.

Dehshir (Naein-Baft) fault
The NW-SE trending, vertical dipping, right-lateral strike slip Dehshir fault has a length of at least 350 km (possible maximum length: 500 km). It extends from the southwestern of Naein city to near Sirjan. The southern termination is not evident but it possibly reaches the Jazmurian embayment and even the Pakistan-Iran border. This fault displaces Upper Cretaceous rocks by about 50 km (Amidi 1975). It also cuts off the Quaternary deposits, indicating the recent movements. No earthquake focus has been recorded associated with this fault. However, earthquake occurrence can presumably be anticipated (Berberian 1976b). The presence of a tract of colored mélange 200 km long along its western part indicates the southwestern boundary of the Central Iran microcontinent.

Sarvestan fault
The NW-SE trending, west-southwest dipping Sarvestan fault is located 75 km southeast of Kerman and has a length of 100 km. It has thrust Paleocene rocks eastward over Quaternary alluviums in the east. The middle part of this fault cuts truncates Quaternary deposits with clayey and salty zones, and acts as an oblique right-lateral strike-slip fault.

The Sarvestan fault is located in the continuation of the Gowk fault zone and, in association with the latter, it forms the seismically-active zone in the Kerman area (Berberian 1976b).

Shahdad fault
The arch-shaped, NW-SE trending, southwest dipping, Shahdad fault which is located 2.5 km south of Shahdad City, approximately defines the southwestern boundary of the Lut desert. The Shahdad thrust fault is a Neogene-Quaternary fault that has west- to southwestwardly thrust Miocene and Neogene conglomerate, marl and red gypsum-bearing sandstone over the Quaternary alluvium deposits of the adjacent plain in the northeast and east (Berberian 1976c). No instrumentally-recorded data exist on this fault.

Kuhbanan fault

The NW-SE trending, northeast dipping Kuhbanan fault, north Kerman has a length of 90 km and brings Jurassic rocks into contact with Devonian strata in the northern Kuhbanan (Huckriede et al. 1962). It also truncates hills of the Recent alluvial sediments.

The activity of this fault is evident in the Cambrian, Paleozoic, Triassic and Pleistocene. The Kuhbanan fault has right-lateral and reverse mechanisms. Owing to its truncation of the Quaternary deposits, it is considered to be an active fault, accompanied by earthquakes (Berberian 1976c).

Jorjafak fault

The NW-SE trending, southwest dipping, reverse Jorjafak fault (length of than 130 km) has thrust Cretaceous rocks to the southwest over the Quaternary alluvium to the northeast. Along the central and southeastern parts of this fault, it brings the upper Precambrian and Paleozoic rocks of the Davaran Mountain into contact with Pliocene conglomerate and Quaternary alluvium (Aghanabati 2004).

The Jorjafak fault is characterized by the occurrence of a highly crushed zone containing fault breccia, springs and fault escarpments. No instrumentally-recorded data were obtained from this fault (Aghanabati 2004).

Golbaf (Guk) fault

The NNW-SSE trending, oblique reverse Golbaf fault with (100 km long) extends from the western Bam to the Western Shahdad in Kerman province. 11 km subsidence has occurred on its western flank compared to 8 km on the eastern one. This is considered to the one of the most active faults in the area, associated with at least 6 medium to large magnitude and destructive earthquakes in the Golbaf area. It also extends on both sides of the Holocene playa. The 26th December 2003 earthquake on the southern part of the Golbaf fault zone can be regarded as a warning of future earthquake in the most southerly parts.

Nayband fault

The N-S trending, right lateral Nayband fault (Wellman 1966) extends from the eastern Shotori Mountains to the Bam area, where it forms the border between the Lut block to the east and Shotori Mountains to the west (Stöcklin and Nabavi 1971). The Nayband fault is therefore thought to be a dextral fault that had a dip-slip component in Quaternary and proceeding geological times (Mohajer-Ashjai et al. 1975). Its most northerly part has caused dropping downthrown the Bajestan and Boshruyeh deserts. The middle part played a significant role in formation of Shotori Mountains, as well as their subsequent uplift (Nabavi 1976). The displacements of the Recent alluvial deposits hints at the recent activity of this fault. The 1976 Tabas earthquake was associated with movement of one of the unknown subordinate branches of the Nayband Fault (Berberian 1976b).

Tabas fault

The NE-SW trending, reverse, Tabas fault is located at the western Shotori Mountains and the eastern border of the Poshtbadam Block. It is one of the structures that has caused 2 km of uplift. The Tabas earthquake was associated with movements of the Tabas fault.

Avaj fault

The NW-SE trending, southwest dipping Avaj fault (length 163 km) extends from the northern Saveh in the southeast to the Ghare-Bulagh in the northwest. Along the Avaj fault, low-grade metamorphic Mesozoic rocks and Eocene Volcanics are thrust over Neogene-Quaternary deposits. The 22nd June 2002 Changureh (Avaj) earthquake (Mw. 5.6) with a thrust mechanism was associated with the reactivity of this fault (Ghasemi and Ghorashi 2004).

Eshtehard fault

The E-W trending, north dipping Eshtehard fault (length of 64 km) extends from Mardabad in the east to Kalleh-Dar in the west. The stratigraphical successions of the two opposite flanks of the fault indicate a reverse mechanism. Along this fault, the Upper Red Formation is thrust over Quaternary deposits. The Eshtehard fault possibly has a strike-slip mechanism (Berberian et al. 1993).

The 1177 AD Buin-Zahra earthquake is presumably associated with either the Eshtahard fault or the Ipak faults (Nazari et al. 2011). Also, most recently the 20th December 2017 Mallard earthquake has been related to the activity of this fault.

Ipak fault

The E-W trending, south dipping Ipak fault (length 106 km) extends from Jarfarabad in the east to the Valijan in the west. The Buin-Zahra earthquake indicated that it has a left-lateral, strike-slip mechanism. Along this fault, the Upper Red Formation and Olio-Miocene volcanics are juxtaposed with Quaternary deposits, indicating a reverse component mechanism. The northwestern continuation of this fault joins the Soltanieh fault (Beberian et al. 1993).

The 3rd millennium BC Buin-Zahra earthquake is among the oldest historically recorded earthquakes in Iran. Archaeological investigations in the Sagezabad Cemetery and hills indicated the destruction of villages of the area and proved the occurrence of an earthquake. Bachmanov et al. (2004) reported a left-lateral offset of 25 m and 30 m of latePleistocene terraces on the southern slopes of the Jaushalu Ridge, across the eastern segment of the Ipak fault.

The proposed epicenter of the 10th December 1119 Ghazvin earthquake matches the Ipak fault (Ambraseys and Melvile 1982). The Buin-Zahra earthquake of 1st September 1962 is presumably also associated with movement of this fault. The earthquake was accompanied by a N50°W trending faulting which extends from the western Ipak to the near Ildarchin (a length of 80 km). The horizontal and vertical displacement were measured at 140 and 60 cm, respectively (Berberian et al. 1993).

Sabzevar fault

The E-W trending, north dipping Sabzevar fault (length 60 km) is located at the southern margin of the Siah-Kuh Mountains, northern Khorasan, where Cretaceous ophiolite and volcanic, and Eocene and Miocene red marls are exposed in its hanging wall. In some parts, alluvial and fluvial sediments are covered by marly deposits (Fattahi et al. 2006). The Sabzevar fault lies at the foot of the range, and has uplifted a band of red Neogene conglomerates, which form a low step in the topography (Hollingsworth et al. 2010). The historical 1052 AD Bihagh (modern Sabzevar), 12th December 2004 (Mb. 4.2) and 17th December 2004 Dec (Mb. 3.4) earthquakes were presumably associated with the action of this fault.

Neyshabur fault

Crossing the northwestern Neyshabur, the NW-SE trending, northeast dipping, compressional Neyshabur fault has thrust Eocene volcanosdeimentry rocks in its middle and northwestern part to the northeast, over alluvial fans to the southwest (Berberian and Yeats 1999).

According to Berberian and Yeats (1999), following earthquakes possibly occurred as the result of Quaternary compressional Neyshabur fault movements:

– The late seventh century AD Neyshabur earthquake
– The 1145 AD Neyshabur earthquake (Ms. 5.3)
– The 1251 AD Shadiakh (Neyshabur) earthquake (Ms. 5.3)
– The 7th October 1270 Neyshabur earthquake (Ms. 7.1)
– The 23rd November 1405 Neyshabur earthquake (Ms. 7.6)
– The 21st August 1928 Neyshabur earthquake (Ms. 5.2).

The instrumentally-recorded earthquakes of 16th August 1977 (Mb. 4.5) and 7th June 1984 (Mb. 4.8) which occurred near Neyshabur were possibly associated with weak movements of this fault.

The distribution of damage suggests that two of these earthquakes (1270 and 1405) ruptured the 55-km-long Neyshabur reverse fault. Historical records show that in 1209 AD, the district of Neyshabur in Neyshabur City in the west to Daneh village in the east (SE Neyshabur in the Zebarkhan district) was totally destroyed, and in 1389 AD, landslides from the Binalud Mountains destroyed several villages in the same region (Berberian and Yeats 1999).

3.6   Eastern and Southeastern Iran

Nehbandan fault

The N-S trending Nehbandan fault, located in the Nehbandan area, 250 km north of Zahedan, consists in the northern part of two fault segments; "East Neh" and "West Neh" (Berberian 1973, unpublished) meeting each other in the southern part of the Nehbandan fault. In the southern part, the Nehbandan fault is divided again into the "Nosratabad" and "Kahurak" faults, of which the former, which separates the ophio-lite of eastern Iran from the Lut block (Berberian 1976a), is the southeastern branch of the Nehbandan Fault, and the latter is the southwestern branch which extends to the northern Bazman volcano. The oldest rocks affected by the movements of the

Nehbandan fault are the Paleozoic-Triassic rocks of the Lut block, so the age of the fault is older than Triassic and it was most probably active since the Precambrian. It was the location of Neo-Tethys rifting and oceanic crust formation during the Mesozoic. The major component is believed to be right-lateral but left-lateral movements have also been recorded along its subordinate branches. Quaternary deposits have been cut by this fault, showing its recent movement.

The Nehbandan earthquake focus of 1928 probably was associated with this fault. The destruction of Nehbandan city, Shurak and Behelabad resulted from the 1991 earthquake associated with the last movement of this fault which borders the eastern Lut desert.

Bashagerd fault
The E-W trending, north dipping Bashagerd fault, located at the southern Jazmurian embayment, extends from the Kahnuj, northeastern Bandar Abbas, to the Pakistan-Iran border. It has played a major role in controlling the distribution of the colored mélange (Aghanabati 2004). Like the other faults having same trend and age, such as the Fahnuj and southern Jazmurian faults, it was formed during the formation of the Makran tectno-sedimentary basin in Mesozoic time (Aghanabati 2004). The initial mechanism has been reported to be normal but, after the subduction of Oman oceanic crust and the formation of the accretionary prism, it changed to a north dipping thrust. The Goharan-Bashagard earthquake of 11th May 2013 (Mw = 6.2) is presumably related to movement of this fault.

Harirud fault
The N-S trending, left-lateral strike slip (Nabavi 1976) Harirud Fault (length 825 km) extends from the Harirud and Tajan rivers, crosses the western Zabol plain and ends in Zahedan. The effects of this fault are very evident in Turkmenistan and the Ural Mountains, where it has formed the disconformity on both sides of the Tajan River. The initial movements of this fault presumably date back to the Pan-African orogeny, from which it continues laterally. It forms the border between eastern Iran and the western Helmand block and, like the Nehbandan fault, it separates the eastern Iran Mountains (Fig. 3.8).

3.7 Tectonic Phases of Iran

Several tectonic phases have been recognized in Iran, resulting in the general features of the country, such as the development of faults, igneous and metamorphic events and sedimentary sequences. The most important tectonic phases of Iran that are recognized are as follows:

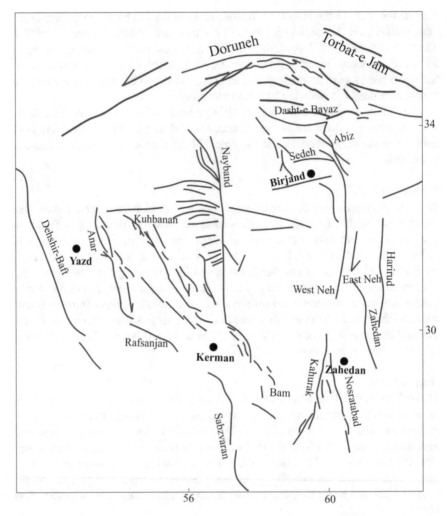

Fig. 3.8 Most faults of the east Central Iran and eastern Iran (modified after Walpersdorf et al. 2014)

3.7.1   Pan-African Orogeny

Being the time-equivalent of the Baikalian and Katangaei orogenic phases, the Pan-African orogeny affected Iran during the Neoproterozoic-early Cambrian interval. Related geological phenomena such as metamorphism, granitization, magmatism and basin development are easily recognized in Iran (Haghipour 1974). They are of such an extent that attribution to the "Iranian Orogeny" instead of the Pan-African orogeny is more logical (Nabavi 1976).

Following an initial extensional mechanism, that resulted in the development of rift basins, oceanic crust and basins in the Takab and Anarak areas, the Pan-African orogeny caused metamorphism, folding (Huckriede et al. 1962; Haghipour 1974), major fault development (Berberian 1981; Berberian and Berberian 1981) and closure of basins and rifts (e.g. A1-Shanti and Mitchell 1976). Some features of the Pan-African orogeny are described below.

Many metamorphic rocks in Iran were previously assigned to the Pan-African orogeny. However, recent studies have shown that they are actually younger than Precambrian (the age of rock or their metamorphic event; Aghanabati 2009; Zanchi et al. 2009; Rosseti et al. 2017). The geological data indicate that metamorphism was one of the effects of this phase. However, its intensity and distribution were always exaggerated; even though some of the metamorphic complexes of Central Iran and Takab developed during the same orogeny (Ghorbani 2013). The metamorphism of the Kahar, Kalmard and Tashk formations also occurred during the Cimmerian orogeny.

In Central Iran, the Azerbaijan and Alborz, the Neoproterozoic-lower Paleozoic rocks that are overlain by the Barut, Zaigun, Heshem, Aghda and Lalun Formations, are locally and regionally metamorphosed, but the overlying formations show no signs of metamorphism. This indicates low-grade metamorphism resulting in slate, phyllite and schist facies, presumably being simultaneous with the closure of marine basins and their subsequent folding. However, during Precambrian, rifting of deep basins has cuased oceanic crust development, high-grade metamorphism to amphibolite facies or even migmatization in several places (Ghorbani 2014).

Faulting

Most of the Iranian faults affecting on the subsequent geological phenomena in Iran were created during this phase (Aghanabati 2004). They are characterized by a parallel N-S trend and echelon geometry. Faults that developed during this phase include the Harirud, Nehbandan, Nayband, Kalmard, Qatar-Kazerun faults and the Main Zagros Thrust. The N-S trending structures of the Iranian Plateau attributed to this phase are located on islands to the south of Iran, such as Bahrain, Qatar and United Arab Emirates (Aghanabati 2004).

Magmatism

This phase is characterized by a large volume of basic and acidic volcanic masses and granitic to gabbroic intrusions. Some of these include volcanic rocks which belong to the Kushk, Rizu and Desu formations (Huckriede et al. 1962), as well as the Hormoz and Gharedash series (Stöcklin 1968). The schist and amphibolite exposures of the Anarak and Takab areas having basic-ultrabasic protoliths can also be considered as magmatic consequences of the Pan-African orogeny. Of the igneous intrusions, the Doran-type granite of Azerbaijan as well as Narigan-Zarigan type of Central Iran can be mentioned (Ghorbani 2013).

3.7.1.1 Evolution of the Igneous Basement in the Neoproterozoic

In the Neoproterozoic, the Iranian basement evolved to the continental crust owing to the several magmatic, metamorphic and folding events. The establishment of a platform territory, persisted to the latest Paleozoic, is considered as one of the consequences of Pan-African Orogeny (Forster et al. 1972).

Development of fractures provides the introductions for establishment of separated basins, which their evidence is visible in subsequent events. Some of the metallic ore deposits of Iran were also developed during this phase (Forster et al. 1972).

Some geologists were of the opinions that the development of latter sedimentary basins such as Zagros were related to the Pan-African consequences and even the framework of modern Alpine orogeny is associated with the Katangaei basement trends (Berberian 1983).

3.7.2 Caledonian Tectonic Phase

The Caledonian orogeny in northwestern Europe originated with the closure of the Caledonian (Iapatus) Ocean. Consequently, the north Atlantic regions were affected by the Caledonian orogeny during the Ordovician—early Devonian time interval.

The obvious geological evidence such as acidic magmatism (dacite to rhyodacite) in Ghaleh-Chai, Ajabshir, and that of the Maku metamorphism are related to Caledonian activity. The metamorphosed Sargaz volcanics and slates were also considered by Sabzehei (1974) to be the effects of same orogeny. However, according to the present author, the evidence is not related to the Caledonian phase and they were actually affected by either the Hercynian or the Eo-Cimmerian orogeny. No folding and faulting was associated with this phase because Iran, being far from the Caledonian collision zone, suffered only epeirogenic movements characterized by regional regression of the Silurian sea, regional disconformity (Berberian and King 1981) and some local unconformities (in the north) at the base of the Middle-Upper Devonian rocks (Stöcklin 1968; Stampfli 1978).

In the late early Cambrian, Caledonian activity caused some geological changes; with the marine Barut and Zaigun Formations grading into the continental Lalun Formation. According to (Aghanabati 2004), this facies change was accompanied by an angular unconformity and even a stratigraphical gap at the Barut-Zaigun boundary but this interpretation is rejected by the present author because in fact, these geological features are related to the latest activities of Pan-African orogeny.

The effects of the Caledonian orogeny began in latest Cambrian time with facies changes, and continued through the Middle Devonian, represented by Eifelian hiatus or erosional phases which are considered to be the prominent consequences of the Caledonian orogeny in Iran (Berberian and Hamdi 1977). Most researchers consider this as an epeirogenic phase (Berberian and King 1981; Aghanabati 2004, 2009; Hairapetian et al. 2017).

It seems that, in some places, the Caledonian event was accompanied by an extensional phase whose climax occurred during Ordovician and Silurian. For example, basalt bodies of considerable size can be observed in the Shahroud area (Khosh-Yeilagh, Soltan-Meydan, and Bastam; Jenny 1977; Stampfli 1978) in the Niur Formation (Ruttner et al. 1968).

This event is considered to have had more effects than previously considered (e.g. Nabavi 1976; Berberian and Hamdi 1977). According to the present author, most of attributed effects, such as the metamorphic rocks of Alborz and Maku were actually produced by subsequent events. The only prominent activity that can be attributed to the Caledonian orogeny is the extensional phase that resulted in the development of Ordovician-Silurian volcanic rocks, especially in northeast, north and east Iran. This can be linked to the opening of the Paleo-Tethys and is especially prominent in northern Iran.

3.7.3 Hercynian Tectonic Phase

Based on the geological evidence, the Hercynian phase began in the early Late Devonian and ended in late Early-middle Triassic time. The role of this phase in geological deformation has been largely neglected owing to the absence of any major magmatism, metamorphism and folding. Extensional events were prominent rather than compressional ones. During this phase, the Paleo-Tethys Ocean developed, also referred to as Tethys II. Actually, the development of Tethys II, its subsequent dwindling and related phenomena such as basaltic volcanics can be regarded as the most important effects of the Hercynian event.

The establishment of marine basins and successive marine transgressions and regressions are assumed to be associated with the events described above. The consequences of these are as follows:

- Main transgression during the Late Devonian-early Carboniferous, facies variation during the Carboniferous and subsequent Late Carboniferous regression;
- Major facies changes and locally-developed Carboniferous stratigraphical gaps.
- Marine transgression of the Early Permian sea and its regression in some parts of Iran.

The establishment of the Tethys rift and its development in north Iran, accompanied by an extensional event, generated a considerable volume of volcanic rocks (Majidi 1978, 1981; Davies et al. 1972) that were metamorphosed by subsequent phases. Some of the metamorphic rocks in this igneous province, especially the mafic type, formerly assigned to the Precambrian, are related to this phase, for example, the protolith of the Gorgan, Shanderman and Gasht schists (Zanchetta et al. 2009), the Fariman ultramafic rocks (Majidi 1981) and the Aghdarband erosional window (Afshar-Harb 1979). However, metamorphism may have occurred during subsequent events (Aghanabati 2004; Zanchi et al. 2009).

- Volcanic rocks of rhyolite, trachyte and andesite in Sanandaj-Sirjan Zone such as Songhor series
- Andesi-basaltic rocks associated with the Permian sedimentary rocks in the Sanandaj-Sirjan Zone and some parts of Alborz (Delavari et al. 2016).

Actually, the Tethys II event occurred during the Hercynian event, i.e. the Tethys II development and its closure, being the most significant event of Iran's geological history should be considered as the consequences of Hercynian orogeny. This event has not yet been yet been considered in the context of Paleo-Tethys discussions and so its triggering mechanism has been largely ignored. It should be mentioned that the Hercynian event is more evident in northern Iran (e.g. Rosseti et al. 2017). Furthermore, recent research has proved that the extensional regime which dominated in the Carboniferous-Permian of the Alborz has resulted in re-activation of inherited basement and synsedimentary faults, causing significant lithostratigraphic, facies and even age variations on the blocks of these faults (Zandkarimi et al. 2016, 2017a, 2017b, 2019).

3.7.4 Eo-Cimmerian Orogeny

The Eo-Cimmerian orogeny (Stöcklin 1974a) is one of the major geological events of the Earth's history. During this phase, Iran experienced several events including metamorphism, folding and faulting, basin development, facies change and magmatism. It is also associated with a compressional phase in the north and an extensional phase in the south. In the south of Iran, during or slightly before the compressional phase (Early Triassic or even older), extension and rifting commenced which resulted in Paleo-Tethys or Tethys II closure. Most probably, the Cimmerian orogeny started during the final stages of the Hercynian orogeny.

The volcanic rocks of the Songhor series include rhyolite, trachyte and andesite and were presumably associated with the extensional phase of the Eo-Cimmerian event. The other related events are as follows:

Sedimentary basin and facies variations
During this event, two large, distinct basins were created, their boundaries corresponding to the Main Zagros Thrust. The south-southwestern basin encompasses the recent Zagros and the northern Arabian plate. The stratigraphic succession on this deep water platform consists mostly of limestone, dolomite, marl and mudstone. In the northern basin which encompasses Sanandaj-Sirjan, Alborz and Central Iran, most of deposits are of continental and shallow provenance, showing major facies difference in comparison to the southern basin. This started with the onset of a continental rift associated with the volcanic activity, and formed the narrow and elongated ocean, namely Neo-Tethys Ocean, here termed Tethys III". With the initial opening of Neo-Tethys, being contemporaneous with the latest Hercynian phase, the southern basin pushed the Iranian plate northward, resulting in the final closure

of Paleo-Tethys, Tethys II, a process that had begun in the early Hercynian. As a result, the basins located between the Iran and Turan were closed as the Iran and Turan plates collided. The same depositional regime is observed in the north Iranian basins, northern Afghanistan, Turkmenistan, and northern Europe.

Metamorphism

The compressional forces that resulted in the closure of Tethys II in the Middle Triassic also caused severe metamorphism of the Iranian plate. As a consequence, its evidence can be observed through most of north Central Iran, Sanandaj-Sirjan, and the suture zone of Iran-Turan plate (Sabzehei and Berberian 1972; Sabzehei 1974). Some of more important metamorphic exposures related to this event are as follows:

Northern Iran (Stocklin 1974; Majidi 1978):

- Shanderman, Gasht and Asalem schists
- Metamorphic rocks of the Taknar area
- Gorgan schist
- Metamorphic rocks of the Aghdarband mafic-ultramafic
- Metamorphic rocks of the Maku area.

Central Iran
Jandagh Schists

Metamorphic rocks of the Biabanak and Bafgh areas, green schist facies (Huckriede et al. 1962).

Sanandaj-Sirjan (Sabzehei 1974)

- Muteh schists
- Hamedan slate and phylite rocks such as those of Aligodarz
- Metamorphic rocks of Songhor series and surroundings of iron ores of Galali-Babali
- Metamorphic rocks, marble and green schists of Saqqez-Divandareh
- Metamorphic rocks of Esfandagheh, and Kuli-Kesh, Surian and Tunak complexes.

Unconformity

An angular unconformity can be observed between Lower -Middle Triassic or older strata and the younger overlying beds. This resulted from the Eo-Cimmerian orogeny in the Middle Triassic (Aghanabati 2004).

Magmatism

Following the Middle Triassic extensional phase with its associated magmatic intrusions, an Upper Triassic-Lower Jurassic extensional phase affected the Alborz, Sanandaj-Sirjan and Central Iran (Alric and Virlogeux 1977; Berberian and Berberian 1981). The results of this are assumed to be the formation of the Shemshak basin and the creation of horsts and grabens, accompanied by a large quantities of volcanic rocks in Central Iran, Alborz, and especially Sanandaj-Sirjan.

During this event, volcanic rocks are dominant; though these have not yet been investigated in detail. However, rare studies have mostly indicated an alkaline-type provenance. These rocks are overlain by the Shemshak Formation, consisting of the Babun basalt of Central Alborz and the Lut Block (Birjand and Abgarm areas; Aghanabati 2004). The volcanic rocks of this phase are more widely distributed in the Sanandaj-Sirjan Zone than elsewhere.

3.7.5 Mid-Cimmerian Orogeny

The Mid-Cimmerian orogeny was firstly introduced by Aghanabati (1998) who presented a substantial body of supporting evidence. The study of Jurassic strata, magmatism and metamorphism evidence shows that this is an orogenic event accompanied by folding, volcanic activity, emplacement of igneous intrusions and even metamorphism. These events indicate that most of the phenomena formerly assigned to the Late Cimmerian actually belong to this event. Contrary to previous interpretations, the Eo-Cimmerian phase is believed to have had only epeirogenic effects. The most significant events of this phase are as follows:

Cessation of sedimentation
Across wide areas of Central and Northern Iran, the sedimentation ceased and erosion occurred during the Middle Jurassic (Bajocian-Bathonian) (Aghanabati 2004; Seyed-Emami et al. 2004; Fürsich et al. 2009).

Emplacement of igneous intrusions
In most of Iran, large and small magmatic intrusions were emplaced during the Late Jurassic such as the Shirkuh Granite (176 ± 8 Ma)m, the Sorkh-Kuh Granite of the Central Lut desert (165–170 Ma) (Tarkian et al. 1984), the Arusan Granite (165 Ma; Reyer and Mohafez 1972), and the Masuleh Granite (175 ± 10 Ma; Crawford 1977) all of which correspond to a Bathonian age (Ghorbani 2014).

Metamorphism
The metamorphism of Lower to Middle Triassic rocks was one of the consequences of the Mid-Cimmerian orogeny. In the middle part of the Sanandaj-Sirjan belt, Eghlid area, the Lower Jurassic rocks were slightly metamorphosed and the upper Jurassic conglomerate disconformably overlies them (Houshmandzadeh and Soheili 1990).

The Hamedan phyllite of Golpayegan, Khansar, Broujerd, Malayer and Hamedan as well as slate and phyllite rocks with Upper Triassic-Jurassic pelitic protoliths were all clearly metamorphosed during this phase, as they are overlain by unmetamorphosed Lower Cretaceous rocks.

3.7.6 Late Cimmerian Orogeny

During the Late Jurassic-Early Cretaceous, about 140 Ma ago, Iran experienced a tectonic event whose evidence can be observed as the folding, faulting, and the facies variations across the basin, unconformity, magmatism and metamorphism. The most important consequences are as follows:

Continental condition
The late Cimmerian compressional phase resulted in the exposure of most of basins. Where marine deposition continued, facies variations occurred. The resulting basin shallowing and clastic-dominated regime can be traced throughout Iran in areas such as Central Iran, Sanandaj-Sirjan, Alborz and even Kopeh-Dagh and Zagros.

The presence of an unconformable contact between the Jurassic and Cretaceous beds can be explained by this event. Occurrences of the refractory mineral deposits in the Sangrud of Loshan and the Kaftarrud of Gonabad, as well as evaporites in the Damavand area indicate the arid environmental conditions that dominated during this time.

Magmatism
This tectonic event was accompanied by intrusive magmatism in the Late Jurassic-Early Cretaceous, resulting in some of the intrusions of Central Iran (Shirkuh, Tut, Zarrin and Jandagh granite intrusions), Alborz (Msulteh and Ghaser-Firuzeh granite) and Sanandaj-Sirjan (Kolah-Ghazi granite). The volcanic activity of Alborz, represented by the Gypsum-Melaphyre Formation (Allenbach 1966; Steiger 1966), underlying the Lar Formationwas also considered to be related to the event.

Extensional phase
Following the initial compressional phase of the Late Cimmerian orogeny, an extensional episode occurred during Early-Middle Cretaceous time, represented by the formation of sedimentary basins, mostly with carbonate deposition, and considerable volcanic activity. Mostly distributed in the northern Sanandaj-Sirjan Zone, the volcanic rocks with andesitic-basaltic composition can be traced through most of this zone, for example along the Divandareh-Saqqez road, in the Piranshahr area, the Sanandaj area, etc. The opening of the Neo-Tethys in eastern Iran (Nehbandan-Khash), the northern Doruneh fault and margins of Central Iran microcontinent are considered to be the most important consequences of this extensional episode.

Metamorphism
Following the compressional and folding episodes, metamorphism of Jurassic and older rocks of Sanandaj-Sirjan also occurred, indicated by the highly-metamorphosed rocks of the Shemshak Group along the Golpayegan-Hamedan road (Houshmandzadeh et al. 1978). This group is overlain by the mid- to Upper Cretaceous rocks.

3.7.7 Laramide Orogeny

This was one of the most important tectonic events in Iran, which resulted in significant geological consequences in the Late Cretaceous to Eocene of the country. A compressional phase was followed by an extensional phase whose activities are considered to have had substantial effects. During the compressional episode, closure of the oceanic basins and Neo-Tethys Ocean commenced. In some areas, oceanic crust was thrust or uplifted, forming the ophiolites and colored mélange of Iran (Dimitrijevic 1973; Setudehnia 1978; Berberian and King 1981).

The compressional episode of this phase was accompanied by magmatic intrusions that are traceable in the Sanandaj-Sirjan and in Central Iran (see Ghorbani 2014). Syn-folding metamorphism, closure of oceanic basins, rifting, continental collision and formation of deep trenches also occurred. The most important rocks are as follows:

The schists and amphibolites of south Birjand; schists and glaucophane schists of Fahnuj and all of the metamorphic rocks associated with the ophiolite complexes. In addition to these rocks, evidence of regional metamorphism can be observed in the northern Sanandaj-Sirjan and eastern Iran flysch zone. As a result, in the Sanandaj-Marivan, Mahabad and Piranshahr areas, lower Cretaceous rocks experienced low and very low pressure-temperature metamorphism to green schist facies (e.g. Eftekharnezhad 1980).

Following the Late Cretaceous-early Paleocene compressional episode, an extensional phase occurred during the Late Paleocene-Eocene reaching its climax in the middle Eocene. This resulted in the formation of sedimentary basins, accompanied by volcanic activity with a range of compositions from basaltic to rhyolitic, though mostly andesitic.

3.7.8 Pyrenean Tectonic Phase

This was an intense compressional event, the evidence for which is variation of sedimentary environment (facies variations), sedimentary disconformity and folding.

The above changes can be traced at the top of the Paleocene-Eocene succession and the base of the upper Eocene-lower Oligocene. During this time, the modern morphology of Iran was formed, with most basins experiencing continental conditions. All rocks older than Oligocene were eroded and a great quantity of clastics was transported to the subsiding areas, fans and inter-montane basins, resulting in the deposition of the clastic lower red formation.

Plutonism

During this event, granite-gabbro intrusions were emplaced over most of Iran, except for Zagros and Kopeh-Dagh. The volume of magma intruded during this period exceeds that of all former periods.

Most of the intrusions previously attributed to the earlier event, are here assigned to the Pyrenean event, for example, the gabbro intrusions of northern Sanandaj-Sirjan

(Piranshahr, Gharreh-Bagh, Qorveh and Sardasht; Sabzehei 1974) and the Alborz Mountains (Alamkuh, Azadkuh, Taleghan and Masuleh; Reyer and Mohafez 1972).

Metamorphism

It seems that the the upward transfer of heat resulted in the emplacement of intrusions and core-complex-type metamorphism. Apparently, most of the metamorphic complexes of Iran can be attributed to core-complex type metamorphisms that was prevalent during this event.

– Metamorphic rocks of Azerbaijan (Azizi 2001)
– Poshtbadam Complex of Central Iran (Houshmandzadeh, pers. comm.)
– Neybaz Complex (Ghorbani 1999)
– Metamorphic rock of Homont in southern Iranshahr (Ghorbani 1999).

A unique interpretation of the formation of the above rocks is not obvious; owing to the fact that the processes that formed them is complex. Indeed, this problem should be addressed in the geology of Iran. However, if the Homont Mountain is studied from a core complexperspective, ideas concerning the metamorphism of other areas can be evaluated.

3.7.9 Savian Tectonic Phase

During this phase (late Oligocene-early Miocene) which was associated with extension and subsidence, basin depth gradually increased, resulting in deposition of calcareous sediments over the continental Lower Red Formation. The extension episode was accompanied by increased depth of marine environments and extensive volcanism continued to the Burdigalian. The major rock components of this time interval are tuffite, tuffaceous rocks, volcanic lava, marlstone and limestone. In some areas such as Takab, Tafresh, some parts of Azerbaijan and eastern Iran, major components are Qom equivalents, tuffite and volcanic lava. The tuffite and tuffaceous rocks are more widely distributed while volcanic lavas are only observed along faults or at the junctions of faults (Aghanabati 2004).

3.7.10 Styrian Tectonic Phase

In Iran, the activities representing this phase are explained by the uplifting of the Qom basin, with associated changes of sedimentary conditions from marine to continental, resulting in changes from calcareous facies to silty sandstone, conglomerate or, in places, claystone. According to the present author, no sedimentary disconformites, either angular or otherwise have been observed between the Qom Formation and the overlying Upper Red Formation that could be associated with this phase. The

lithological variation of the upper lower to middle Miocene succession can be linked
to the rapid fall in global sea level at this time.

3.7.11 Atican Tectonic Phase

The Atican phase (late Miocene-early Pliocene) is the most evident phase of in the
Cenozoic of Alborz, eastern and Central Iran. Its consequences include folding and
deformation of older rocks, especially those of Oligo-Miocene age. The present day
morphology of Oligo-Miocene rocks at surface was formed during this phase. The
disconformity between the Oligo-Miocene (especially the Upper Red Formation)
and the overlying Pliocene conglomerate is observed everywhere this contact crops
out, such as in the Kheyrabad, Moshampa and Mahneshan areas. During this phase,
the Zagros zone was exposed and experienced continental conditions.

In the most of Iran, such as Azerbaijan (Ahar area), Sabalan, the Urmia-Dokhtar
belt and eastern Iran, this phase can be traced by associated severe magmatism.
Most of small intrusions of granite, tonalite and diorite bodies were intruded into
the Iranian crust at this time, especially those in eastern and central Iran. They have
been accounted to be indications of Tertiary magmatism in Iran. According to the
present author, these intrusions are of metallogenic importance, because most of the
epithermal gold and porphyric copper reservoirs of Iran are associated with the acidic
to medium composition bodies of this phase.

3.7.12 Pasadanian Orogeny

Being the youngest orogenic phase of Iran, this most important tectonic phase that
formed the modern morphology of Iran is assumed to represent the Pasadanian phase
(late Pliocene-early Pleistocene). It is characterized by:

Folding and uplifting of Zagros and the increase of topographical dip between
northern Zagros and southern Zagros (Khuzestan Plain, Fars and southern coasts of
Iran).

The relocation of Iran between the Arabian plate in the west, Indian plate in the
east and Turan plate in the north; the change of mechanism to strike-slip movements;
compressional mechanisms associated with crustal shortening, and thickening and
earthquake events with compressional mechanism (Aghanabati 2004).

The beginning of new but scatteed episode of volcanism in Iran and its termination
except for some areas such as Damavand and Taftan presumably are not related to
the Cenozoic volcanism.

Some of the volcanic activities related to this event are described below:

Volcanic activity in Central Alborz, Damavand, Alamut, Taleghan, though their
initial activities are presumably related to the earlier event, and primary eruption
occurred during Pasadanian event.

- Old Sabalan collapses and there is a subsequent decrease of height; development of a collapse caldera and young Sabalan.
- Young basaltic activities in eastern Iran, Central Iran and Azerbaijan (however some of these presumably began before this phase).
- Onset of volcanic activities and its termination in the northern Sanandaj-Sirjan Zone.
- Development of volcanoes in the Taftan-Bazman belt (Taftan, Kuh-e-shahi, Bazman and Kuh-e-Soltan of Pakistan) and subsequent cessation of volcanic activity.
- Major earthquake activity in most of Iran with reduction at the end of this phase. The modern earthquakes are assumed to be the continuation of faulting, having low magnitude and frequency in comparison to those of the Early Quaternary.
- Formation of deep valleys and their subsequent destruction caused by dip variations in the late stages of this phase.

An extensional mechanism in the early stages of this phase (Early Tertiary to late Miocene) later was changed to a compressional phase. Some evidence of thrusting and crushing are evident, such as the fractures on the Zanjan-Tabriz road (Fig. 3.9), which have resulted from the activities of this phase,

Fig. 3.9 A view of the Recent faults along the Zanjan-Tabriz Road

3.8 Seismotectonic of Iran

As already mentioned, Iran is situated in the middle part of Alpine-Himalayan belt (Stöcklin 1968; Berberian and King 1981; Alavi 1991). The stable cratons such as northern Africa, Arabia and India are located in the southern part of this belt, whereas southern parts of Eastern, middle Asia and China are in its northern part. The Alpine-Himalayan belt is a seismic belt that most global earthquakes are associated with it (Fig. 1.7; Berberian 1981, 2014, 1995; Berberian et al., 1996).

The geological and tectonic history of the Iranian plateau indicates an intracontinental plate whose sedimentation processes are controlled by several major basement faults with different mechanisms.

The recent collisions of the Iranian plate with neighboring plates have resulted in crustal thickening and shortening associated with reverse faulting and folding, as well as subsequent exposure of the areas (Berberian and King 1981; Aghanabati 2004). The Iranian plateau is composed of several microcontinents, mechanically delimited by old, major faults (Stöcklin 1968; Berberian and King 1981; Berberian 2014). These differ seismically from the immediately adjacent plates.

The old faults are those that affected both old and young sedimentation and the caused thickness variations of formations on both their sides (Berberian 1976b).

The following account builds on the previous descriptions of the faults and discusses the tectonoseismic activity. The records of Iranian earthquakes are then listed.

3.8.1 Zagros Active Fold and Thrust Belt

Located at the northeastern margin of the Arabian Plate, the Zagros Fold and Thrust belt is characterized by a Precambrian basement and a relatively simple geological history, indicating continuous sedimentation in a basin which was subsiding during the Precambrian-Miocene interval. Although the collision was distributed through the Zagros fold and thrust belt, Alavi (1980) believed that it extended further into the Sanandaj-Sirjan Zone. From the Precambrian to the Permian-Triassic, it was dominated by a shallow platform, subjected to several marine regression recorded in the lithological units (James and Wind 1965; Setudehnia 1975; Alavi 1980). This belt was subsiding during the Late Permian-Triassic, and a thick sedimentary succession accumulated, indicating the opening of Neo-Tethys Ocean between Central Iran and the Arabian plate. Sedimentation in the Neo-Tethys Ocean continued until Miocene time, with the sedimentation type (mainly carbonates) suggesting a passive continental margin with contemporaneous Mesozoic crustal thinning and subsidence. This subsidence occurred along the Main Zagros Fault due to the inter-continental rifting, initiated in the Late Paleozoic that subsequently lead to crustal thinning caused by normal faults.

During Miocene-Pliocene time, tectonic activity was initiated that resulted in the formation of some NW-SE structures, although the High Zagros had already been affected by the same compressional mechanisms in the Late Cretaceous (Setudehnia 1975; Alavi 2007).

The collision that occurred between the Arabian and Iranian plates resulted in re-activation of pre-existing normal faults that had originated during the opening of the Neo-Tethys. The main deformation of the Zagros area is characterized by the reversal of movement direction of pre-existing normal faults. The folding intensity decreases significantly from north to south. The southern and southeastern flanks of the Zagros folds are mostly reversed, indicating the over-thrusting mechanism of the faults at depth (Berberian 1976b).

At the present time, the Zagros fold and thrust belt is experiencing shortening in its northwestern areas, where it is ca. 200 km wide compared to 350–400 km in the southeastern regions. Counter maps of Zagros suggest a 50–55 km thickness for the Zagros crust, indicating the Arabian-Iranian collision zone or even the subduction of northern edge of Arabian plate beneath the Central Iran Microcontinent (Berberian 1976c).

In this wide belt, compressional movements were accompanied by shallow earthquakes having less than 30 km focal center. The number of surface centers of earthquakes increases towards the northeast and the focal depth slightly increases (Berberian 1976a, b). The seismicity of this belt terminates with the main Zagros reverse fault and in southeastern part to the folds. The Minab-Zendan Fault is regarded as the southeastern limit of Zagros belt. However, as already mentioned, the seismicity of this belt decreases beside this fault. The earthquakes of Zagros belt are of low-medium magnitude, i.e. less than 7.

Analysis of fault planes in the twentieth century indicates a high-dip thrusting mechanism. These faults, which trend NW-SE are buried by recent sediments in the folded Zagros (Sepehr and Cosgrave 2007).

The thickness of the alluvial terrace together with geophysical evidence such as the young age of the rivers indicates the rapid rate of uplift and high seismicity of this area. The minimum rate of uplift of the folded Zagros was measured to be 1 mm/year during the Pliocene to Recent interval. Geological sketch sections indicate ca. 20% shortening of the Zagros Ranges by ca. 250 km. Considering the syn-tectonic sedimentation of the Bakhtiari Formation that dates back to 5 Ma ago, it can be inferred that the beginning of this shortening and uplift is of the same age as Bakhtiari Formation.

It is also of note that the northern part of the belt was exposed earlier than the southern part, and, therefore, the Bakhtiari Formation is of older age.

The Zagros Range is the widest mountain range in Iran. It is unique in this area and is divided into several subzones. The boundaries of this zone are generally defined by deep thrust faults.

3.8.2 Alborz

Located at the southern end of the Caspian Sea, the Alborz Mountains are an elongated E-W trending structure which has been considered to be part of the Alpine -Himalayan belt. Geographically they extend from the Lesser Caucasus and Talesh mountains in the northwest and west to the Kopet-Dagh and Binaloud in the northeast and east of the Iranian Plateau, respectively. In terms of geological characterization, the Alborz extends from the southern coast of the Caspian Sea, from Talesh in the west to their junction with the Kopeh-Dagh in the east (Ghorbani 2013; Berberian 2014). Except for Kopeh-Dagh, all of these areas are seismically and structurally similar.

The Alborz Mountains include Neoproterozoic to Quaternary rocks in complex structures of folds and thrusts formed during multiple orogenic events such as the Pan-African, Hercynian and Alpine orogenies (Stöcklin 1974b; Berberian and King 1981; Alavi 1996). However, it is believed that the most important tectonic event in the formation of the Alborz was the Cimmerian orogeny (Zanchi et al. 2009).

Uplift of the old structures associated with the Pliocene to Recent geological movements, are mostly controlled by E-W trending basement faults, such as the Roudbar, Taleghan, Mosha, Firuzkuh, Astaneh and Jajarm faults in the interior part, the Caspian and North Tehran faults in the northern areas, and the Damghan, Garmsar and northern Gazvin faults in the southern Alborz (Nazari and Shahidi 2011).

The presence of abundant Quaternary faults in the Alborz, especially on its southern flanks, the shape and the high frequency of earthquake occurrence during the last century (see the list of Iranian earthquakes, Table 3.1) indicates the active nature of these faults. The occurrences of historically-recorded large earthquakes along the old geological structures within this zone indicate reactivation of some basement faults along this northern edge of the Iranian Plateau. Due to having no relation to the old faults and the relatively thin crust, the earthquakes of the Alborz zone have almost been of large magnitude with relatively high destructive effects (e.g. Berberian and Yeats 2016). The earthquakes recorded in the southern part have been more intense than in the northern part.

3.8.3 Kopeh-Dagh Belt

This belt with its Hercynian basement is located at the most southerly margin of the Turan plate and the northeastern margin of Iran (Stöcklin 1968; Afshar-Harb 1979; Berberian and King 1981; Ghorbani 2013). The sedimentary cover on this basement is 80 km thick and consists of Mesozoic to Recent rocks which were deposited in a gradually subsiding basin (Afshar-Harb 1979). Although, the extent of the basin was not widespread, the thick sedimentary cover hints at a significantly thinned crust. This crustal thinning was associated with several normal faults (Afshar-Harb 1979).

The NW-SE trending mountains were folded during the last episode of the Alpine orogeny. The minimum crustal shortening in the Iranian part of these mountains is 15% and the present width was measured to be 70 km. In the western Kopeh-Dagh, the folds are characterized by low dips on their northern flank and high dips in their southern flank, associated with thrust faulting. In the eastern Kopeh-Dagh, symmetrical folds are crossed by a number of shear faults (Afshar-Harb 1994). These later faults, such as the Baghan-Garmab fault, have caused folding and longitude displacement associated with reverse faulting.

The Kopeh-Dagh belt is limited to the north by the major Kopeh-Dagh reverse fault and to the south; it is separated from Central Iran by a great number of reverse faults (Afshar-Harb 1979). The northern Mashhad Mountains and Hezarmasjed faults mark the southern part of this belt.

The salt layers and evaporative sediments, which are characteristic of the Zagros Zone, cannot be seen here. Meanwhile, the subordinate formations such as the Shurijeh and Pesteligh formations have plastic properties. The latter, with an age of Late Jurassic-Early Cretaceous includes gypsum layers, and the former, of Paleocene age, is composed of gypsiferous mudstone (Afshar-Harb 1979, 1994). The thickness of these formations is high in the Kurkhud quadrant but in the Mashhad area it is thin or absent. This caused the differential deformation in the western and eastern parts. As a result, in the western part, most of the folds are inverted and their formation was related to the deep thrust faults, while the deformation and folds in the eastern part were associated with strike-slip faults.

The belt is also characterized by north-dipping thrusts, such as the Main Kopeh-Dagh reverse fault in the north and south-dipping thrusts in the south. However, the north-dipping thrusts are more abundant.

The Kopeh-Dagh belt is associated with widespread distribution of shallow, large magnitude earthquakes. The right-lateral displacement of fold axes caused by the activity of the Baghan-Garamab fault was measured to be 9 km. If we consider the movement initiation in the Pliocene period based on the syn-tectonic sediments of foreland regions, then the displacement rate has been 4–5 cm/year (NIOC 2012) from the Pliocene to Recent. The Moho-level curves indicate a thickness of 45 km for the northern part.

Apparently, the crustal thickness of southern part is more than that of the northern part, reaching 50 km, owing to the presence of numerous reverse faults (Afshar-Harb 1979, 1994; Robert et al. 2014).

3.8.4 Central Iran

The Central Iran Zone is located between Zagros, Alborz and Kopeh-Dagh. It was affected by several tectonic events and is characterized by several syn-tectonic metamorphic events, especially during the late Paleozoic (Haghipour 1974; Berberian and King 1981), Middle Triassic (Wilmsen et al. 2009), Jurassic and Late Cretaceous (e.g. Wilmsen et al. 2015) with intense metamorphism and multiple eruptions in its southern part. During the Precambrian to Late Paleozoic, it was a part of the Gondwanaland and was separated from Eurasian plate by the Hercynian Ocean (Berberian 1983). The late Paleozoic rifting between the Iranian and Arabian plates was caused the separation of Central Iran from the Arabian plate, with the subsequent opening of the Neo-Tethys Ocean and final collision of the Central Iran and Turan plates during the Cretaceous to Miocene interval (Stöcklin 1968; Berberian 1983; Muttoni et al. 2009).

The Moho depth curves for Central Iran show a maximum depth of 50 km in the Sanandaj-Sirjan Zone and 45 km at its contact with Kopeh-Dagh.

The continuation of the Andian tectonic regime and subduction of oceanic plate beneath Central Iran was initiated between170 and 210 Ma and was associated with major intrusions that caused the occurrence of hot spots in the upper mantle beneath Central Iran. The presence of many active volcanos confirms that the upper mantle of Central Iran has a higher thermal gradient than Zagros and Kopeh-Dagh. Therefore, the brittle-plastic zone should be shallower, as the focal centers of Central Iranian earthquakes mostly occur at shallow depths.

Central Iran is characterized seismically by the occurrence of rare and statistically non-continuous earthquakes. These are of high magnitude but shallow, and occur along pre-existing reverse and strike-slip faults.

Central Iranian earthquakes all occur along strike-slip faults, most of which were formed along the normal faults created during the subsidence of Central Iran during extension and associated crustal thinning. Therefore, the recent active faults are deep, have a long geological history, and controlled the old sedimentary basins.

Some researchers are of the opinion that the absence of earthquakes along some faults is presumably due to the seismic silence that may occur after severe seismic activity. In Central Iran, large magnitude earthquakes have occurred along some faults, followed by a period of inactivity ranging from 11 years to 11 centuries (Berberian 1976a) before the re-occurrence of severe earthquakes.

This is probably because the energy that thrust and reverse faults need in order to reactivate them.

Some large-magnitude earthquakes in Central Iran have not any fore- and after shocks. Seismic activity along the faults, analysis of fault planes and fold axis of Neogene and later sediments indicate the movement pattern of Central Iran. There is an apparent eastward widening in this area, probably owing to the eastward wedge movement of Iran, caused by the collision of the Arabian and Eurasian plates.

The Shotori Mountains of Central Iran

The Shotori Mountains trend N-S and are located in the middle part of Central Iran. The western boundary of this range is formed by the Tabas fault in the north and the Kuhbanan fault in south. The Nayband fault forms the eastern limit of this range. Having a Precambrian basement, the Shotori range is surrounded by active major faults crossing the Quaternary sediments. The Kalmard Mountains are located in the western part of the Shotori range and generally are of lower elevation. The Paleozoic sediments covering the flanks of the Tabas faults differ, so that in the eastern part, the Paleozoic-Triassic sediments are 12.6 km thick, and in the western part 1.5 km. The

Tabas fault can therefore be assumed to be an old lineation that affected sedimentary processes. Geological studies carried out on this range indicate deformation stages as following:

1. Precambrian basement thinning associated with extensional movements, development of crossed spoon faults and shallow water sedimentation during subsidence of the area during the Mesozoic and Cenozoic.
2. Reversal of movements owing to compressional mechanisms and crustal shortening in the Late Cretaceous to Quaternary.
3. Separation and rupture of Paleozoic and other sediments from the basement by tectonic movements and fold formation associated with the occurrence of reverse imbricate faults.

Studies on aftershocks and focal mechanisms of the earthquakes indicate that the deformation and changes in depths of the western part of the Shotori Mountains are related to thick skinned tectonics. However, this deformation results in a decrease of dip of steeply-dipping faults at or close to the surface, forming imbricate thrusts.

3.8.5 Accretionary Wedge Zones of the Makran Mountains and the Zabol-Balouch Subzone (Eastern Iranian Flysch Zone)

These zones are the Post-Cretaceous flysch belts which meet in the southeast of Iran and continue to the Baluch- Pakistani Mountains. The flysch sediments are probably Late Cretaceous in age and are associated with the ophiolite mélange derived from continental slope deposits.

The E-W trending Makran range is characterized by southward thrusts and extends from the Minab-Zendan fault in Iran to the Arnach-Nai and Chaman faults in Pakistan. The direction of forces in the belt have a N-W trend and contour level curves of the Moho indicate a crustal thickness of 50 km in the northern Makran, decreasing southward to 25 km in the coastal area.

The Makran zone is an aggregation of accretionary wedges formed by subduction, at a rate of 5 mm/year, of the Oman Ocean beneath Central Iran. It has been a continental margin since Cretaceous time. The subducted oceanic crust is 6.7 km thick and 7–12 km of sediments have been deposited over it. Some of these sediments are obducted over the continental crust as the accretionary wedges are driven beneath the continental crust with their subsequent melting resulting in the formation of the active volcanos of Taftan-Bazman (Ghorbani 2014).

The uplift of convergent margins is probably due to the occurrences of high magnitude earthquakes. Studies carried out on C14 and U-Th values in the uplifted crust of the Iranian coastal area indicate a rate of uplift of 0.01–0.02 cm/year during the Holocene, which increases eastward.

This zone is characterized by low frequency earthquakes compared to Zagros. The focal depths of earthquakes occurring in this belt are very shallow in the coastal part, gradually increasing to 80 km to north. Fault plane studies indicate that the area is subjected to two tectonic-seismic regimes.

1.   Deep-focus earthquakes with thrusting mechanism occurred in accretionary wedge due to the re-activation of surface faults.
2.   medium-deep focus earthquakes occurred, owing to deformation of the subducted zone.

No normal faulting is observed in the Makran zone. The Jazmurian depression, located in the northern part of this range has high-angled reverse faults whose traces are observed at surface. During recent centuries, two earthquakes with large magnitude (more than 7) were occurred in this area. These occurrences of very recent, shallow-focus earthquakes suggest that this area may be hazadeously the location of further large-magnitude earthquakes in the future.

The active faults of Iran can be considered to be direct evidence of recent deformation of the Iranian crust. Movements along these faults have generally resulted from the continental collision of the Arabian and Eurasian plates. A number of these faults that control the morphological units of Iran are old, active faults re-activated by younger tectonic activity. Some of these faults, such as the Kalmard, Zagros and Nayband faults have affected sedimentary processes and geological movements during different geological periods (Huckriede et al. 1962; Huckriede 1962; Stöcklin 1968; Aghanabati 2004; Ghorbani 2013).

3.9   Active Faults and Tectonic Block Movements in Iran

In studies of the active faults of Iran, geomorphological methods and quaternary stratigraphy have rarely been used. Active faults are established based on studies of seismic data. Based on these data, the focus and movement mechanism of faults are determined (Hesamiazar et al. 2003). However, (Fig. 3.1), the extensive Quaternary cover, as well as structural complexity, especially in the areas of continental collision, makes the investigation of active Iranian faults very difficult. A contributory factor that compounds this difficulty is that the seismicity of such areas is not limited to single, distinct faults but has resulted from the activity of fault zones more than several kilometers wide. As a consequence, in areas with a high density of faults and rare seismic data, earthquakes have sometimes been interpreted to be related to more than one fault. Additionally, providing the large-scale maps and determining the high resolution focal centers of faults are very difficult for the following reasons:

1.   Most of earthquakes, such as those of Zagros have resulted from the activity of faults that are buried by sediments and do not reach the surface.
2.   Gradual movements of the earth in the in the two fault blocks are not associated with large-magnitude earthquakes over long periods of time.

In the recent map of active Iranian major faults, the measured rates were calculated based on GPS data, relation of slip vector and compression axis derived from focal mechanism solution.

The he active faults of Zagros are buried by sediments but the focal mechanism solution (FMS) implies the presence of reverse and strike-slip faults in the basement. In other parts of Iran, the faults reach the surface. However, the FMSs of eastern and central Iran indicate the dominance of strike-slip movement in a shear compressional regime (Hesami et al. 2003).

The active faults of the Azerbaijan area mostly are mostly of strike-slip type and formed during a shear extensional regime (Hesami et al. 2003). In the Alborz and Kopeh-Dagh area, the fault systems are relatively related to each other; so that the introduction of single and distinct fault is difficult. Although there are abraded terraces along the coastal area bordering the Oman Sea, there is little data concerning active faults in the Makran.

The N-S movements of the Iranian plateau resulting from northward movement of the Arabian plate have been accounted for most of the seismic activity. Most deformation has occurred in the Zagros, Alborz, Kopeh-Dagh and some movements are recorded along the major strike-slip faults in Central Iran, the Lut block and Southern Khazar.

Mapping using a GPS network including 38 benchmarks was completed in September 1999 and October 2001. The studies of Vernant et al. (2004) carried out on movement rate based on these GPS points indicates that:

- A N-S movement at a rate of 22 ± 2 mm/year has been recorded.
- The benchmarks located in eastern Sistan do not indicate a higher movement rate compared to Eurasian. However, Central Iran is moving northward in relation to Eurasia, at a rate of 14 ± 2 mm/year.
- The shortening measured along the Zagros range is such that the movement rate is 4 ± 2 mm/year in the northwest and 9 ± 2 mm/year in the southeast.
- The measured movement of northeast Iran is 12 mm/year with towards N350E, and 14 mm/year with azimuth N26E in the Kora basin. These values can be explained by strike-slip movement along the Tabriz fault (Hesami et al. 2003).
- The Alborz Mountains are located between Central Iran in the south and the south Caspian basin depression in the north. Here there is a shortening rate of 5 ± 2 mm/year (Vernant et al. 2004). The mountains are presumably deformed more than Zagros owing to its narrowness.
- The subduction rate of the Oman plate beneath Central Iran is measured to be 18 mm/year but in the western Makran Mountains, the movement rate is 15 mm/year (Vernant et al. 2004).
- The right-lateral strike-slip movement between the Helmand plate, Afghanistan, and the Lut plate is 14 ± 2 mm/year.
- The right-lateral strike-slip movement between Zagros and the Makran Mountains observed along the Minab-Zendan fault is 11 ± 2 mm/year.

The seismic data collected during 1964–2006 indicate that seismic activity in Iran is related to plasticity recorded by the GPS network. The Zagros Zone is of high seismic activity, with low-magnitude and high-frequency earthquakes, whereas the Alborz and Kopeh-Dagh zones are characterized by the low-frequency and high-magnitude earthquakes. Jackson and McKenzie (1988) stated that deformation of Zagros is mostly non-seismic and is associated with rock deformation. In contrast, deformation in Alborz and Kopeh-Dagh is mostly associated with earthquakes.

The values obtained from the GPS network indicate geodesic deformation of the earth. The comparison of seismic activity with earth geodesic deformation implies that:

- In Zagros, seismic activity is strongly associated with severe geodesic deformation. As a result, Zagros is an aseismic zone.
- Central Iran has weak seismic activity but geodesic deformation is not observed. Consequently, major deformation is associated with fault development.
- The Lut block, Alborz and Kopeh-Dagh exhibit medium to low seismic activity but with severe geodesic deformation.
- The ratio of deformation/earthquake frequency in northwest Iran and Alborz is higher than that of Lut and Kopeh-Dagh. Based on measuring benchmark points of Zagros, Tatar et al. (2002) stated that the total movement of Saudi Arabia in relation to Iran is 21 mm/year, of which 10 mm/year occurs in the Zagros Mountains. As a result, half of deformations occur in this area.

3.10   List of Iranian Earthquakes from 3000 Years Ago to Recent

In this list, about 3000 earthquakes recorded by national and international nets as well as from historical records are presented in table form. All of these data were provided by Dr. Mehdi Zar'e and contain data such as date, coordinate, focal depth (FD), body-wave magnitude (Mb), surface wave magnitude; moment magnitude (Mw); local magnitude (ML) and occurrence area.

Figure 3.10 shows the position of Iran in the Alpine-Himalayan Belt and Figs. 3.11 and 3.12 respectively show the distribution of Iranian earthquakes having a magnitude less than 6 and those with magnitude equal to, or more than 6.

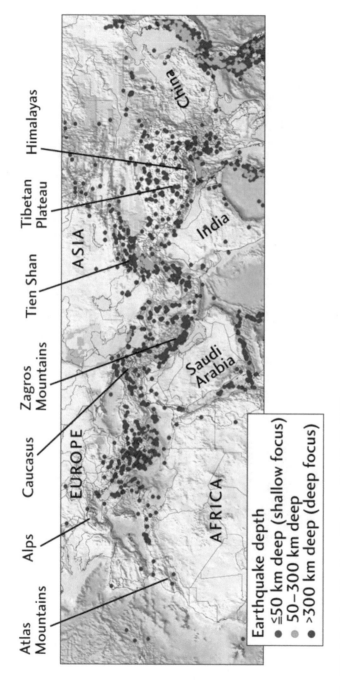

Fig. 3.10 Seismicity of the Alpine-Himalayan Belt (Keskin 2005)

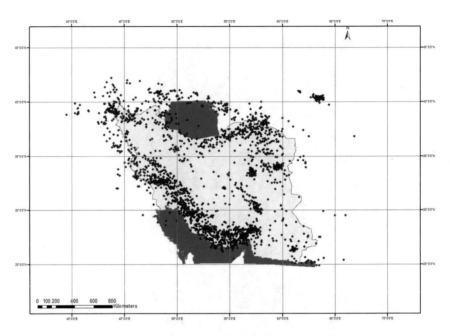

Fig. 3.11 Distribututution of earthquakes with magnitude less than 6 Richter ocurred in Iran an surroundings since 3000 ago to recent

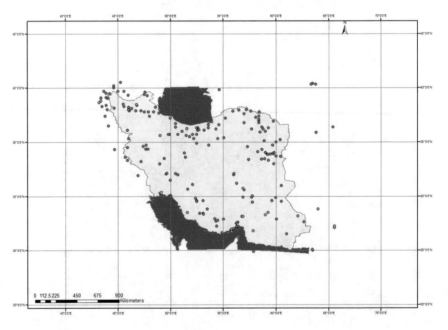

Fig. 3.12 Distribution of earthquakes with magnitude 6 ora more occurred in Iran and surroundings

References

Afshar-Harb A (1979) The stratigraphy, tectonics and petroleum geology of Kopet-Dagh region. Unpublished Ph. D. thesis, Imperial College of Sciences and Technology, University of London, p 316

Afshar-Harb A (1994) Geology of Kopet Dagh, in treatise on the geology of Iran. In: Hushmandzadeh A (ed), Tehran: Geological Survey of Iran, p 275

Aghanabati A (1975) Etude geologique de la region de Kalmard these Doctorate thesis, No AD 11/623 Gronoble, France

Aghanabati A (1998) The Jurassic stratigraphy of Iran, Geological Survey of Iran

Aghanabati A (2004) Geology of Iran. Geological Survey of Iran, Tehran

Aghanabati A (2009) Stratigraphic lexicon of Iran, V. 1–3. National Geosciences Directory of Iran, Geological Survey of Iran

Alavi M (1980) Tectonostratigraphic evolution of the Zagrosides of Iran. Geology 8(3):144–149

Alavi M (1991) Sedimentary and structural characteristics of the Paleo-Tethys remnants in northeastern Iran. Geol Soc Am Bull 103(8):983–992

Alavi M (1996) Tectonostratigraphic synthesis and structural style of the Alborz mountain system in northern Iran. J Geodyn 21:1–33

Alavi M (2007) Structures of the Zagros fold-thrust belt in Iran. Am J Sci 307:1064–1095. https://doi.org/10.2475/09.2007.02

Alavi-Naeini M (1972) Etude géologique de la région de Djam, Geological Survey of Iran, Report no. 23, p 288

Allenbach P (1966) Geologie und Petrographie des Damavand und seiner Umgebung (Zentral-Elburz), Iran. Abhandlung zur Erlangung der Wurde eines Doktors der Naturwissenschaften der Eidgenossischen Technischen Hochschule Zurich, p 145

Alric G, Virlogeux D (1977) Pétrographie et géochimie des roches métamorphiques et magmatiques de la région de Deh-Bid. Bawanat: Chaîne de Sanandaj-Sirjan. Iran, Universite scientifique et medicale de Grenoble

Al-Shanti AMS, Mitchell AHG (1976) Late Precambrian subduction and collision in the Al-Amer-Idsas region, Arabian Shield, Kingdom of Saudi Arabia. Tectonophysics 31:41–47

Ambraseys N, Melville CP (1982) A history of Persian earthquakes. Cambridge University Press, New York

Amidi M (1975) Contribution à l'étude stratigraphique, pétrologique et pétrochimique des roches magmatiques de la région Natanz-Nain-Surk (Iran central) (Doctoral dissertation)

Andalibi MJ, Oveisi B, Yousefi T (1998) The report of Kelestan sheet 1:100000, Ministry of Industry & Mine Geological Survey of Iran Publications, Iran

Annells RN, Arthurton RS, Bazley RAB, Davies RG, Hamedi MAR, Rahimzadeh F (1977) The Geology map of SHAKRAN, 1: 100000, No. 6162. Geological Survey of IRAN

Assereto R (1966) Geological map of Upper Djadjerund and Lar valleys (Central Alborz, Iran). Scale 1:50,000 with explanatory notes. Geol. Inst. Univ. Milano. Ser. G., publ. 232, p 86

Authemayou C, Bellier O, Chardon D, Malekzade Z, Abbassi M (2005) Role of the 857 kazerun fault system in active deformation of the Zagros fold and- thrust belt 858 (Iran). Compt Rendus Geosci 337:539–545

Aziz H (2001) Petrography, petrology and geochemistry of the Khoy Metamorphic Rocks. PhD Thesis (in Farsi), University of Tarbiat Moalem, Tehran, Iran, p 255

Bachmanov DM, Trifonov VG, Hessami KT, Kozhurin AI, Ivanova TP, Rogozhin EA, Hademi MC, Jamali FH (2004) Active faults in the Zagros and central Iran. Tectonophysics 380:221–241

Barzegar P (1981) Razak fault. Third Geology of Iran Symposium, Iranian Petroleum Society, Abstract Volume (In Iranian)

Barzegari A, Ghorashi M, Nazari H, Fontugne M, Shokri MA, Pourkermani M (2017) Paleoseismological analysis along the Astara fault system (Talesh Mountain, North Iran). Acta Geol Sinica-English Ed 91(5):1553–1572

Berberian F Berberian MJ (1981) Tectono-plutonic episodes in Iran. Zagros Hindu Kush Himalaya Geodynamic Evolution. 1981 January 1, 3:5–32

Berberian M (1973) The seismicity of Iran: preliminary map of epicentres and focal depths. Geological survey of Iran. Seismotectonic group, 1973

Berberian M (1976a) an explanatory note on the fi rst seismotectonics map of Iran, a seismotectonic review of the country, in contribution to the seismotectonic of Iran (Part II). Geological of Iran, report no. 43, p 518

Berberian M (1976b) Documented earthquake faults in Iran. Geol Surv Iran 39:143–186

Berberian M (1976c) Quaternary faults in Iran. Geol Surv Iran 39:187–258

Berberian M (1981) Active faulting and tectonics of Iran. Zagros Hindu Kush Himalaya Geodynamic Evolution 3:33–69

Berberian M (1983) structural evolution of the Iranian Plateau: contribution to the seismotectonics of Iran, Part IV: continental deformation in the Iranian plateau. Geol Surv Iran, Report 52, 19–68

Berberian M (1995) Master blind thrust faults hidden under the Zagros folds: active basement tectonics and surface morphotectonics. Tectonophysics 241:193–224

Berberian M (1997) Seismic sources of Trans-Caucasian historical earthquakes. In Historical and Prehistorical Earthquakes in the Caucasus, Vol. 28 of NATO ASI Ser., Ser. 2, Eds. by D. Giardini and S. Balassanian (Kluwer, Dordrecht, 1997), pp 233–311

Berberian M (2005) The 2003 Bam urban earthquake: a predictable seismotectonic pattern along the western margin of the rigid Lut Block, southeast Iran. Earthquake Spectra. 21:35–99

Berberian M (2014) Earthquakes and coseismic surface faulting on the Iranian Plateau. Dev Earth Surf Process 17:2–714

Berberian M (2017) Development of geological perceptions and explorations on the Iranian Plateau: from Zoroastrian cosmogony to plate tectonics (ca. 1200 BCE to 1980 CE). Tectonic evolution, collision, and seismicity of Southwest Asia: In Honor of Manuel Berberian's Forty-Five Years of Research Contributions, 525, p 25

Berberian M, Arshadi S (1976) On the evidence of the youngest activity of the North Tabriz fault and the seismicity of Tabriz City. Geol Surv Iran 39:397–418

Berberian M, Ghorashi M, Arjangravesh B, Mohajer Ashjaie A (1993) Seismotectonic and earthquake-fault hazard investigations in the great Ghazvin Region. Geological Survey of Iran, Report, (61), p 197

Berberian M, Ghoreshi M, ArjangRavesh B, Mohajer Ashjai A (1992) Seismotectonic and earthquake—Fault hazard investigation in the Tehran region. Geological survey of Iran

Berberian M, Ghoreshi M, Talebian M, Shojae-Taheri J (1996) Research on Neo-tectonic, Geoseismic and hazards of earthquakefaulting in Semnan area, Report no. 63, p 266

Berberian M, Hamdi B (1977) First discovery of Ordovician beds and conodonts in the slightly metamorphosed rocks of Kuh-e-Agh-Baba, Maku Quadrangle, Azarbaijan. Geol Surv Iran, Int Rep, p 7

Berberian M, King GCP (1981) Towards a paleogeography and tectonic evolution of Iran. Can J Earth Sci 18(2):210–265

Berberian M, Ghorashi M (1986) General geology and seismotectonics of the Chahar Mahal Bakhtiari Dam Site. Chahar Mahal Bakhtiari Dev. Proj., Plan and Budget Org (in Persian)

Berberian M, Ghorashi M, Arzhang-Ravesh B, Mohajer-Ashjai A (1985) Recent tectonics, seismo-tectonics and earthquake fault hazard investigations in the Greater Tehran region: contribution to the seismotectonics of Iran, part V. Geol Surv Iran 56:316

Berberian M, Ghorashi M (1989) Seismotectonic and earthquake-fault hazard study of the Neyshabur Power Plant Site, Moshanir Consulting Engineers, Internal Report, Ministry of Power, Tehran, p 145 (in Persian)

Berberian M, Yeats R (1999) Patterns of historical earthquake rupture in the Iranian Plateau. Bull Seismolo Soc Am 89. https://doi.org/10.1016/b978-0-444-63292-0.00016-8

Berberian M, Yeats R (2001) Contribution of archaeological data to studies of earthquake history in the Iranian Plateau. Paul Hancock Memorial Issue. J Struct Geol 23:563–584. https://doi.org/10.1016/S0191-8141(00)00115-2

Berberian M, Yeats RS (2016) Tehran: an earthquake time bomb; In tectonic evolution, collision, and seismicity of Southwest Asia. In: Honor of manuel berberian's forty-five years of research contributions, Geological Society of America. Special Paper p 525

Braud J, Ricou LE (1971) The Zagros fault or main thrust, an overthrust and a strike-slip fault (in French): Acad Sci Comptes Rendus, ser. D, 272(2), pp 203–206

Crawford MA (1977) A summary of isotopic age data for Iran, Pakistan and India. In: Libre a la memoire del A.F. de Lapparent. Mémoire hors-serie 8. Societé Géologique de France, pp 251–260

Davies RG, Jones CR, Hamzepour B, Clark GC (1972) Geology of the Masuleh, Sheet 1:100,000; Northwest Iran, Gel Surv Iran, Rep 24, p 110

Dedual E (1967) Zur geologie des mittleren und unteren Karaj-Tales, zentral-Elburz (Iran). Mitt Geol Inst ETH Univ Zurich 79:45–75

Delavari M, Dolati A, Mohammadi A, Rostami F (2016) The permian volcanics of central alborz: implications for passive continental margin along The southern border of paleotethys. Ofioliti 41(2):59–74. https://doi.org/10.4454/ofioliti.v41i2.442

Dellenbach J (1964) Contribution a l'étude de la région située à l'est de Teheran (Iran)

Dimitrijevic MD (1973) Geology of Kerman Region. Gel Surv Iran, Report Number Yu/52, p 334

Eftekharnezhad J (1980) Subdivision of Iran into different structural realms with relation to sedimentary basins (in Farsi). Bull Iran Petrol Inst 82:19–28

Falcon NL (1967) The geology of the north-east margin of the Arabian basement shield. Adv Sci 24:31–42

Falcon N (1969) Problems of the relationship between surface structure and deep displacements illustrated by the Zagros range. Geol Soc Lond Spec Publ, 3/1, pp 9–21

Falcon NL (1974) Southern Iran: Zagros Mountains. Geol Soc, Lond Spec Publ 4(1):199–211

Fattahi M, Walker R, Hollingsworth J, Bahroudi A, Nazari H, Talebian M, Armitage S, Stokes S (2006) Holocene slip-rate on the Sabzevar thrust fault, NE Iran, determined using optically stimulated luminescence (OSL). Earth Planet Sci Lett 245(3–4):673–684

Favre G (1975) Structures in the Zagros Orogenic Belt OSCO (No. 1233E52). Report

Forster H, Fesefeldt K, Kiirsten M (1972) Magmatic and orogenic evolution of the Central Iranian volcanic belt. Proc. 24th Int. Geol Congr 2:198–210

Fürsich FT, Wilmsen M, Seyed-Emami K, Majidifard MR (2009) The Mid-Cimmerian tectonic event (Bajocian) in the Alborz Mountains, Northern Iran: evidence of the break- up unconformity of the South Caspian Basin. In Brunet M-F, Wilmsen M, Granath JW (eds) South Caspian to central Iran Basins: London, Geol Soc Lond Spec Publ 312:189–203. https://doi.org/10.1144/SP312.9

Ghasemi MR, Ghorashi M (2004) Regional study of the main seismic faults in the Alborz mountain, report of the Scientific Research Committee, Geological Survey of Iran, Tehran, Iran, p 58

Ghorbani M (1999) Petrological investigations of Tertiary-Quaternary magmatic rocks and their metallogeny in Takab area. PhD thesis, Shahid Beheshti University

Ghorbani M (2013) Economic geology of Iran, vol 581, Mineral deposits and natural resources: Springer

Ghorbani M (2014) Geology of Iran, The magmatism and metamorphism, vol 3, Arian zamin publication

Gidon M, Berthier F, Billiault JP, Halbronn B, Maurizot P (1974) Sur les caractères et l'ampleur du coulissement de la "Main Fault" dans la région de Borudjerd-Dorud (Zagros oriental, Iran). Comptes Rendus de l'Académie des Sci 278:701–704

Haghipour A (1974) Etude géologique de la région de Biabanak-Bafgh (Iran central), Pétrologie et tectonique du socle precambrien et de sa couverture, France, PhD thesis, Sci Nat, Gronoble University, p 403

Hairapetian V, Ghobadipour M, Popov LE, Männik P, Miller CG (2017) Silurian stratigraphy of Central Iran-an update. Acta Geologica Polonica, p 67

Hesamiazar KH, Jamali F, Tabasi H (2003) Map of active faults in Iran, producers of, colleagues of, scale 1, 2,500,000, International Institute of Earthquake Engineering and seismology

Hessami K, Pantosti D, Tabassi H, Shabanian E, Abbassi MR, Feghhi K, Solaymani S (2003) Paleoearthquakes and slip rates of the North Tabriz Fault, NW Iran: preliminary results. Ann Geophys 46(5):903–915

Hollingsworth J, Nazari H, Ritz JF, Salamati R, Talebian M, Bahroudi A, Walker RT, Rizza M, Jackson J (2010) Active tectonics of the east Alborz mountains, NE Iran: rupture of the left-lateral Astaneh fault system during the great 856 AD Qumis earthquake. J Geophys Res: Solid Earth, 115(B12)

Houshmandzadeh A, Alavi-Naeini M, Haghipour A (1978) The evolution of geological phenomena in the Torud region (from Precambrian to the Present Age), Geological Survey of Iran, pp 138

Houshmandzadeh A, Soheili M (1990) Explanatory text of the Eqhlid quadrangle map, scale of 1: 250,000, Gel Surv Iran, p 157

Huckriede R (1962) Zur geologie des gebietes zwischen Kerman und Sagand (Iran). Beihefte zum Geologischen Jahrbuch 51:1–197

Huckriede R, Kürsten M, Venzlaff H (1962) Geotektonische Kartenskizze des Gebietes zwischen Kerman und Sagand, (Iran), Bundesanstalt für Bodenforschung

Iranpanah A (1986) Structural characteristics of Razak Lineament (Zagros Fold Belt) and Tyrone-Mount Union Lineament (Appalachian Fold Belt)–a Comparison. Geol Soc Am Abstracts with Programs 18(1):25

Jackson J, McKenzie D (1984) Active tectonics of the Alpine-Himalayan belt between western Turkey and Pakistan. Geophys J R Astron Soc 77:185–264

Jackson J, McKenzie D (1988) The relationship between plate motions and seismic moment tensors, and the rates of active deformation in the Mediterranean and Middle East. Geophys J Int 93(1):45–73

Jackson J, Priestley K, Allen MB, Berberian M (2002) Active tectonics of the South Caspian Basin. Geophys J Int 148:214–245

Jahani S (1998) Investigation of fragile deformity and continuity of Doroud fault in Rudbar area of Lorestan (70 km south of Aligudarz); MSc thesis

Jahani S, Callot J, Letouzey J, Frizon de Lamotte D (2009) The eastern termination of the Zagros Fold-and-Thrust Belt, Iran: Structures, evolution, and relationships between salt plugs, folding, and faulting, J Tectonics, vol 28

Jahani S, Nogol-e Sadat AA (1999) Study of fragile transformation in the direction of "Bakhtiari Section" from the main Zagros earthquake zone (70 km south of Aligudarz), 3th Conference of Geological Society of Iran, Shiraz University

James GA, Wind JG (1965) Stratigraphic nomenclature of Iranian oil consortium agreement area. AAPG Bull 12:2182–2245

Jenny JG (1977) Géologie et stratigraphie de l'Elbourz oriental entre Aliabad et Sharud, Iran: these, Imprimerie nationale

Kazemi-Safa A (2011) Structural analysis of fracture arrays south of Angelo fault south of Damghan, MSc thesis, Faculty of Earth Sciences, Damghan University

Kent PE (1958) Recent studies of south Persian salt diapirs. AAPG Bull 42:2951–2972

Keskin M (2005) Domal uplift and volcanism in a collision zone without a mantle plume: evidence from Eastern Anatolia. http://wwwmantleplumes.org/Anatolia.html

Majidi B (1978) Etude pétrostructural de la région du Mashhad (Iran), Les problemes des méta-morphites, serpentinites et granitoides Hercyniens, Thése Universite Scientifique et Medical de Granobel France, p 277

Majidi B (1981) The ultrabasic lava flows of Mashhad. North East Iran. Geol Mag 118(1):49–58

Masson F, Djamour Y, Van Gorp S, Chéry J, Tatar M, Tavakoli F, Nankali H, Vernant P (2006) Extension in NW Iran driven by the motion of the South Caspian Basin. Earth Planet Sci Lett 252(1–2):180–188

Meyer SP (1967) Die Geologie des Gebietes Velian-Kechiré (Zentral-Elburz, Iran), vol 79, Offsetdruck P. Schmidberger

Mohajer-Ashjai A, Behzadi H, Berberian M (1975) Reflections on the rigidity of the Lut block and recent crustal deformation in eastern Iran. Tectonophysics 25(3–4):281–301

Motiei H (1995) Petroleum geology of Zagros. Geological Survey of Iran (in Farsi), p 589

Mousavi Z, Pathier E, Walker RT, Walpersdorf A, Tavakoli F, Nankali H, Sedighi M, Doin MP (2015) Interseismic deformation of the Shahroud fault system (NE Iran) from space-borne radar interferometry measurements. Geophys Res Lett 42(14):5753–5761

Muttoni G, Gaetani M, Kent DV, Sciunnach D, Angiolini L, Berra F, Garzanti E, Mattei M, Zanchi A (2009) Opening of the Neo-Tethys Ocean and the Pangea B to Pangea a transformation during the Permian. GeoArabia 14(4):17–48

Nabavi M (1976) An introduction to geology of Iran, Geological Survev of Iran, Tehran, p 109

Nazari H (2006) Analyse de la tectonique recente et active dans l'AlborzCentral et la region de Teheran: approche morphotectonique et paleo-seismologique, PhD thesis. University of Montpellier II, Montpellier, France

Nazari H, Ritz JF (2008) Neotectonic in central Alborz. Geosci, GSI, 17(1)

Nazari H, Ritz JF, Ghassemi A, Bahar-Firouzi K, Salamati R, Shafei A, Fonoudi M (2011) Paleoearthquakes Determination of Magnitude ~ 6.5 on the North Tehran Fault, Iran, JSEE, 13(1):159–166

Nazari H, Ritz JF, Salamati R, Shafei A, Ghassemi A, Michelot JL, Massault M, Ghorashi M (2009) Morphological and palaeoseismological analysis along the Taleghan fault (Central Alborz, Iran). Geophys J Int 178(2):1028–1041

Nazari H, Ritz JF, Salamati R, Shahidi A, Habibi H, Ghorashi M, Bavandpur AK (2010) Distinguishing between fault scarps and shorelines: the question of the nature of the Kahrizak, North Rey and South Rey features in the Tehran plain (Iran). Terra Nova 22(3):227–237

Nazari H, Salamati R (1998) Geological map of Rudbar, scale 1: 100000, Geological Survey of Iran

Nazari H, Shahidi A (2011) Tectonic of Iran «Alborz». Gological Survey and Mineral Exploration of Iran, Research institute for Earth Science

NIOC (1977) Tectonic map of southwest Iran, 1: 2500000 Scale. NIOC, Tehran, Iran

NIOC (2012) Fault and lineament demarcation and maping in Coastal Fars Province (Between Kazeroon and Razak Faults), Pars Geological research center (Client: Exploration Directorate of National Iranian Oil Company)

Omidi P, Nogolsadat M, Ghorashi M (2002) Damghan system fault setting in the Astaneh-Attary convergent shear zone, Earth science quarterly, pp 39–40 (in Persian)

Pattinson R, Takin M (1971) Geological significance of the Dezful embayment boundaries. National Iranian Oil Company, Report, p 1166

Pedrami M (1981) Pasadenian orogeny and geology of last 700,000 years of Iran. Geological Survey of Iran, p 273

Reyer D, Mohafez SA (1972) First contribution of the NIOC-ERAP agreements to the knowledge of Iranian Geology. Edition Techniqs, Paris, p 58

Ricou LE (1974) L'étude géologique de la région de Neyriz (Zagros iranien) et l'évolution structurale des Zagrides (Doctoral dissertation)

Ritz JF, Nazari H, Ghassemi A, Salamati R, Shafei A, Solaymani S, Vernant P (2006) Active transtension inside central Alborz: a new insight into northern Iran–southern Caspian geodynamics. Geology 34(6):477–480

Ritz JR, Rizza M, Vernant P, Peyret M, Nazari H, Nankali H, Djamour Y, Mahan S, Salamati R, Tavakoli F (2011) Morphotectonics and geodetic evidences for a constant slip-rate along the Tabriz Fault (Iran) during the past 45 kyr. AGUFM, 2011, pp T44A-07

Rizza M, Vernant P, Ritz JF, Peyret M, Nankali H, Nazari H, Djamour Y, Salamati R, Tavakoli F, Chery J, Mahan SA (2013) Morphotectonic and geodetic evidence for a constant slip-rate over the last 45 kyr along the Tabriz fault (Iran). Geophys J Int 193(3):1083–1094

Robert AM, Letouzey J, Kavoosi MA, Sherkati S, Müller C, Vergés J, Aghababaei A (2014) Structural evolution of the Kopeh Dagh fold-and-thrust belt (NE Iran) and interactions with the South Caspian Sea Basin and Amu Darya Basin. Mar Pet Geol 57:68–87

Rossetti F, Monié P, Nasrabady M, Theye T, Lucci F, Saadat M (2017) Early Carboniferous subduction-zone metamorphism preserved within the Palaeo-Tethyan Rasht ophiolites (western Alborz, Iran). J Geol Soc 174:741–758

Ruttner A, Nabavi M, Hajian J (1968) Geology of the Shirgesht area (Tabas area, East Iran). Tehran: Geol Surv Iran. Report 4, p 133

Sabzehei M (1974) Les Mélanges ophiolitiques de la région d'Esfandagheh (Iran méridional): étude pétrologique et structurale, interprétation dans le cadre iranien, Université Scientifique et Médicale de Grenoble

Sabzehei M, Berberian M (1972) Preliminary note on the structural and metamorphic history of the area between Dowlatabad and Esfandagheh, south-east Central Iran, Geol Surv Iran, Int Rep 30; and 1st Iranian Geol Symp

Safari H, Qoreshi M, Abbasi MR (1999) Deformation analysis of cross-sectional Sabzpoushan areas- south Shiraz. Geosci J Geol Surv Iran, no. 31–32

Samadian M (1984) Investigation of neotectonic, Seimicity and Seismotectonics of Shiraz region, Zagros, Iran. Geological Survey of Iran publication

Sepehr M (2001) The tectonic significance of the Kazerun Fault Zone, Zagros fold-thrust belt, Iran. PhD thesis, Imperial College, University of London

Sepehr M, Cosgrove JW (2004) Structural framework of the Zagros fold–thrust belt, Iran. Mar Pet Geol 21(7):829–843

Sepehr M, Cosgrove JW (2007) The role of major fault zones in controlling the geometry and spatial organization of structures in the Zagros Fold-Thrust Belt. Geol Soc Lond Spec Publ 272(1):419–436

Setudehnia A (1975) The Paleozoic sequence at Zard Kuh and Kuh-e-Dinar. Bull Iran Petrol Inst 60:16–33

Setudehnia A (1978) The Mesozoic sequence in South-West Iran and adjacent areas. J Pet Geol 1:3–42

Seyed-Emami K, Fürsich FT, Wilmsen M (2004) Documentation and significance of tectonic events in the northern Tabas Block (east-central Iran) during the Middle and Late Jurassic. Rivista Italiana di Paleontologia e Stratigrafia, 110(1)

Sherkati S, Letouzey J (2004) Variation of structural style and basin evolution in the central Zagros (Izeh zone and Dezful Embayment), Iran. Mar Pet Geol 21(5):535–554

Stampfli GM (1978) Etude géologique generale de l'Elbourz oriental au sud de Gonbad-e-Qabus (Iran NE) (Doctoral dissertation, Université de Genève)

Steiger R (1966) Die Geologie der West-Firuzkuh-Area (Zentralelburs, Iran). Mitt. Geol. Inst. ETH Univ. Zürich [N.F.] 68

Stöcklin J (1968) Structural history and tectonics of Iran: a review. Am Assoc Pet Geol Bull 52(7):1229–1258

Stöcklin J (1974a) Northern Iran: Alborz Mountains. Geol Soc Lond Spec Publ 4(1):213–234

Stöcklin J (1974b) Possible ancient continental margins in Iran. In The geology of continental margins. Springer, Berlin, Heidelberg, pp 873–887

Stöcklin J, Nabavi MH (1971) Explanatory text of the Boshruyeh Quadrangle map 1: 250,000. Geol Surv Iran Geol Quad J 7:1–50

Stocklin J, Nabavi MH (1973) Tectonic map of Iran. Geol Surv Iran 1:5

Talebian M, Jackson J (2004) A reappraisal of earthquake focal mechanisms and active shortening in the Zagros mountains of Iran. Geophys J Int 156(3):506–526

Tarkian M, Lotfi M, Baumann A (1984) Magmatic copper and lead-zinc ore deposits in the Central Lut, East Iran. Neues Jahrbuch für Geologie und Paläontologie-Abhandlungen, pp 497–523

Tchalenko JS (1975) Seismicity and structure of the Kopet Dagh (Iran, USSR). Philos Trans R Soc Lond Ser A, Math Phys Sci 278(1275):1–28

Tchalenko JS, Braud J (1974) Seismicity and structure of the Zagros (Iran): the main recent fault between 33 and 35 N. Philos Trans R Soc Lond Ser A, Math Phys Sci 277(1262):1–25

Tirrul R, Bell IR, Griffis RJ, Camp VE (1983) Thesistan suture zone of eastern Iran. Geol Soc Am Bull 94:134–150

Vahdati Daneshmand F, Nadim H (2001) Geological map of Marzan abad, 1: 100000 Geol. Surv., Iran

Vernant P, Nilforoushan F, Hatzfeld D, Abbassi MR, Vigny C, Masson F, Nankali H, Martinod J, Ashtiani A, Bayer R, Tavakoli F (2004) Present-day crustal deformation and plate kinematics in the Middle East constrained by GPS measurements in Iran and northern Oman. Geophys J Int 157(1):381–398

Walpersdorf A, Manighetti I, Mousavi Z, Tavakoli F, Vergnolle M, Jadidi A, Hatzfeld D, Aghamohammadi A, Bigot A, Djamour Y, Nankali H (2014) Present-day kinematics and fault slip rates in eastern Iran, derived from 11 years of GPS data. J Geophys Res Solid Earth 119(2):1359–1383

Wellman HW (1966) Active wrench faults of Iran, Afghanistan and Pakistan. Geol Rundsch 55(3):716–735

Wilmsen M, Fürsich FT, Majidifard MR (2015) An overview of the cretaceous stratigraphy and facies development of the Yazd Block, western Central Iran. J Asian Earth Sci 102:73–91

Wilmsen M, Fürsich FT, Seyed-Emami K, Majidifard MR, Taheri J (2009) The Cimmerian orogeny in northern Iran: Tectono- stratigraphic evidence from the foreland. Terra Nova 21(3):211–218

Zanchetta S, Zanchi A, Villa I, Poli S, Muttoni G (2009) The Shanderman eclogites: a Late Carboniferous high-pressure event in the NW Talesh Mountains (NW Iran). Geol Soc Lond Spec Publ 312(1):57–78

Zanchi A, Zanchetta S, Berra F, Mattei M, Garzanti E, Molyneux S, Nawab A, Sabouri J (2009) The Eo-Cimmerian (Late? Triassic) orogeny in North Iran. Geol Soc Lond Spec Publ 312(1):31–55

Zandkarimi K, Najafian B, Vachard D, Bahrammanesh M, Vaziri SH (2016) Latest Tournaisian–late Viséan foraminiferal biozonation (MFZ8–MFZ14) of the Valiabad area, northwestern Alborz (Iran): geological implications. Geol J 51(1):125–142

Zandkarimi K, Vachard D, Cózar P, Najafian B, Hamdi B, Mosaddegh H (2017a) New data on the Late Viséan–Late Serpukhovian foraminifers of northern Alborz, Iran (biostratigraphic implications). Rev Micropaléontol 60(2):257–278

Zandkarimi K, Vachard D, Najafian B, Hamdi B, Mosaddegh H (2017b) Viséan-Serpukhovian (Mississippian) archaediscoid foraminifers of the northern Alborz, Iran. Neues Jahrb Geol Paläontol Abh 286(1):105–123

Zandkarimi K, Vachard D, Najafian B, Mosaddegh H, Ehteshami-Moinabadi M (2019) Mississippian lithofacies and foraminiferal biozonation of the Alborz Mountains, Iran: Implications for regional geology. Geol J. Article first published online

Chapter 4
Magmatic Phases and Their Distribution in Iran

Abstract In this chapter, the Precambrian to Pliocene magmatic rocks of Iran have been described. The distribution of igneous rocks in all parts of Iran is demonstrated in a map. Magmatic rocks have been described and analyzed in all structural zones of Iran, while the regions of their occurrence have been described and analyzed separately. Magmatic rocks of Iran except ophiolites and Quaternary intrusive and volcanic masses have been discussed in this chapter in terms of petrography, magmatic series, tectonic environment and the geodynamic model of igneous rocks have been analyzed for each geologic period and region.

Keywords Magmatism of Iran · Magmatic phases of Iran · Petrogenesis of magmatic rocks of Iran · Volcanoes of Iran · Magmatic belts of Iran · Urmia-Dokhtar · Torud · Arasbaran

4.1 Magmatism of Iran

The volcanic rocks and intrusive bodies of Iran can be used to trace the magmatic activities from Precambrian to Quaternary. During some intervals of this long period, both magmatic and volcanic activities have affected other geological events, and in contrast, the lack of volcanic activities also existed in many periods.

4.2 Late Precambrian-Early Cambrian Magmatism

During the time interval of 530 to 620 Ma., volcanic and magmatic activities have dramatically occurred in many parts of Iran. Their associated evidence is traceable particularly in the Central Iran, Azerbaijan and Zagros.

The volcanic and magmatic activities at this time are most possibly related to the Pan-African tectonomagmatic phases (Hushmandzadeh et al. 1988). The following igneous rocks are associated with these activities.

Doran-type intrusive masses in Azerbaijan: The Doran-type granites are hololeucratic intrusions which often consist of two minerals, quartz and sodic-alkali feldspar and lack of ferromagnesian minerals. These intrusions are mostly composed of some inclusions of metamorphic sedimentary rocks.

The Doran-type granites do not form the large batholiths. They are mostly intruded into the Kahar and Bayandor formations (Proterozoic in age) (Ghorbani 2015), but never cut the Soltanieh Formation. Several age determinations indicate different ages for the Doran granite. Nevertheless, most of the results considered it to be about 620 Ma. (Stocklin and Setudehnia 1971).

The Doran-type granites are mostly "A" and "S" alkaline types. The geochemical studies show that they have formed in an extensional phase (Valizadeh and Esmaili 1996). Some of them are as follow: Doran-Granite, Sarv-e Jehan, Moghanloo, Mah Neshan, Incheh, Ugh Kand, Alam Kandi, Salim Khan, and Yeiylaghi (Fig. 4.1).

Intrusive rocks of Narigan-Zarigan Type: This type of intrusions is traceable from the Anarak to Bafgh areas (Haghipour 1974) as well as Kuhbanan, Behabad and Kalmard in Tabas Region. They are intruded into the volcanic rocks which are equivalent to the Rizu and Desu 'Series' in Central Iran. Their petrogenesis highly differs from the Doran Granite and is composed of ferromagnesian minerals such as: biotite, hornblende; with accessory minerals like zircon and apatite, and fewer inclusions. They have formed in a higher temperature environment than that of Doran-Granite. Geochemically, they are more likely alkaline and calc-alkaline (Mousavi Makavi 1998).

Their geological evidence indicates an extensional zone of I type granite which contains plenty of iron mineralization. The following granitic series can be attributed to this group:

Narigan, Zarigan, Behabad, Robat-e khan (Kuh-e Lakharbakhshy) in Tabas area.

The intrusions of Eastern Central Iran: From southern Shahroud to northern Bardaskan there are several intrusions showing Late Proterozoic ages such as Shotor Kuh, Bornaward, Khartouran, etc. (refer to Chap. 5, i.e. Intrusive rocks).

Rhyolitic volcanic rocks of the Qareh-Dash 'Series' (Azerbaijan): The Qareh-Dash 'Series' extends from Zanjan to Takab, Shahin Dezh and Mahabad. Their petrographic composition indicates rich content of quartz phenocrysts and other minerals (Shahbazi 1999; Ghorbani 1999) which forms the matrix of the rhyolitic volcanic rocks with albite feldespate.

Like the Doran granite, the Qareh-Dash 'Series' does not contain the ferromagnesian minerals. The Qareh-Dash 'Series' is located on the top of the Kahar Formation and below the Bayandor Formation or with interfingering contact with the latter (Ghorbani 2019). The overlying formation, however, is reported to be the Soltanieh Formation. Their composition is most likely alkaline, and they are extrusive equivalent rocks of Doran-Granite (Fig. 4.2).

Volcanic rocks in Central Iran: The volcanic rocks in Central Iran are associated with the Kushk 'Series' and the Rizu and Desu formations. They are widely distributed in Central Iran particularly in Bafgh-Saqand as well as in the Shirgesht and Tabas areas.

Fig. 4.1 Two views of the Doran-Granite in the Doran village

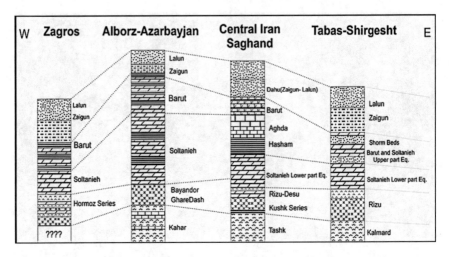

Fig. 4.2 Correlating the Late Precambrian volcanic rocks of the Azerbaijan, Central Iran, Eastern Central Iran and Zagros areas

These volcanic rocks show wide range of petrology and geochemical composition. Accordingly, they petrologically consist of a wide range of volcanic rocks such as: basalt, rhyolite, andesite, and acidic tuffs, basic to intermediate dikes and sills. Nevertheless, acidic tuffs, rhyolite, and a plenty of basalts form the majority of rocks.

Based on the geochemical characteristics of acidic tuffs and rhyolites, they are originated from an extensional zone of a continental crust; but the alkaline nature of basic and basaltic rocks indicates oceanic rifted zones (Huckriede et al. 1962).

The volcanic rocks in Central Iran formed during a long time interval of late Proterozoic and Pre-Cambrian. They are associated with the Kushk, Tashk and Rizu-Desu 'series'. The Lalun sandstones have been cut by igneous sills and dikes in the Shirgesht area. This confirms that they are genetically related to the volcanism of Central Iran. Please refer to magmatism of Yazd area (Figs. 4.3, 4.4, 4.5, and 4.6).

Volcanic rocks in the Zagros (Hormuz 'Series'): Igneous rocks of the late Precambrian-early Cambrian in Zagros Zone are associated with the Hormuz 'Series' which consist of the extensive and significant parts of Hormuz Formation. According to their petrological composition they are divided into two different kinds of rocks (Ahmadzadeh-Heravi et al. 1990; Nouraei 1997; Figs. 4.7 and 4.8).

- Acidic volcanic and pyroclastic rocks (rhyolite and rhyolitic tuff in type) mostly observed in Bandar Abbas area accompanied by salt domes.
- Basic volcanic rocks and dikes (basaltic, diabasic and occassionaly gabbroidic) that are mostly observed in Fars area accompanied by salt domes.

The geochemical specification of Hormuz 'Series' igneous rocks are as follows:

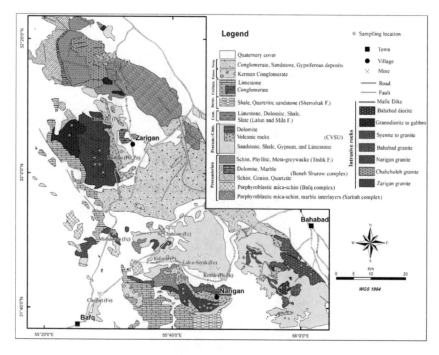

Fig. 4.3 Geological map of Bafgh area (Mehdipour 2019)

- Acidic volcanic rocks are more widely distributed in Bandar Abbas, and decreased in Fars area. These rocks have been originated from melting of a continental crust (Ghafari 2010).
- Basic rocks are mostly alkaline type. The tectonic evidence show that they have formed in a continental rift zone, but during the latest stages of developing-rift, in Fars region (associated with Salt Domes in the Lar Area), their chemical composition tends to tholeiitic rocks, which is a significant transfer from a continental rift to an oceanic zone (Ghafari 2010; Nokhbeh 2016).

Metamorphosed igneous rocks of the late Precambrian-early Cambrian: Many of the late Precambrian-early Cambrian metamorphic rocks which exhibits metamorphism from greenschist to lower amphibolite facies are originally volcanic lava, tuff and tuffite that are extended in Takab and Anarak, although their protoliths have been formed during late Precambrian (Ghorbani 1999).

Detailed studies indicate that their protoliths have a wide variation of petrological composition, but most of amphibolites or schists are associated with the igneous mafic and ultramafic protoliths. The characteristics of these metamorphic rocks indicate that they are originated from an oceanic crust which was already confirmed in Takab and Anarak (Ghorbani 1999) (Figs. 4.9 and 4.10).

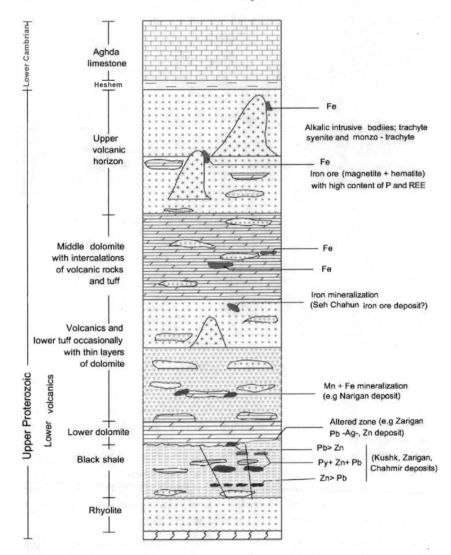

Fig. 4.4 The schematic view of the volcanism in Bafgh area

The Takab schist in Zarshuran and Anguran mines as well as Alamkandi area, the Chah Gorbeh schist and Patyar in the Anarak have been originally ultramafic rocks which have been metamorphosed later.

Fig. 4.5 A view of the volcanosedimentary rock units of Rizu Fm. in Kushk and Bafgh area

Fig. 4.6 A view of the Aghda limestone and granite-syenite and rhyolithic tuffs of Rizu Fm.in Yazd area

Fig. 4.7   Rhyolites of the Hormuz Island

Fig. 4.8   The Hormuz rhyolites associated with salt and dolomite

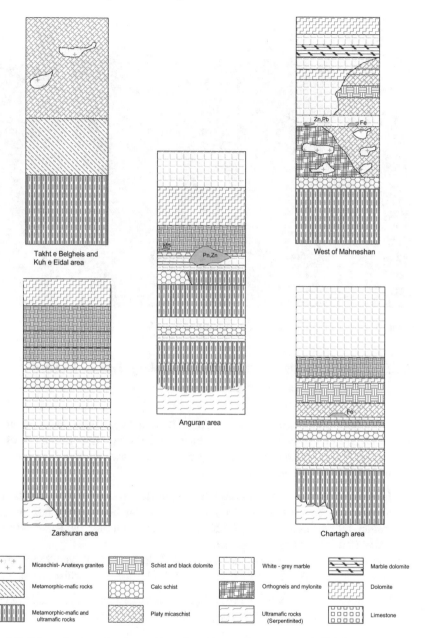

Fig. 4.9 Schematic view of metamorphic rocks with oceanic crust source in Takab area

Fig. 4.10   A view of metamorphic rocks in Takab area

4.2.1   Geodynamics and Mechanism of Magmatism in Late Precambrian-Early Cambrian

To clarify the mechanism of magmatism during the late Precambrian-early Cambrian, we should consider some facts about the Precambrian-Cambrian magmatic rocks:

The late Precambrian-early Cambrian magmatic rocks are classified into two groups (e.g. Berberian and Berberian 1981):

1.   The acidic magmatic rocks, consisting of the Doran–Granite, Zarigan-Narigan Granite and the related volcanic rocks such as Qareh-Dash rhyolites in Azerbaijan, all volcanic rocks of the Kushk 'Series', including the Rizu and Desu formations and Hormuz formation.
2.   Mafic-ultramafic rocks associated with tuffs in the Anarak and Takab areas. Some of them are metamorphosed inamphibolite and greenschist phases.

The volume of metamorphosed volcanic rocks is greater than that of metamorphic plutonic rocks. Also, some sedimentary rocks such as carbonate (as interfinger and interlayer forms) and pelitic rocks are associated with the basic and acidic tuffs (Fig. 4.11) (Ghorbani 1999).

Regarding the relationship between the interfinger structure of pelitic and interlayer carbonate rocks, the depth of deposition was deep in Takab and shallow at the margin. The facies of mafic rocks and their associated tuffs, at the marginal basin, have been changed to the clastic rocks. As a result, the depth of the basin in Bafgh

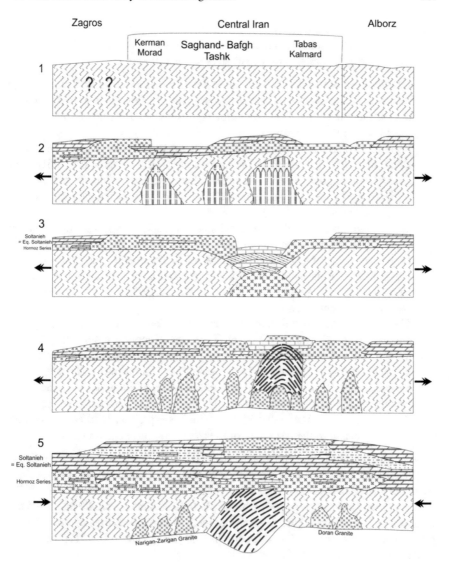

Fig. 4.11 Geodynamic model of magmatism in late Precambrian-early Cambrian

and Saqand was less than those of Anarak and Takab. Whenever the ultramafic rocks are exposed, the deep zone represented as a strip (Kani and Ghorbani 2016).

It is noteworthy that the Takab mafic and ultramafic rocks have the MORB geochemical characteristics (Farrokhmanesh 1998; Ghorbani 1999).

According to what mentioned above and taking into the account all the geological factors of Central Iran such as stratigraphy, magmatism, mineralization and lithology, probably a narrow rift zone has been formed during the late Precambrian- Cambrian.

It extended from the Takab to Anarak and Saqand in Central Iran. This rift is similar to the African and Arabian rifts that are associated with Pan-African Orogeny.

Some of the events related to the rift zone are as follows:

1. Development of the Qareh-Dash 'Series' acidic rocks and Doran Granite might be formed at an extensional zone, and probably derived from an early melting of Central Iran's crust. These rocks are equivalent to the Kushk 'Series' and the Rizu Formation in Central Iran.
2. Formation of mafic and ultramafic rocks in Takab and Anarak could be associated with the mafic and ultramafic magma of the mid-Oceanic rift, which built a new oceanic crust. The lithological and stratigraphic evidence and also the sedimentary facies with the mafic and ultramafic complex indicate that the new oceanic crust was not widely developed.
3. The stratigraphic and petrological evidence of the Rizu and Desu formations in Central Iran signifies that this rift zone in northwest of Takab and Anarak was deeper and wider than Saqand-Kerman area. Consequently, the rift zone is not recognizable in the mentioned area because the oceanic crust is not exposed along the rift axis. In other words, the tectonic condition of the area could be varied at different zones.
4. The volcanic activities during these eras provided a rich source of evaporite deposits in the Central Iran's basin. Then, the evaporate sediments produced a huge salt and gypsum deposits in the basin. The evidence of the evaporite deposits are observed in the in Desu 'Series', in Central Iran, and the 'Hormuz Series' in the Zagros. There are also traces of evaporite deposits in the Barut Formation. Eftekharnezhad (1975a, b) believes that the primary cavities of the Barut Formation in the Mahabad area, which was previously filled by gypsum, could be related to the last phase of the rift's activities in central Iran.

Accordingly, from the geodynamic point of view, during the late Precambrian (600–650 Ma) the continental crust in Iran diverted and ridged. As the result, a continental rift developed. In some areas, the rift even extended form an oceanic rift, then, it gradually closed to provide a broad platform; which eventually deposited the Zaigun and Lalun formations. Converging and diverging activities of the rift took place as the same time as Pan African Orogeny, discussed in the "Tectonics and structural geology of Iran" (Fig. 4.11).

4.3 Paleozoic Magmatism

4.3.1 Early-Middle Paleozoic Magmatism

During 419 to 458 Ma., the middle Ordovician- Silurian, magmatic activities severely affected Iran (Houshmandzadeh and Nabavi 1986). This wide magmatic rocks crop out in many parts of northern Iran (Alborz and northern Central Iran) (Jamei et al.

2020). For example, the basaltic rocks in the Shahroud and Khosh-Yeiylagh areas and basaltic-andesitic rocks of the Niur Formation in Central Iran.

Generally, in Alborz and the eastern Central Iran, the early Paleozoic basic volcanic activities are as followed.

4.3.1.1 Volcanic Rocks of Alborz

There are some basaltic layers on the upper part of the Ordovician Abarsaj Formation, Shahroud area. These layers gradually increase in thickness in the area. As a result, Abarsaj Formation's sedimentary rocks disappear and basaltic layers finally dominate and even reach to 700 m in thickness.

The basaltic rocks extend in Shahroud, Khosh-Yeiylagh, east of Firuzkuh and some other areas. In the Alborz zone, these rocks are known as the Soltan-Meydan Basalts which are of a Silurian age (Jenny 1977). Based on detail studies, wherever the basaltic layers convert into a massive form, they indicate the characteristic of a submarine and sometimes continental basalts (Jenny 1977).

Apparently, the eruption of basalts in oceanic basin had such a severity that volcanic eruptions exposed out of sea level. As a result, the volcanic rocks formed as continental basalt. The chemical composition of these basalts have been attributed to the continental tholeiitic basalt (Zohur Ghorbani 1995).

4.3.1.2 Volcanic Rocks of Central Iran

The Silurian of Central Iran contains some basaltic rocks which are represented by the Niur Formation. Though, these rocks are usually situated at the base of Niur Formation, in Shirgesht area, they normally form two distinct horizons within the formation. The geochemical composition of the basalts, signify the continental tholeiitic magma (Ruttner et al. 1968).

Actually, most of these volcanic rocks are reported from the Alborz and eastern Central Iran which were geographically closer to the Paleo-Tethyan region compared to any other areas in Iran. The geodynamics of the rocks at the last stage of early Paleozoic-early middle Paleozoic (late Ordovician-Silurian) show that the affinity of the basalts belongs to continental tholeiitic series.

The volcanic rocks in Central Iran could be an indication of preliminary stages of the Paleo-Tethys rifting in northern Iran. Accordingly, most of the rocks are exposed in the northeastern Central Iran and eastern Alborz and are of less exposure in western Alborz and western Central Iran. A series of metamorphic volcano-sedimentary rocks are observed in the Sanandaj-Sirjan Zone which can be probably attributed to the Silurian. However, no detail studies have been yet presented in terms of their accurate ages. There is a controversial discussion between researchers; some of them believe that the rocks belong to the early Paleozoic (Hosseini 2011) and some other point it out a late Paleozoic. However, certainly their ages are related to the Paleozoic.

4.3.2 Middle-Late Paleozoic Magmatism

According to the many geological evidence, there are certain magmatic activities which occurred in late Paleozoic (Late Devonian) and early-Mesozoic in Iran. Occasionally, the volcanic activities of basalt-andesite rocks were more dominant than the others during the late Paleozoic. Some of these rocks are as follows.

4.3.2.1 Volcanic Rocks of Northern Iran

The Geirud Formation of late Devonian is associated with basalts which is the best known volcanic phase in the Alborz zone, northern Iran (Assereto 1963). Their equivalent rocks in northern Iran are exposed in Amol, Geirud (type section), eastern Tehran and Talesh. These rocks have geochemical characteristics of inter-Continental tholeiitic basalts (Shirmohammadi 2014) (Fig. 4.12).

The middle-late Paleozoic volcanic rocks are exposed in some regions in Central Iran, such as Abdolhossien Mountain with the Devonian rocks in Anarak and Abadeh area close to the Esteghlal mine and south Damghan area (Figs. 4.13 and 4.14).

Carboniferous volcanic rocks: Notwithstanding lack of detailed reports about the Carboniferous volcanic rocks in Iran, some Carboniferous rock formations are probably originated from tuffs such as Sardar Formation between Shish-Angoshti Mountain and Bagh-e Vang area in Tabas. Also, the Carboniferous rocks associated with tuffaceous shale are similar to Sardar Formation in Abar Kuh- Bekheirkheng Valley (Sharkovski et al. 1984).

The northwestern part of Iran is structurally similar to Central Iran during the Paleozoic. The Carboniferous volcanics and intrusions (basalt to trachyte and gabbro, syenite as well as granite that all show alkaline affinity) can be observed in southern Julfa and Mishu mountains, close to the Urmia city (Asadpour and Heuss 2018; Jamei et al. 2020).

Devonian-Carboniferous: A complex of different metamorphic ultramafic, mafic and volcano-sedimentary rocks of middle-late Paleozoic is identified in north eastern and northern Iran. The complex is extended in the Fariman, Gorgan, Talesh mountains as well as the north of Meshgin Shahr. Probably, it confirms the activities of opening and expanding of Paleo-Tethys oceanic crust in the northern Iran.

The volcanic rocks in the Fariman area have characteristics of komatiite series (Dakhili 1995; Moaf-pourian 2010). Apparently, these rocks signify the Late Devonian-Permian continental rifting zone in northern Iran. Then, a wide ocean, developed in the Permian which was closed in early Mesozoic. The complex of the oceanic rocks was metamorphosed at Greenschist-Amphibolite phases. The Devonian basalts which formed in Alborz zone and Central Iran are related to the extensional pulses in the southern part of the Permian ocean.

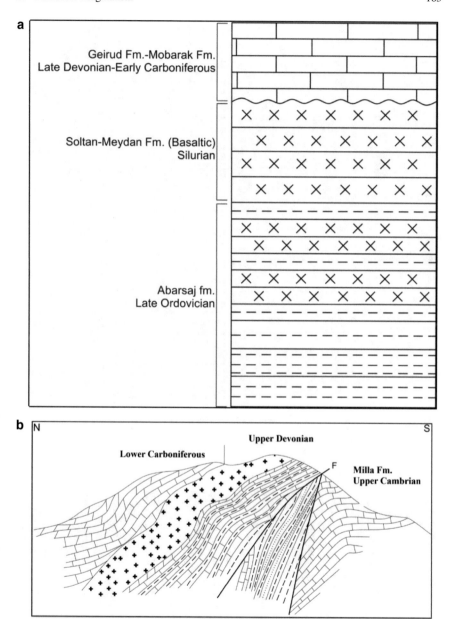

Fig. 4.12 A schematic view of basaltic rocks in Alborz area (Negarman and Abarsanj, NW Shahrud, near Bastam (**a**) and Paleozoic formations of the Geirud section, northern (**b**)

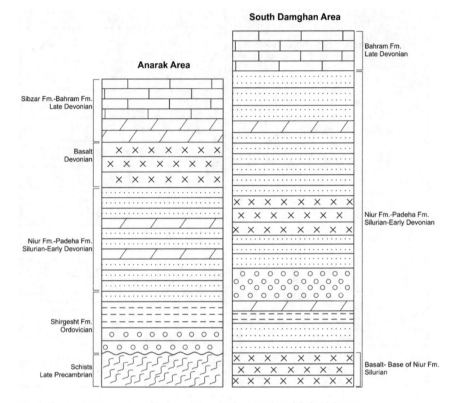

Fig. 4.13 A schematic view of Paleozoic basalts in Anarak and Soltanieh area

Fig. 4.14 A view of volcanics and Sandstone of Padeha Fm., Ghusheh area

4.3.2.2 Volcanic Rocks of the Sanandaj-Sirjan Zone

In addition to the volcanic rocks of northern Iran, a wide range of basalt, andesite and trachyte (mostly rhyolite associated with tuff, and volcanic rocks) are exposed in the Sanandaj-Sirjan Zone, southern Iran. Some small diorite to granite bodies have intruded in the volcanic rocks. It is believed that they belong to the middle Paleozoic, but some geologists attribute them to late Paleozoic-Early Triassic (Ahmadi Khalaj et al. 2007).

The present author has attributed these rocks to upper Paleozoic-Triassic based on his investigations on the associated iron mineralizations. As noted before, these rocks are volcano-sedimentary in origin which metamorphosed in greenschist and rarely amphibolite phases. Some of them are as follows:

– Sirjan Metamorphic Zone.
– Tutak and Surian Complex Metamorphic Rocks
– Kuli Kosh Complex Metamorphic Rocks in Deh Bid Zone
– Daran and Saman Metamorphic rocks
– Aligudarz-Azna Metamorphic rocks
– Northern Ali baba-Golali Metamorphic rocks
– Hamedan-Qorveh Metamorphic rocks
– Northern Divandareh-Zarineh Obatu Metamorphic rocks, Kurdistan province (Abdi 1996).

These rocks are referred to Pre-Cambrian in some geological reports and maps. But recently, it is strongly believed to be formed in Paleozoic (refer to Chap. 8, i.e. Metamorphism rocks). The volcanic rocks accompanied by the small intrusive bodies are associated with the intercontinental alkaline rocks which are composed in an extensional zone (Ghorbani 2007; Hosseini 2011; Hadizadeh 2010).

According to the geodynamic characteristics of Sanandaj-Sirjan zone, it represents an intercontinental extensional system during Carboniferous to Early Triassic and could be attributed to the beginning of the Neo-Tethyan rifting at its southern marginal basin (Fig. 4.15).

4.4 Mesozoic Magmatism

The Mesozoic magmatic activities which are related to the Laramide and Cimmerian orogenies have formed some continental and oceanic rifts in Iran. Gradually, the rifts were closed and subducted beneath a large area of Iran. The occurrence of such widespread magmatism can be classified into three categories:

1. **Magmatic rocks in ophiolite complexes**: the magmatic rocks of Jurassic and Cretaceous to Paleocene associated with ophiolite complexes are formed in an oceanic basin. These rocks have previously introduced as the "Tethys III". They will be discussed in the chapter of ophiolites of Iran.

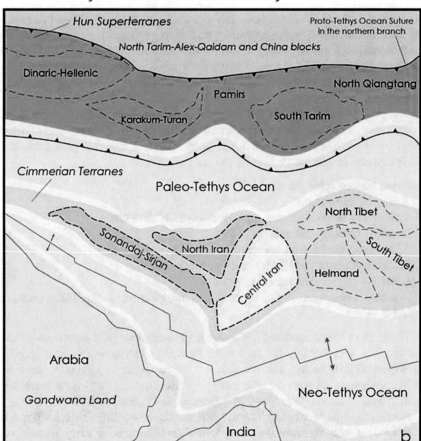

Fig. 4.15 A schematic view of the Magmatic rocks of south Jolfa and Mishu mount (Jamei et al. 2020)

2. **Volcanic rocks**: the volcanic rocks were formed in an extensional basin, by rifting and partly subduction of the oceanic crust under the continental crust of Iran. They are commonly found in Iran especially those of Early Triassic and Late Jurassic-Cretaceous.

3. **Intrusive rocks**: Many varieties of gabbroic to granitic rocks with different ages formed in Mesozoic. These rocks have following characteristics:

   – In Triassic-Early Jurassic, the volcanic rocks are more dominant compared to intrusive bodies. The alkaline volcanic rocks are widely extended in the Sanandaj-Sirjan zone. In many areas, they are covered by the younger rocks.

– In Jurassic and Cretaceous, particularly in Middle Jurassic and Late Cretaceous, the intrusive bodies and large batholiths are more dominant in
comparison to volcanic rocks.

The magmatic rocks of Triassic, Jurassic and Cretaceous are briefly described.

4.4.1 Triassic

Some magmatic rocks were formed in late Paleozoic-Early Triassic. They are associated with the Late Hercynian orogeny (Palatian phase) and some of them have been
formed in the Middle Triassic to Late Triassic related to early Cimmerian orogeny.
The Triassic magmatic rocks are mostly intrusive bodies in northern Iran while they
occur as volcanics in central and southern Iran (Allenbach 1966).

Sanandaj-Sirjan Zone: The Triassic magmatic activities appeared as volcanic
rocks and occurred in Sanandaj-Sirjan Zone. These rocks are mainly exposed in
the southern Sirjan, Eghlid and northern basin of Sanandaj-Sirjan Zone (Songhor,
Kamyaran and surroundings). The Triassic volcanic rocks are the oldest rock units
in the zone and have been cut by diabasic dikes. These rocks are usually covered by
the younger rocks. The mid-Jurassic thick shales from central Eghlid to Hamedan
are the main part of the young sediments. Accordingly, the volcanic rocks are of less
exposure (Alric and Virlogeux 1977).

The volcanic rocks are mostly alkaline lava and composed of trachyte-rhyolite
volcanic rocks. Also, the gabbro-ophiolite and ortho-amphibolite complexes occur in
the east southern Sanandaj-Sirjan zone. In some areas, these igneous sequences are
covered by the Early Jurassic sediments. According to Sabzehei (1974), the intrusive
layered rocks vary from ultrabasic to granite and are segregated from tholeiitic basalts
in Sykhouran and Esfandagheh. This magma has a rich calcium oxide composition
with a poor content of alkalis.

The intrusive complexes which are of different ages can be classified in three
categories: ultramafic, isotropic gabbro, and diabasic dikes. The isotropic gabbro cut
the huge ultramafic Komatiite unit, and as a result a massive contact metamorphism
occurred in Paleozoic rocks. Also, the scattered dibasic dikes cut the old ultramafic
and gabbroic rocks. This complex might be related to an extensional zone, in which,
Neo-Tethys Ocean was opened and developed.

Several basaltic sills in early Triassic intruded the base of Triassic rocks. The
Triassic basal rocks are equivalent to the Elika Formation in Dehbid, Shahreza
(Shahzadeh Ali Akbar) areas (Ghorbani 2019). In southern Julfa (Zal section) the
basaltic rocks are of wide exposure at the basal parts of the Elika Formation. Indeed,
the Elika Formation in Alborz is accompanied by volcanic rocks.

Central Iran Zone: It is a controversial discussion that biotite-granite and also,
basic rocks of Abadeh are attributed to Triassic. Crawford (1977) estimated 240 Ma.
by Rb-Sr method for the Ismail Abad granite's age (late Middle Triassic). According
to the investigations of the present author such age estimation cannot be fully correct.

Anyway, it is included in late Paleozoic-Triassic intrusions of Iran and shows the magmatic activity of this period of time in northern Iran. During the late Paleozoic-Early Triassic the Paleo Tethys (Tethys II) was completely closed and syn-collision magmatism occurred.

Alborz-Azerbaijan Zone: In the Alborz-Azerbaijan Zone, the magmatic activities can be traced as intrusive rocks. The largest intrusion is Lahijan Granitoid which has metamorphosed the Carboniferous sediments. The Jurassic conglomerate contains some pebbles of the metamorphosed Carboniferous sediments in northern Alborz. So, this granite has been formed later than the Carboniferous and before Jurassic. However, Moussavi (1994) doubtfully attributed it to the final stages of Hercynian orogeny.

Taking into account the tectonics of the region, the meta-aluminous granitoid is "I type" which originated from a calc-alkaline magma, at the post orogeny stage.

There is a kind of tourmaline-bearing granite in Gasht, Masuleh that based on the geological evidence it is formed within the Permian to Triassic interval. The Rb-Sr age determination indicated 180 ± 5 Ma. for the mentioned intrusion (Emami quotes from Crawford 1972, 1977).

The huge intrusive bodies composed of porphyry granite, biotitic-granite, granodiorite and biotitic-hornblende tonalite have intruded in Khorasan; north and northeastern, south and south western Mashhad (Aghanabati and Shahrabi 1987). These rocks were attributed to the late Paleozoic-Triassic. Though, French B.R.G.M experts reported the intrusive rocks in eastern Qaen which are formed in Proterozoic, likely they belong to Permian-Early Triassic. Based on K-Ar age determination, on biotite of some of the porphyry granites, the age estimation for these rocks are 256 ± 10 and 215 ± 9 Ma., Permian Triassic (Emami 2000). Age determination analyze on biotites of leucogranite showed 245 ± 10 Ma. Consequently, the intrusions have apparently intruded between Permian and Triassic and could be emplaced as multiphase intrusive bodies. The magmatic activities seem to be related to the Hercynian and Early-Cimmerian orogenies. Based on geochemical composition, these rocks have probably developed after Hercynian orogeny and mixed with the continental crust rocks.

The numerous basaltic dikes in northeastern Iran (Aghdarband area) cut the lower part of Anisian limestone (Ruttner 1984). These dikes can be related to the shallow depth Triassic intrusive bodies. Also, as mentioned before, a series of basaltic rocks have formed within the Elika Formation in northern Iran, Alborz zone.

4.4.1.1 Magmatism, Mechanism and Geodynamics

As previously discussed, the Triassic magmatic rocks are classified into two groups:

1. Volcanic rocks which are mostly alkaline and can be found over a large area in Iran. However, they are of less distribution in individual zones. Based on geochemical studies most of them were formed during extensional phases.

2. Intrusive rocks have mostly formed as scattered igneous bodies. They are more distributed in northern and northeastern Iran. These intrusive rocks are related to the orogeny phases. According to the geological evidence and stratigraphy of the nearby rocks as well as their various age determinations; they have formed at Late Paleozoic and Early Mesozoic, Permian to Triassic periods (refer to Chap. 5, i.e. Intrusive rocks).

According to what mentioned above, the Triassic intrusive bodies have been formed in compressional phases, when the Paleo-Tethys Ocean (Tethys II) was closing in the northern and east northern Iran. However, the volcanic rocks of northwestern Iran, Central Iran, and the Sanandaj-Sirjan zone have formed in extensional phases. Such environments could be found in several parts of the world. Generally, the late Paleozoic-Carboniferous-Early Triassic magmatism should be investigated two distinct parts: northern Iran which is related to the closure of the Paleo-Tethys ocean (Tethys II) and southern and central Iran as well as Sanandaj-Sirjan zone in relation to the opening of the Neo-Tethys ocean (Tethys III).

4.4.2 Jurassic

This era precedes and succeeds the early Cimmerian orogeny phase in the Late Triassic and late Cimmerian orogeny in the early Cretaceous, respectively.

It was previously thought that most of the magmatism of Jurassic is related to the late Cimmerian orogeny. However, new findings indicate that they are caused by the middle Cimmerian tectonic event (Aghanabati 2004). On the other hand, the volcanics of the base of Shemshak Formation which have previously attributed to Jurassic, is now dated as late Triassic (Fursich et al. 2009).

Some of the basic and intermediate igneous rocks formed in Jurassic are:

– Andesitic rocks in Torud, Isfahan, Shahrekord, Mahabad and Sanandaj.
– Andesitic to basaltic rocks in Ardabil
– Spilite basalt in Sanandaj-Sirjan, between Esfandagheh to Isfahan.

The volcanic rocks widely occur within the sedimentary layers of the Shemshak Formation. The Laramide orogeny has folded, faulted and deformed some parts of the Jurassic rocks.

4.4.2.1 Volcanic Rocks

Both intrusive and volcanic rocks were formed during Jurassic. It seems that the well distributed volcanism continued in an extensional zone in Jurassic. The volcanic rocks extend in different areas as follows:

The vast alkaline basalts, small gabbroic and gabbro-foid intrusions occurred in northern Alborz- Baladeh and Siah-Bisheh. Based on stratigraphic sequence, these rocks were formed in different periods and covered the older sedimentary rocks such as the Shemshak, Dalichai and Lar Formations (Doroozi et al. 2017). Regarding to the stratigraphic evidence, the magmatic rocks formed in early to late Jurassic.

The foid-bearing gabbro has intruded into the Permian rocks in northern Iran, near Siah-Bisheh Valley. The composition of these beds are very similar to basaltic rocks and occur at the top and within the layers of Shemshak, Lar, and Dalichai formations. Also, a series of evaporites are associated with basalt and diabase formation that can be found at top of the Lar and Tiz-Kuh formations (Sussli 1976). These volcanic and evaporitic complexes constitute the Gypsum-Melaphyre Formation of Damavand, eastern Tehran (Aghanabati 2004). The volcanic and diabasic rocks are basanit, which indicating their development in an extensional basin. The inter-bedded basaltic lava occurs in the Shemshak Formation of the northern Iran, Roudbar; which is probably formed in Jurassic. Nevertheless, probably this interlayered andesitic-basaltic lava could be Cenozoic basaltic sills.

There are a plenty of volcanic rocks occurred in Central Iran, in Damghan and Torud, Sanandaj-Sirjan Zone, Gol-Gohar and Esfandagheh which were most likely formed in Jurassic. Nevertheless, there is no detailed geochemical studies about their ages.

4.4.2.2 Intrusive Rocks

During Jurassic, the main phases of granitoid magmatism were formed by intruding magma into the continental crust of Iran. This activity is related to the tectonic events of middle Cimmerian orogeny, extending over a wide area in different parts of Iran like granite-granodiorite of Shir-Kuh (Yazd), Bafgh, Sabzevaran, Shah Kuh- Deh Salm, Absard, Brujerd, Jajarm, Miami and Jandagh; and Granite-Granodiorite in Alvand and Diorite in Almogholagh (refer to Chap. 5, i.e. Intrusive rocks).

The granodiorite-diorite and gabbro-diorite intrusive rocks in some regions like Sabzevaran were assigned to the Jurassic and some to Jurassic-Cretaceous and Late Cretaceous. For example, Atyrakan granite's age determination by the Rb-Sr and K-AR methods indicates 168 ± 8 Ma. (Reyer and Mohafez 1972) and 113 ± 9 Ma. (Aghanabati 1998), respectively. The Golpayegan granite was probably formed during the Pre-Cambrian and late Cretaceous. Some of the Jurassic intrusive rocks discuss as follows:

Shir-Kuh Granite: Shir-Kuh intrusion has cut shales and sandstones of the Shemshak Formation and has vastly metamorphosed them in the southwestern Yazd. The clastic sediments overlay the Shir-Kuh granitic batholith by a nonconformity at southwestern Taft.

In the Lut block, large granitic batholiths cut the Jurassic shales and sandstones and a sequence of Cretaceous clastic rocks covers them. Shah-Kuh Granite with 45 km length has formed a contact metamorphism within the Jurassic rocks in the northern and southern margins. In the southern part of the same batholith, Deh-Salm area, metamorphic rocks are of hornfels affinity. Also, an intrusion with 40 km length and 2 km width injected into the layers of Jurassic sedimentary rocks including shale and sandstone of Chahar Farsakh. The elongated shape of the Chahar Farsakh granite-granodiorite and pegmatite indicates its emplacement along the major longitudinal fractures of the region. The Chahar Farsakh intrusion is more or less similar to Shah-Kuh batholith. Although, these granitic batholiths are attributed to the late Jurassic; it is most likely formed in the middle Jurassic related to the middle Cimmerian orogeny phase (Aghanabati 1998).

Gonabad intrusive sills (rhyolite-dacite): A series of metamorphosed porphyry sills (initially composed of rhyolite to dacite) have injected into the Shemshak Group, in southern Ab-Kuh, Gonabad area. In the Gonabad to Ferdows, southern Kuhsar Hesar, similar plutonic stock exposes that never cut any Early and Late Cretaceous sedimentary rocks in the region. It is expected that they have formed during Jurassic.

In the Sargaz area, coarse-grained granites have injected into the Dogger volcano-sedimentary rocks. The radiometric age determination of biotite granite and gneiss of Chah-Dozdan area indicates 164 ± 4 Ma. for them (Sabzehei et al. 1970).

The Cheshmeh Ghassaban gabbro clearly crosscut the Hamedan shales and the Shemshak sandstones. The basic portion (gabbro-diorite) of the Almogholagh intrusion, 30 km along the Hamedan-Sanandaj road, has probably differentiated from the upper mantle magma, with the age of 144 ± 7 Ma. (Emami 2000). This intrusion is attributed to Late Jurassic, but the age determination and regional evidence indicate a middle Jurassic which is associated with the middle Cimmerian orogeny phase. Recently, Shahbazi et al. (2010) has determined the age of Hamedan batholith to 170 Ma.

The basic and acidic intrusions are attributed to the Jurassic and Late Cretaceous-Paleocene, in Sanandaj-Sirjan zone, respectively. Both types of the intrusions are intertwined and in fact, regarding to available data, their age, tectonic setting, magmatic type and composition cannot be distinguished. Probably, the basic intrusions are more extended than the acidic intrusions in this zone (refer to Chap. 5, i.e. Intrusive rocks).

Kolah Ghazi granodiorite: Kolah Ghazi granodiorite has intruded into the Shemshak shales 50 km southwestern Isfahan as several separable magmatic bodies and has been covered by Barremian-Aptian basal conglomerate. Zahedi believes that their age is Late Jurassic. However, they could be middle Jurassic (Zahedi 1976).

A gigantic granitic batholith has emplaced into the Pre-Cambrian metamorphic, Permian igneous and fossiliferous rocks in western mountains of Posht-e Badam north of Sefid Donbeh and Khoshumi Mountains. This granitic body is apparently covered by the Cretaceous conglomerates. Hence, it seems that it is formed in late Triassic to Oligocene. Referring to the Iran's Intrusions table, which is depicted in the next chapter. Many of them have formed in Jurassic and mostly have intruded in Sanandaj-Sirjan Zone.

4.4.2.3 Geodynamics of Jurassic Volcanics

As noted earlier, it seems that the Jurassic volcanic rocks correspond to a local extensional zone which is associated with the faults of the deep zone and have taken place at the faults crosscutting. Nevertheless, regarding to those of Sanandaj-Sirjan, it should be mentioned that these intrusions have characteristics of calc-alkaline series mostly generated in the subduction and orogeny zones.

These intrusions are mainly emplaced in Sanandaj-Sirjan zone. Regarding the previous situation of the Neo-Tethys which was integrating into a great ocean during the Triassic to Early Jurassic, it seems that the oceanic plate has subducted beneath the Iranian continental plate in the Sanandaj-Sirjan zone during the Early Jurassic. The emplacement of intrusive bodies in Sanandaj-Sirjan zone has taken place later in early Middle Jurassic.

4.4.3 Cretaceous

4.4.3.1 Cretaceous Magmatism

The Cretaceous magmatic rocks are classified into three categories.

A. Volcanic,
B. Intrusive rocks and
C. Ophiolite complex

These groups of rocks are relatively independent. The first two groups are described here and the third group will be discussed in Chap. 7, i.e. ophiolites.

(A) Volcanic rocks: There is no considerable detailed study about the Cretaceous volcanic rocks. However, according to their petrography and geochemical specifications well as their dispersions, they are widely extended in Iran. The specifications of Cretaceous volcanic rocks are as follows:

Alborz-Azerbaijan Zone: Widespread and thick volcanics are exposed in the northern slopes of the Javaherdeh (Sothern Amol) and Dohezar valley, Alborz mountains (Clark et al. 1975) and in Azerbaijan (Majidabad, Kaleybar and Arasbaran).

The Cretaceous volcanic rocks of the northern Alborz (Javaherdeh and Dohezar valley) were formed submarine alkaline basalts which gradually show variations to a continental alkaline to tholeiitic basalts. It is considered that these basalts have occurred along with the gabbroic magma which have the same origin and has formed in Early Cretaceous (Hagh-Nazar 2009).

The alkaline and foid-bearing basalts were formed as spilite, pillow and vacuolated in Azerbaijan, Majid Abad and northern Kaleybar (Fardaei 2009). Based on the stratigraphic characters, these rocks have occurred between Middle to Late Cretaceous; and according to their geochemical composition, they are related to the potassic alkaline series (Geological map of Ahar, 1:250,000 in scale Fardaei 2009).

More than 1000 m basalts and andesites were formed during the Middle to Late Cretaceous in both sides of Aras River of Iran, Azerbaijan and Armenia Republic. On the other side of Aras River in Armenia, these rocks extend up to the Caucasus region (Geological map of Siah Rud, 1:100,000 in scale Mehrpartou et al. 1997). The geochemical composition of them is varied from the Calc-alkaline to alkaline.

However, they are mostly composed of calc-alkaline type (Kazmin et al. 1986). Tectonically, they mostly formed at a continental subduction zone. However, they could be also attributed to an extensional zone.

There is no detailed geodynamics and mechanical studies about the origin of these volcanic rocks. However, most recently, Doroozi et al. (2015) indicated that based on low $^{87}Sr/^{86}Sr$, high $^{143}Nd/^{144}Nd$ and HFSE/REE ratios, south Marzanabad volcanic rocks show the similar source characteristics and derivation from the depleted mantle source with regard to bulk earth composition.

Sanandaj-Sirjan Zone: till now, the volcanic rocks were of less importance compared to the Cretaceous intrusive rocks in this zone. However, detailed field observations indicate that they are more dominant in northern Sanandaj-Sirjan zone in comparison to intrusions, in terms of volume. Certainly, the petrological interpretation of the mentioned volcanics is a useful tool in better understanding of the Sanandaj-Sirjan geodynamics (Eftekharnezhad 1980).

The Cretaceous volcanic rocks in the northern Sanandaj-Sirjan Zone are traceable in following locations:

– In Sanandaj, particularly around the Gheshlagh Dam, the majority of which is composed of andesite (Geological map of Dehgolan, 1:100,000 in scale).
– A wide range of basalt to basaltic andesite rocks were formed along Divandareh-Saqqez road (Iran-Shah area), and Western Saqqez. They have occasionally been cut by medium-sized intrusive stocks (Geological map of Iran-Shah, 1:100,000 in scale).
– Andesite and rhyolite are outcropped around the Piranshahr batholith, along the Piranshahr to Mahabad road (geological map of Mahabad, 1:250,000 in scale).

The geochemical compositions of the Cretaceous volcanic rocks in Sanandaj-Sirjan zone are associated with the calk alkaline series (Kazmin et al. 1986). These rocks should be formed in a subduction zone where the Neo-Tethyan oceanic crust subducted beneath the Sanandaj-Sirjan continental crust zone (Ghorbani 1999; Tarkhani 2010).

Other parts of Iran: Except the volcanic rock related to the ophiolites, the Cretaceous volcanics are not widely exposed in Central and western Iran, compared to Alborz-Azerbaijan (western Alborz) and Sanandaj-Sirjan Zones. However, they can be scatteredly found in the mentioned areas such as Khar-e Touran in Central Iran, Birjand and occasionally at northern Zahedan, eastern Iran.

(B) Intrusive bodies: In addition to the intrusions formed during the Jurassic to Cretaceous, some other intrusive bodies were formed in Early Cretaceous (90–100 Ma).

Many Late Cretaceous plutonic granitoids injected into the continental crust of Iran, were considered to be related to the Laramide Orogeny. It is believed that a gabbroic magma moved up and injected into the siliceous continental crust and melted to form the major plutonic bodies up to the end of Cretaceous. Apparently, the age of last plutonic activities could be younger than Cretaceous.

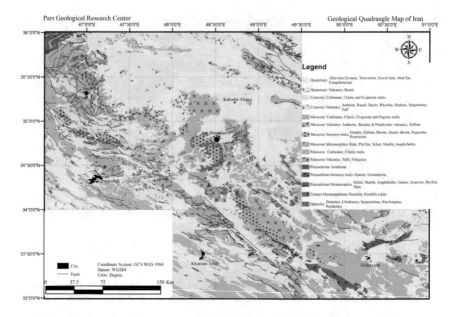

Fig. 4.16 The intrusive rocks in northern and southern part of the Sanandaj-Sirjan zone

Granite and granodiorite intrusions have cut the Early Jurassic shale and sand-stones in Golpayegan and southwestern Arak (refer to Chap. 5, i.e. Intrusive rocks).

The Cretaceous magmatic intrusions are considerable in Sanandaj-Sirjan zone, but they are not distinguishable in other parts of Iran such as the Alborz and Central Iran. However, there are several intrusive bodies in Qaenand Hajiabad in eastern Iran which could be related to the Cretaceous magmatic phases (Geological map of Shahrakht, 1:250,000 in scale).

The Cretaceous intrusive rocks are mostly exposed in northern (associated with Jurassic intrusive rocks) and in a minor extent in the southern part of the Sanandaj-Sirjan zone. They have a wide range of chemical compositions and as a result it would be complicated to distinguish them from the Jurassic intrusive rocks (Fig. 4.16). More detailed age determinations and geodynamics' studies are needed to identify their ages. More information is given in Chap. 5, i.e. Intrusive rocks.

4.5 Cenozoic Magmatism

The Cenozoic Era is a short geological time, approximately 65 Ma. Indeed, it is shorter than earlier eras. Nevertheless, the Cenozoic magmatism is significance in terms of their extent and volume.

The Cenozoic sedimentary rocks overlay the Cretaceous or older rocks by an angular unconformity. However, the K/Pg boundary is associated with the facies changes in Kopeh-Dagh and some areas in Zagros.

Generally, the Cenozoic must be known as "the time of serious magmatic activities" that have affected Iran except the Zagros and Kopeh-Dagh areas. Regarding the "Iran intrusions distribution map", the Cenozoic intrusive bodies are more than the previous ones (Fig. 4.17).

Various studies conducted in different parts of Iran such as Sahand, Sabalan, eastern and the north eastern Azerbaijan as well as the southern Birjand indicate that the volcanic activities and the injection of the intrusive bodies have initiated in the Late Cretaceous and reached its climax during the Eocene. In many areas, accompanied by some short gaps, it continued till the Quaternary.

Although, the majority of intrusive activities occurred and extended during the Cenozoic to the Quaternary, but four periods of Cenozoic, the magmatic activities soared:

Middle Eocene: Volcanic activities are much more dominant compared to intrusive bodies in middle Eocene and apparently there is no sign of plutonism. But the small intrusions are likely a source of the middle Eocene volcanoes.

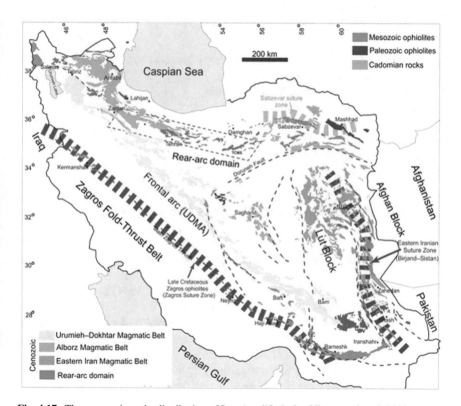

Fig. 4.17 The magmatic rocks distribution of Iran (modified after Niroomand et al. 2020)

Late Eocene-early Oligocene: During the late to early Oligocene, the intrusive bodies are more dominant than the volcanic ones. Most of the Cenozoic large batholiths have been formed in this period.

Late Oligocene-early Miocene: The magmatism started as volcanic activities in late Oligocene. At the end of the early Miocene (Bourdigalian), the volcanic activity which was accompanied by small intrusions gradually decreased.

Middle Miocene-Pliocene: During the middle Miocene-Pliocene, the Magmatism started by strong volcanic eruptions and ended by the injection of acidic-intermediate magmatic domes. Most of the Cenozoic's small intrusions of Iran are formed in this period.

Quaternary: Damavand, Sahand, and Sabalan volcanoes are resulted from the Quaternary volcanic activities. Also, the dispersed young basalts formed in Azerbaijan and eastern Iran are of same origin. The young volcanism in Iran is depicted in the next chapter.

4.5.1 Distribution

The Cenozoic magmatic activities, particularly volcanism are widespread in Iran which are listed as following:

1. Volcanic Zone of Urmia-Dokhtar or Sahand-Bazman zone
2. Southern margins of Central Iran,
3. Western Alborz including Tarom-Hashtjin,
4. Sabalan, Ahar-Jolfa axis and Mianeh area,
5. Sabzevar-Torud-Moalleman axis,
6. Torbat-e Jam-Kashmar-Bardeskan axis,
7. Margins of the Lut and Central Iran zones,
8. Bazman-Taftan axis,
9. Eastern Iran.

The magmatic rocks distribution of Iran is depicted in Fig. 4.17.

4.5.1.1 Urmia-Dokhtar Volcanic Zone (Sahand–Bazman)

Various lithological studies (e.g. Emami 1981; Amidi et al. 1984) in several parts of Iran indicate that volcanic rocks are mostly as follows:

Andesite, andesite basalt, basaltic andesite, basalt, rhyolite, dacite and rhyodacite along with acidic to intermediate tuffs, and sometimes, latite andesite, latite trachyte, alkali-trachyte, and lucite phonolites.

Intrusive rocks: the intrusive rocks particularly have a compositional range from diorite to granite which have been formed during the Eocene to Pliocene and occasionally in the Quaternary. There is not general agreement regarding the origin of the mentioned rocks. Some researchers assigned the provenience to volcanic arcs and some to Island arcs, nevertheless, the majority confirms that they are formed as a result of oceanic-continental subduction and subsequent development of volcanic arcs.

The Urmia-Dokhtar volcanic belt is not only confined to Iran, but also it begins from the Serbia to Pakistan. The belt extends from Urmia in northwestern part of country to the Dokhtar area of southeastern part or the Bazman volcanic area (western Neyriz; southern Iranshahr area, Sistan and Baluchestan province). Generally, it has the following geological characteristics in Iran:

1. NW-SE trend,
2. Composing of volcanic to volcano- sedimentary series,
3. They are composed of the various lavas and pyroclastic rocks and are compositionally varied from basalt to rhyolite,
4. Sedimentary and volcanic rocks formed as alternating series, but the clastic rocks form the majority of them,
5. Many Eocene-Pliocene intrusions with the composition of gabbro to granite have cross-cut this volcanic belt,
6. More exposures of the Precambrian-Cambrian basement can be observed in the northern parts compared to the southern parts of the volcanic belt and
7. The intensity of volcanic activities was different during the Cenozoic. The volcanic activities increased during some periods such as Middle Eocene, Late Eocene-Early Oligocene, Late Oligocene-Early Miocene and Late Miocene-Pliocene.

As it was mentioned, numerous studies have been conducted on the Urmia-Dokhtar zone, but most of them cover a small part of the area.

Studies carried out by Emami (1981) in the Qom-Aram, Ghorbani (1999) in the Takab, Amidi et al. (1984) in the Natanz-Surk, Moradian (2005) and Hassanzadeh (1993) in the Shahr-e Babak, and Omrani et al. (2007) in Zagros indicate that the magmatism of the Urmia-Dokhtar belt were active during the Cenozoic.

Contrary to the wide exposure and abundance, the Cenozoic magmatism has not been well studied. There is a controversial discussion between Iranian and foreign geologists regarding the mechanism of Cenozoic magmatism. Two main approaches are discussed as below:

1. Continental rift theory and aulacogens (Sabzehei 1974; Emami 2000): There is not enough evidence to substantiate the continental rifting theory and there are not so many researchers who agree with this theory.
2. There is plenty of evidence which confirm the Subduction Zone Hypothesis, based on which the Neo-Tethys Oceanic crust has subducted below the Iranian continental plate. Therefore, the two plates of Iran and Arabia have collided during the Late Cretaceous and Paleocene. Thus, the Neo-Tethys Oceanic crust was closed (Berberian and King 1981; Alavi 1991).

Immediately after the advent of Plate Tectonic Theory, Subduction hypothesis was proposed for better understanding of how magmatic rocks have been formed in Iran, particularly, the formation of the Urmia-Dokhtar volcanic rocks (Stocklin

1968; Alavi 1980, 2004). Many geological characteristics and evidence confirm the Subduction Theory for Cenozoic Magmatism in Urmia-Dokhtar volcanic zone as follows:

1. The Cenozoic Urmia- Dokhtar volcanic zone with linear extension more than 1500 km could be originated from magmatic activities resulted from the subduction of Neo-Tethys beneath the Iranian plate.
2. Regarding facies variations and tectonic system, the Zagros Mesozoic and Cenozoic rocks are the northern extension of the Arabian Plate which highly differs from the Sanandaj-Sirjan and Central Iran zones.
3. The separation boundary between the Zagros and Sanandaj-Sirjan zones which is known as the Main Zagros Thrust is nearly a straight line along NW–SE, the same trend of Urmia-Dokhtar volcanic zone. It is believed that the Arabian and Iranian plates collided and formed the Main Zagros Thrust (e.g. Alavi 2004).
4. Some ophiolitic complexes, a composition of ultramafics along with deep oceanic sedimentary rocks are formed along the Zagros Thrust Zone. These complexes are remnants of the Neo-Tethyan oceanic crust between Iran and Arabian plates.
5. High-thick crust in the Sanandaj-Sirjan Zone at the collision border with Zagros Thrust, comparing to the other parts of Iran, indicates that the NE Arabian plate has been subducted beneath the Iranian Plate.
6. The Kerman's copper porphyry deposits in the Urmia-Dokhtar volcanic zone confirms the subduction theory. It is believed that this type of deposits in Kerman, Pacific Ocean margin and other parts of the world are formed as a result of the subduction mechanism.

The enormous calk alkaline type volcanic rocks are much more dominant than the other types of volcanic and plutonic rocks in the Urmia-Dokhtar zone. Some of the lithology and petrography evidence in Iran indicate that the presence of alkaline rocks does not fully explained by the known subduction zones in other parts of the world (Ghorbani 1999) such as:

1. Alkaline rocks, particularly alkaline potassic rocks formed widely in many parts of Iran such as Taleghan, Tarom, Azerbaijan, Kerman and other areas.
2. Calc-alkaline and alkaline volcanic rocks occurred at the same location. They are not coordinated both in terms of the nature and the time of magmatic events. These rocks seem to be related to the rift not to subduction zones.
3. The Sanandaj-Sirjan metamorphic rocks regarding their metamorphic phases, and ages do not fully comply with the subduction theory. These contradictions should be taken into account the geological characteristics of Iran, the specification of subducted plate, the angle of subducting plate, crust thickness and other factors.

Based on the Cenozoic magmatic phases, the volcanic belt could be classified into three subzones:

1. **Southern Urmia-Dokhtar zone**: Dehaj-Sardouieyeh, Jebal-e Barez axis in Kerman region.

2. **Middle Urmia-Dokhtar zone**: Taft- Kashan-Qom.
3. **Northern Urmia-Dokhtar zone**: Axis of Tafresh-Razan-Takab-Sahand.

All Cenozoic magmatic phases can be traced in the southern zone (e.g. middle Eocene, Eocene-Oligocene, Oligocene-Miocene, Miocene-Pliocene and Quaternary). The most widespread magmatic activities in terms of duration and extent have occurred in the Kerman region.

In the middle Urmia-Dokhtar zone, two phases of magmatism such as the volcanic phase in the Eocene especially in middle Eocene and plutonic phase in Eocene-Oligocene are traceable in Taft-Anarak-Ardestan, and Kashan-Qom (Abdolnshad Mamaghani 2007). The younger magmatism in the middle Urmia-Dokhtar zone is of less widely-distributed than the Kerman region.

The Eocene volcanism and Eocene-Oligocene plutonism are less significant in the northern Urmia-Dokhtar zone. These rocks are traceable in the Tafresh-Razan-Takab and Shahin Dezh, and in terms of volume and variety are less considerable. The volcanic rocks particularly pyroclastics are more similar to Karaj Tuffs. Also, the Eocene-Oligocene intrusive rocks are of fewer occurrences. Nevertheless, the younger Oligo-Miocene and Mio-Pliocene magmatism are remarkable (Amidi et al. 1984).

Southern Urmia-Dokhtar zone (Kerman): This is a part of the southern Urmia-Dokhtar volcanic zone which has about 450 km length and 80 km width. It extends to the Rafsanjan, Kerman and the Bam basin in the north side and to the Meiduk-Sirjan basin in the southern side. The southern zone is cross-cut by some NS faults such as the Anar and Sabzevaran faults at south and Gudak fault at north. Also, the NE-SW faults cross-cut the area (Nedimovic 1973).

The tremendous magmatic activities were expanded during the early Eocene to Quaternary in the Kerman region. The main part of the magmatic rocks was formed in Eocene. In Kerman, the Eocene magmatic rocks are classified into three complexes that are Bahr Asman, Razakand Hezar.

Although, these complexes have been defined in terms of age and lithological characteristics. However, sometimes, the upper part of underlying complex is considered to be the lower part of the overlying one.

Bahr Asman Complex: The name of this complex is derived from the Bahr-Asman (means the See of the Sky) Mountain or Darb-e Behesht (means Paradise Gate) in the south western of Sardouieyeh. This complex with 7 km thickness was formed during early to lowermost middle Eocene. The lowest unit of the complex has various compositions in different areas, but it is generally composed of flysch sediments and in some parts is of limestone affinity. They occurred in many areas of the Kerman belt, such as: Sardouieyeh, Delfard, Bahr Asman Mountain, Sirjan, Shahr-e Babak, and northern Meiduk (Ghorbani 2010).

The lower subdivision of the complex, except for the sedimentary part, is mainly basaltic andesite and trachyandesite to dacite. The basic rocks mainly consist of coarse crystals of plagioclase and clinopyroxene in fine grain matrix. The acidic rocks are not too thick (Dimitrijevic 1973).

The middle subdivision of the complex is more acidic including acidic ignimbrite and acidic tuffs, with high content of zeolite, calcite and quartz. The thickness of the acidic lava is varied from 450 to 4800 m.

Some green andesitic lava is found within the acidic lava. In some areas, up to 95% of the volcanic rocks are tuff, breccia and ignimbrite, which represent a very strong explosions and volcanic eruptions. They mostly occurred as a sequence of thick breccia and agglomerate in Mozahem Mountain and Cheshmeh Firouzi (Ahmadi Pourfarsangi 2004).

The upper subdivision of the complex is composed of basic rocks, dark andesitic basalts, red trachyandesite, basaltic lava, tuffs and basaltic agglomerates and also small amount of rhyolite and rhyolitic tuffs. The thickness of lava is estimated to be less than 1500 m.

The upper Complex is traceable in the southwestern Bam, the volcanic area of Hezar Kuh to Mozahem Mountain, and the northern Shahr-e Babak in Meiduk mine and around the Sara Kuh. This complex is the same as the Razak Complex, which forms the uppermost part of Eocene series (Ahmadi Pourfarsangi 2004).

Middle-upper Eocene volcanic Razak Complex: The name of the complex is derived from the Razak Mountain in southwestern Khaneh-Khatoon. It covers the middle Eocene sedimentary complex and in some other areas disconformably overlies the older sedimentary rocks (Dimitrijevic 1973).

The Razak complex is introduced as the most widespread geological unit in the Kerman region and extends from the Dehaj area to the northern Sardouieyeh, Khaneh-Khatoon-Sabzevaran (Jiroft), Jabal-e Barez mountains and Hana.

The petrological studies of the volcanic complex have indicated three major lithological units:

The lower unit, with more than 2000 m thick has mainly basic and intermediate composition. They originated from trachyandesite and andesitic basalt lava. Also, tuffaceous sandstone, agglomerate, tuff and sandstone formed as interlays within the volcanic rocks.

The basaltic middle unit, with a small amount of rhyolite, is the most widespread unit in Chahar Gonbad and Bardsir.

The basaltic Upper Unit, with a thickness about 1500 m, is more dominant than the other Eocene Units in Jebal-e Barez Mountains (Alae-Taha 2003).

Upper Eocene Volcanic Hezar Complex: This complex represents the last stage of the Eocene volcanic activities. The volcanic rocks indicate high lithological variation in vertical cross section of the Hezar volcanic rocks (Shahabpour 2007). The Hezar volcanic units are mostly exposed in the form of crests along the Dehaj-Sardouieyeh axis. The rocks composition of the complex is mainly acidic. It broadly extends from the NW Mozahem to the Ahoorak Mountains and the Hezar Mount and includes the rhyolite acidic tuffs and rhyodacite of the Dehaj area. The small patches of these rocks expose in Shahr-e Babak and Sardouieyeh area. The complex with more than 1400 m includes basic and acidic rocks in western area of Hezar.

Many researchers including the Yugoslav group have studied the Hezar complex volcanic activities throughout the Kerman volcanic belt (Dimitrijevic 1973). Based on the field studies, the volcanic activities continued into Oligocene along the Sardouieyeh-Dehaj axis, gradually decreased, and later in Miocene the porphyry intrusive magmatic activities increased. Thus, most of porphyry copper deposits of the Kerman belt are originated from the Miocene intrusive bodies. As a result, the volcanic rocks which have been formed in Dehaj-Sardouieyeh changed to the large intrusive activities in Jebal-e Barez. The main rocks of the complex form a sequence of alternating andesite-basalt, andesite and rhyolite lava with agglomerate and related tuffs.

The huge Eocene complexes with an approximate thickness of 15 km could be identified as a belt (with 450 km length, 60 km width and 10 km thickness) in the Kerman volcanic belt. Accordingly, the volume of the complexes is approximately 270,000 km^3 indicates the significance of the magmatic activities in a short period of the time; 16 Ma.

Oligocene: The upper Eocene volcanic complex is covered by the Lower Red Formation in the Dehaj-Sardouieyeh subzone (Dimitrijevic 1973). The later formation is mainly composed of marl, tuffaceous sandstone, conglomerate, sandstone and minor dacite-andesite pyroclastics as well as andesite-basalt lava. At the southeastern part of the belt where the Lower Red Formation is not exposed, the late Eocene volcanic Hezar complex is covered by the Qom Formation which contains Oligocene's fossils.

The Chahar Gonbad Formation (equivalent to the Qom Formation) with the thick sequence of clastic and carbonate deposits, located at the west of Chahar Gonbad Mine, is Oligocene in age. Not only the Oligocene volcanic rocks, but also the sedimentary rocks equivalent to the Lower Red Formation have not been reported in the Jebal-e Barez area. Most probably, they have not been formed. However, large batholiths were developed at the same time in the area (Rasouli et al. 2016).

Miocene: The Miocene volcanic and volcano-sedimentary rocks are rare and not remarkable in the Dehaj-Sardouieyeh subzone. In despite, the porphyry intrusions are considerably exposed (Alizadeh et al. 2015).

The Upper Red Formation composed of red silty rocks, cross-bedding sandstones, and fewer tuffaceous layers is the main geological unit throughout the Miocene period. This formation mostly exposes in Lalezar Mountains (Baft) and Panj Kuh area (Dimitrijevic 1973).

The intrusive magmatic activities and mineralization are the most significant Oligo-Miocene geological event which caused the formation of large copper porphyry, skarn and vein deposits in Kerman belt.

It seems that the significant Oligocene volcanic activities have not occurred in the Miocene at Jebal-e Barez mountains; however, some Miocene intrusions related to the Oligocene batholiths have emplaced in the area (Jahangiri 2007).

Late Miocene-Pliocene: The magmatic activities of the Late Miocene-Pliocene were of more violence and have resulted in the development of acidic domes that form individual high peaks of Kerman belt such as Amiralmomenin Mountain in Sarcheshmeh area, Mardvar mount close to Shahr-e Babak, Narkuh mount along the Shahr-e Babak-Jusam road, Tazraj and Sara mountains near the Meiduk Mine as well as several peaks at Dehaj area (at least 6 peaks including Upper and Lower Aj mountains) that are most probably Pliocene to Pleistocene in age (Ghorbani 2010).

Pliocene-Quaternary: Late Miocene-early Pliocene to Quaternary could be considered as a period of acidic sub-volcanic intrusions in the Kerman region. However, this activity persists in Dehaj-Sardouieyeh, but not in the Jabal-e Barez subzone. The number of subvolcanic intrusions decreases from Dehaj to Sardouieyeh. However, compared to the Jebal-e Barez area, a decrease of the formation of large intrusive and porphyry bodies as well as decreasing the volcanism can be observed at Sardouieyeh Subzone during Miocene-Pliocene.

In terms of Cenozoic magmatic phases and containing porphyry copper deposits in Kerman belt, it could be classified into two subzones, Dehaj-Sardouieyeh and Jebal-e Barez which are described in below.

It is remarkable to know that the Eocene volcanic activities in Jebal-e Barez are more or less similar to Dehaj-Sardouieyeh subzone.

Geological characteristics of the Dehaj-Sardouieyeh subzone: Based on petrographic, mineralogical, geochemical, remote sensing and field studies, this subzone is classified into five magmatic phases, since Eocene to Quaternary, as follows:

Early Eocene-late Eocene: During this period a variety of volcanic rocks have developed including the following rock units:

– Andesite, andesite-basalt to trachyte along with small portion of acidic rocks,
– Breccia, agglomerates, and tuffaceous rocks which have predominantly composed of andesitic rocks; and often along with the acidic tuffs and
– Andesite, andesite-basalt to trachyte rocks.

Eocene-Oligocene: The Eocene-Oligocene rocks are composed of mostly acidic and ultra-acidic units and overly the Eocene andesite-basalt unit.

Miocene: Regarding the regional geological evidence and age determinations, the Miocene magmatic activities form porphyry bodies including Chah Firouzi, Meiduk and Iju. Most of them are composed of porphyry quartz monzonite to porphyry tonalite (Fig. 4.18 and Table 4.1).

According to the remote sensing studies, the alteration zones significantly occurred in this subvolcanic zone. The alteration zones can be categorized by the field geology and mineralogical studies.

Miocene-Pliocene: The formation of Miocene-Pliocene volcanics and subvolcanics is the most important magmatic phase at Dehaj-Sardouieyeh subzone. Most of the rough morphology and individual peaks such as Ayoub mount, Mardvar mount and other peaks along the Dehaj-Keder road are formed during this phase. These are adakitic intrusions (Berahmand 2017) (Fig. 4.19).

Plio-Quaternary: Formation of the Plio-Quaternary volcanic and subvolcanic activities is the youngest magmatic phase in the area which have resulted in several intrusive domes and craters such as Aj, Mozahem, Hezar and Amiralmomenin mountain Sarcheshmeh area.

The magmatic phase in Dehaj-Sardouieyeh axis is similar to Jebal-e Barez area. Nevertheless, the only difference is the more violent Miocene-Neogene-Quaternary magmatic activities at Dehaj-Sardouieyeh (Fig. 4.20).

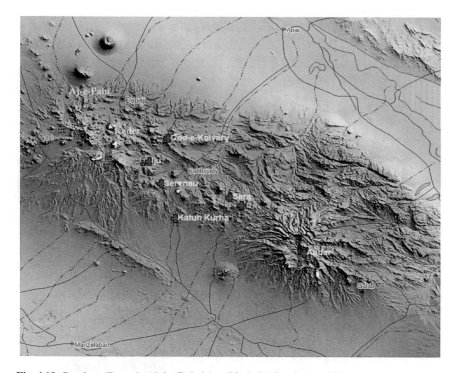

Fig. 4.18 Porphyry Zones in Abdar-Dehaj (modified after Parvinpour 2007)

Jebal-e Barez Subzone
Magmatism

The early, middle and late Eocene magmatism in the Jebal-e Barez is similar to other areas in Kerman copper belt. However, the acidic portions such as breccia agglomerate and acidic tuffs and Oligo-Miocene to Pliocene volcanic rocks are not so significance in the area (Rasouli et al. 2016). In contrary, the Oligocene batholiths are notable (Ghorbani 2010).

Three large granitoid batholiths were formed in the Jebal-e Barez subzone including southwestern Bam, northeastern Jiroft and Delfard. They are Rigan, Jebale-e Barez and northeastern Delfard batholiths where extends from southeastern to northwestern parts of area.

The lithology, petrography and geochemistry of these batholiths are as follows:

These batholiths are originated from a same magmatic complex due to similar lithological and chemical trend. They are very similar to each other, accordingly are known as Jebel-e Barez complex. However, based on their similar characteristics of petrographic, chemical composition, and emplacement environments, all of them could be introduced as Jebal-e Barez Complex.

The petrographic composition of Jabal-e Barez complex changes from diorite to granite. Table 4.2 show the petrography of the Jebal-e Barez batholith (Rasouli et al. 2016).

Table 4.1 The Lithological and Chemical Characteristics of the some Porphyry Intrusions of Dehaj-Sardouieyeh (Parvinpour 2007)

Sr. No.	Name	Petrography and age	Remarks
1	Sarcheshmeh	Porphyritic dioirte/14–16 Ma	Porphyry intrusion has formed at the conjunction of two faults. The other porphyry bodies near the Sarcheshmeh don't have important copper mineralization
2	Abdar	Porphyritic quartz diorite, porphyritic tonalite and porphyritic micro diorite/7.5 Ma., by U-He method	The mineralization might have been destroyed by another magmatic phase
3	Meiduk	Porphyritic diorite/12.5 Ma., by U-He and U-Pb methods	The propylitic alteration zone is widespread around Meiduk, and likely, several other porphyry intrusions exist
4	Pargam	Porphyritic quartz diorite and porphyritic tonalite	It is a small intrusion
5	Goud-e Kalvar	Porphyritic tonalite /Miocene	It seems that, intrusion has occurred in the fault zone, and is locally mineralized
6	Saghino (Sarno)	Porphyritic quartz monzonite and diorite/Miocene	The surficial evidence indicates at least a poor copper mineralization at surface
7	Iju	Porphyritic tonalite, Porphyritic micro diorite- porphyritic quartz diorite/8.4 to 20 Ma	It has a low grade mineralization due to its high volume. It seems there is no important mineralization at depth
8	Chah Firouzi	Porphyritic tonalite, Porphyritic quartz diorite/16 Ma., by U-He and 13 Ma., by U-Pb methods	Porphyry mineralization is observed at fault zone, but is not significant
9	Keder	There is a large porphyritic quartz diorite body at this area/Miocene	–

The geological characteristics of the Jebal-e Barez complex (Alae-Taha 2003) are as follows:

- The Jebal-e Barez complex have intruded into the thick Eocene volcanic and pyroclastic rocks,
- Occasionally forms some contact metamorphism in volcanic host (Ghorbani 2010), and
- The Jebal-e Barez batholith has been formed at the first magmatic phase, and later many intrusive bodies have injected rapidly into the batholith. These intrusions known as "porphyries" have more uniform composition than the batholith. A wide range of exploration has been done in the area.

The rock composition of the late intrusions within the Jebal-e Barez batholith is:

Fig. 4.19 Two views of dacitic dome with adakitic attitude in Shahr-e Babak area

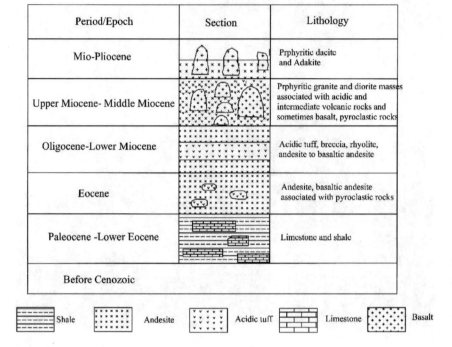

Period/Epoch	Section	Lithology
Mio-Pliocene		Prphyritic dacite and Adakite
Upper Miocene- Middle Miocene		Prphyritic granite and diorite masses associated with acidic and intermediate volcanic rocks and sometimes basalt, pyroclastic rocks
Oligocene-Lower Miocene		Acidic tuff, breccia, rhyolite, andesite to basaltic andesite
Eocene		Andesite, basaltic andesite associated with pyroclastic rocks
Paleocene -Lower Eocene		Limestone and shale
Before Cenozoic		

Shale   Andesite   Acidic tuff   Limestone   Basalt

Fig. 4.20 Magmatic phases of the Kerman area

Table 4.2 Petrography of the intrusions

Batholith Name	Petrography
Reagan	Granite to grano diorite
Central Jebal-e Barez	Micro diorite -tonalite, and grano diorite- granite
Northeastern Delfard	Alkali granite-grano diorite

Alkali granite and light colored granites, mostly in Mijan area. These were formed during the second intrusive phase of the area.

– The leucocratic acidic porphyry intrusions are mostly formed as alkali-granite and alkali syenogranite. Nevertheless, those rocks with different petrographic composition are mostly light in color due to their less content of mafic minerals. The porphyry rocks are characteristically fine-grained and rich in quartz. The field studies indicate that these rocks have semi-stockwork structure formed by later silicified veins and veinlets.

– The structural evidence of some late intrusions indicates that they have mostly intruded in the fault zones and occasionally in the old calderas in some areas such as Mijan and Darreh Hamzeh. The silicic, phyllic (quartz-sericite), argillic and sometimes potassic alterations can be found.

The porphyry intrusions in the Jebal-e Barez subzone are as follows:

Rigan, Korour, Bagh Golan, Darreh Hamzeh, Mijan, Tanaroyeh, Madin, Zavrak, Hieshin, and Gerdoo Chahar Shanbeh. Chahoon intrusion has some characteristics of porphyry onesand shows some copper mineralization in north of Jebal-e Barez Batholith. But, it is not a part of Jebal-e Barez batholith (Table 4.3 and Fig. 4.21).

Differences between Jebal-e Barez subzone and Dehaj-Sardouieyeh magmatic belt:

In fact, the Jebal-e Barez subzone is a southern extensional part of the Dehaj-Sardouieyeh magmatic belt. The thickness and volume of volcanic rocks in Jebal-e Barez is similar to that of Dehaj-Sardouieyeh belt. The significant differences in terms of magmatism and porphyry characteristics of them are as follow:

In Jebal-e Barez subzone the thickness and volume of volcanic rocks are similar to those of the Dehaj-Sardouieyeh belt. But there are some differences as follows.

- The lithological variation of volcanic rocks in Dehaj-Sardouieyeh is more than Jebal-e Barez subzone. The basement of Jebal-e Barez subzone is composed of sedimentary and pyroclastic rocks and is covered by lava with more or less similar composition (andesite to andesite-basalt). The pyroclastic rocks are mostly intermediate with fewer acidic compositions. The volcanic rocks are dominantly formed in the Eocene while there are no notable volcanic rocks of other ages.
- The Eocene to Quaternary volcanic rocks with wide variations in lithological and geochemical compositions are extended in the Dehaj-Sardouieyeh belt. Those of Oligo-Miocene are of more extent.
- Many large batholiths have formed the Jebal-e Barez intrusion complexes. Nevertheless, these types of complexes are not developed in the Dehaj-Sardouieyeh belt.

Table 4.3 Main features of the late porphyry intrusions in the Jebal-e Barez subzone

Name	Petrographic Composition
Korour	Alkali granite, syenogranite and monzodiorite
Tenaroyeh	Diorite, tonalite and granodiorite
Madin	Tonalite, diorite, granodiorite and granite
Gerdoo Chaharshanbeh	Granite to granodiorite, alkali granite and tonalite
Mijan	Alkali granite, syenogranite, monzogranite, alkali syenite, granodiorite and quartz monzodiorite
Zavrak	Alkali granite, granodiorite and monzodiorite
Hieshin	Alkali granite, monzogranite and diorite
DarrehHamzeh	Alkali granite and quartz alkali syenite
BaghGolan	Granite, granodiorite and quartz diorite
Chahoon	Tonalite to granodiorite and alkali granite to granodiorite
Nesa	Granite, granodiorite, tonalite and diorite
Rigan	Granite to granodiorite

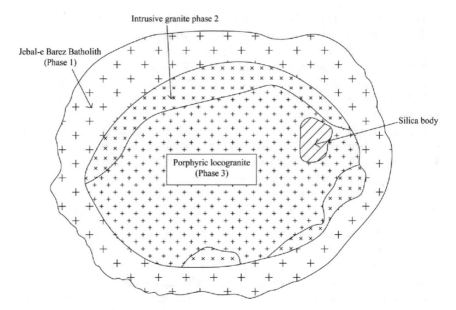

Fig. 4.21 Porphyry intrusions in the Jebal-e Barez, Mijan area

- The Dehaj-Sardouieyeh intrusions often are small and relatively similar to each other. These bodies could be developed during the different intrusive phases, However, the late magmatic phases were not formed in the area.
- In the Jebal-e Barez subzone, porphyry intrusions are associated with the late significant intrusive activities. In fact, the late magmatic intrusions were originated from the differentiating of the Jebal-e Barez batholith. Accordingly, gradually the acidity of the magmatic phases increased. Thus, the first (batholith) and the last magmatic phase (late alkali granites) show an increase in the acidity. The final phase has accompanied by the silicic veins and veinlets forming a stockwork texture exposed very well at Mijan and Hieshin.

Mid. Urmia-Dokhtar Zone

This zone extends from the Taft to Qom. The magmatic activities have started since Lutetian, when the Eocene Sea has started to transgression and the mostly acidic volcano-sedimentary sequences equivalent to the Green Series formed. The activities were of phreatomagmatic affinity and were highly explosive.

The sequences of these volcanic rocks are classified as follows:

Late Paleocene-early Eocene: The lower volcanic sequences are composed of the alkaline and intermediate rocks such as trachy-andesite, trachy-basalt, trachyte and shoshonite. The upper rhyolitic rocks are rhyolitic lava, pumice tuff, breccia tuff and breccia lava (Amidi et al. 1984).

Middle Eocene: The middle Eocene rocks are sedimentary green series, acidic tuffs, tuffit, marble, limestone, and conglomerate. They are more exposed in the Alborz zone.

Late Eocene: The lower parts of volcanic rocks including rhyodacite and ignimbrite belong to the 39 Ma. They exposed within the rhyolitic and rhyodacitic tuffs, pumice and pyroclastic rocks of the upper part of the Eocene (Amidi et al. 1984).

The Pyrenean Orogeny and its consequent intrusive bodies (such as Natanz, Taft, Vashmeh, Kahak, Kuh-e Dome, Calkafy and Talehsiyah) have caused to raise the middle Urmia-Dokhtar Zone up to the sea level.

Different stages of magmatic activities widespread in the Natanz-Naein area during the late Eocene Period (Amidi 1977) and the volcanic activities identified by the very thick pyroclastic sedimentary rocks including andesite, rhyodacite and shoshonitic rocks are as follows:

1. The rhyolitic rocks formed as tuff and ignimbrite in a very shallow marine basin along with conglomerate and clastic limestone.
2. The lower andesitic rocks mostly potassic alkaline in type.
3. The acidic volcanic activities have formed rhyodacite in several parts of Iran. These rocks often are exposed as domes; representing a continental magmatic activity. These rocks sometimes occurred as ignimbrite. The upper eroded surface of the rhyodacitic units indicates a major activity happened significantly in the Eocene.
4. A plentiful of andesite was formed as pyroclastic and andesite lava flow at the middle part. The continental volcanic activities have mainly formed rhyolite as spherulitic- and ignimbritic-tuffs at this stage.
5. The shoshonitic rocks including absarokite, shoshonite, trachy-andesite, tuscanite, rhyodacite to rhyolite with a high content of the primary sodic analcime.
6. The upper andesitic rocks are rich in aluminum and in some parts have caused some small acidic activities.

Oligocene: An eruption of rhyolite caused to form by the continental volcanic activities indicates a widespread epeirogenic movement in Late Eocene-early Oligocene. Consequently a sequence formed by rhyolite lava flows and pyroclastic rocks can be found from lower to upper parts of the mentioned succession.

The main characteristic of the rhyolites of the middle Urmia-Dokhtar volcanic belt is the presence of garnets (similar to spessartine) which chemically indicates that the rhyolitic magma is the result of partial melting of the upper parts of mantle or lower parts of the crust.

Oligo-Miocene: The volcanic activities as submarine andesites formed in this stage. The succession starts with the marls and fossiliferous limestones and the volcanic rocks cover the mentioned unit. The volcanic eruptions are mostly explosive and in some areas the magma has flowed away as lava.

The Oligo-Miocene andesites and basalts contain sodic analcime. The existence of sodic analcime indicates the second series of such rocks in the Oligo-Miocene volcanic rocks at Urmia-Dokhtar belt.

Middle Miocene: The large and small intrusions with the absolute age of 17–19 Ma. were formed in the area. The intrusions most likely are related to the older volcanic rocks. The results of the studies carried on the intrusions and volcanic rocks show a genetic relationship in some areas.

Late Miocene-Pliocene: The calk alkaline continental magmatic activities occurred as a continuous and steady differentiation process in Naein to Natanz. These activities are different from the magmatic activities in Paleogene-early Neogene. Magmatic activities of these periods are classified as follows:

The trend of the volcanic rocks is basaltic-andesite to andesite, dacite and rhyodacite, which are formed in form of scattered domes at the middle Urmia-Dokhtar zone. These rocks originated during differentiation of the calc-alkaline magma and are significance in terms of quality and quantity in magmatism of Iran.

The Eocene and Oligocene volcanic and plutonic rocks are exposed in Ardestan at the middle Urmia-Dokhtar zone. It seems the main magmatic activities started in the Middle Eocene. As a result, the volcanic rocks have been initiated by volcanic activities in the Middle Eocene-Late Eocene.

The Middle-Late Eocene rock sequence is a variety of volcanic rocks in which rhyolite units are older than andesite, basaltic trachyte, trachyandesite units. The ignimbrite and trachyte units are the youngest units in the zone. The volcanic activities mainly occurred in the continental or shallow coastal basins (Emami et al. 1993).

The Oligocene volcanic rocks have been mainly formed as rhyolite and are scattered as white patches.

It should be noted that the diverse pyroclastic rocks in the zone are including: basalt trachyte-andesite trachyte and andesitic tuffs and rhyolitic breccia. Various rock types such as olivine basalt, trachy-basalt, andesite, quartz andesite, pyroxene andesite and mega porphyritic andesite, trachyte, dacite, rhyodacite and rhyolite are formed in this stage. The intrusive activities have formed the small monzogranite to granite intrusions, in Zafarghand and Doorjin of Ardestan area.

According to Ghalamghash and Babakhani (1996), the Cenozoic volcanic rocks in Kahak area (southern Qom) are classified as below:

Eocene: The Eocene volcanic rocks include andesite, basaltic andesite, acidic tuff and ignimbrite with sedimentary interlayers.

Ignimbritic layers are composed of dacitic tuff and lava flows in the interactive parts of the Eocene coastal and continental margins. The sedimentary interlayers, particularly, fossiliferous tuffic limestone is occurred in the Kahak green series.

Oligo-Miocene: the intermediate to basic volcanic rocks are located in the middle part of the Qom Formation and occasionally cover it.

Pliocene: Generally, the volcanic rocks form diverse intermediate pyroclastic rocks.

The intrusive rocks in a relatively wide range of time occurred in two magmatic phases: The Oligocene and Oligo-Miocene intrusive and the Pliocene semi-acidic to acidic semi-volcanic phases in Kahak area.

Northern Urmia-Dokhtar Zone

The Northern Urmia-Dokhtar zone can be traced from Tafrash to Razan, southern Zanjan, Takab, Shahin Dej, Urmia and Turkey as its westernmost border.

In the northern part of the Urmia-Dokhtar zone, the Cenozoic Magmatism specifically Eocene volcanism and Eocene-Oligocene intrusions occur in a fewer volume than the southern and central zones (Alizadeh et al. 2015).

In general, the unique magmatism characteristics of the northern zone which distinguish it from the other two mentioned zones are as follows (Ghorbani 2007):

Eocene: Eocene volcanic rocks in terms of volume, thickness and composition differ from those of other areas; the volume and thickness of the Eocene volcanic rocks are less and rare, and their composition is often acidic pyroclastic rocks and tuffs. They are similar to the Karaj Formation of the Alborz zone.

Eocene-Oligocene: The Eocene-Early Oligocene intrusive rocks are not so considerable and there is not any large batholith in the northern Urmia-Dokhtar zone. However, the basic to intermediate scattered volcanic rocks and small intrusions are occurred conspicuously in many areas such as Mahneshan, the basic to intermediate intrusive rocks of northern parts, Shahin Dej and Mianeh.

Late Oligocene-Miocene: The acidic tuffs developed significantly during the Late Oligocene to the late Miocene. Therefore, the lower part of Qom Formation is composed of tuffs in southern Zanjan, Halab, Mahneshan and Takht-e Soleyman of northern Urmia-Dokhtar zone (Babakhani and Ghalamghash 1995) (Figs. 4.22 and 4.23).

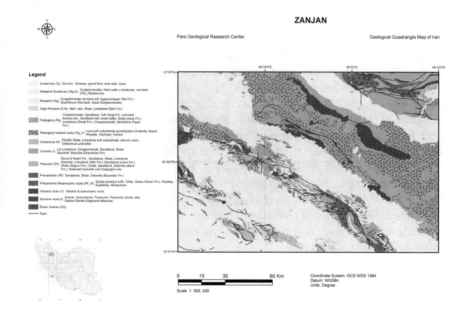

Fig. 4.22 Geological map of Zanjan area

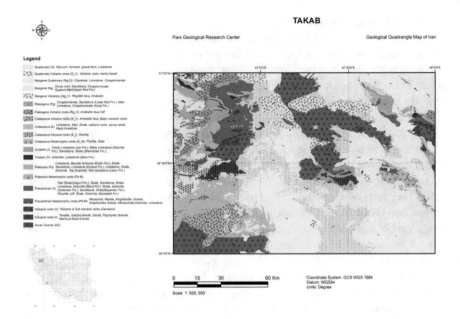

Fig. 4.23 Geological map of Takab area

Generally, the Eocene and Eocene-Early Oligocene magmatism are less considerable than Oligo-Miocene volcanic activities. It is the main difference between the
northern and Middle Urmia-Dokhtar zone.

Geodynamic of the Urmia-Dokhtar Zone

The Neo-Tethys Ocean started to subducting beneath the continental plate during the
beginning of the Early Jurassic. Later, during the Early Cretaceous, it began to become
smaller and were eventually closed. Accordingly, during the Late Cretaceous-
Paleogene the collision occurred between two continents. Therefore, the subduction
zone which initiated in Early Jurassic extended throughout the Cretaceous. Thus the
molten magma derived from the subducting oceanic lithosphere penetrates into the
continental crust (Dimitrijevic 1973; Aghanabati 1998).

The volcanic rocks, particularly andesite, basalt as well as batholiths could be
considered as the results of the onset of a subduction process in the Sanandaj-Sirjan
zone. The subduction activities have continued during the Cenozoic till the Miocene-
Oligocene in the Urmia-Dokhtar zone and west Central Iran, parallel to the northern
side of the Sanandaj-Sirjan zone.

The result of the Cenozoic subduction zone is the development of various types
of volcanic rocks in Urmia-Dokhtar. Omrani et al. (2007) believes that the Sanandaj-
Sirjan magmatism was initiated in the Mesozoic and continued till the beginning
of the Cenozoic which is associated with subduction in Urmia-Dokhtar zone. There
is a gap of magmatic activities, about 20 Ma. It means that the magmatism related
to the subduction of Sanandaj-Sirjan is reduced during Paleocene, and later, the

magmatic activities transferred from Sanandaj-Sirjan to the Urmia-Dokhtar zone and west Central Iran (Omrani et al. 2007).

The tectonic setting and magma types of mixed Cenozoic magmatic rocks in Urmia-Dokhtar zone could be related to the continental subduction plate which has been broken during the subducting in the northern band of the Sanandaj-Sirjan and south and southwestern Central Iran (Urmia-Dokhtar zone). Breaking down of the subducting oceanic lithosphere could be accompanied by a local tension, creating an irregular rift in one hand and on the other hand, the involvement of oceanic lithosphere in the fractured zone could justify the occurrence of the mixed magmatic rocks.

This hypothesis could be proved by the seismic tests providing several profiles perpendicular to the Urmia-Dokhtar zone. If the oceanic lithosphere has been broken during the subducting, the melted segments would have different reaction during cooling and solidifying. Accordingly, seismic technique could recognize them from the peripheral rock units.

4.5.1.2   Alborz Zone

The Alborz mountain range geographically extends from Hindu Kush in Afghanistan to Azerbaijan. Its northern boundary is between the northern slopes, and the Caspian coastal plain. The Southern boundary is not still fully identified. Alborz is exposed as low height hills in southern Tehran and Garmsar-Semnan with the geological characteristics of Central Iran. The similarity to Central Iran is visible in the most parts of Alborz southern boundary. However, stratigraphically the eastern Alborz is more similar to the Central Iran.

Stöcklin (1974) based on the structural characteristics of Alborz divided it into six zones, mostly from north to south as follows:

– Gorgan spore;
– Northern Neogene Zone;
– Central Northern Zone;
– Central Southern Zone;
– Cenozoic Southern Zone; and
– Southern front spore zone

Alborz is divided into several units as well as:

– Eastern Alborz, Kopeh-Dagh or Binaloud Zone; which is the transitional zone between Central Iran and Alborz (Nabavi 1976).
– Central Alborz; and
– Western Alborz and Azerbaijan.

The Cenozoic volcanism in Central Alborz is of a high importance which will be discussed in below.

Central Alborz

The central Alborz begins from the Firuzkuh valley in the east to the Sefidrud fault to the northwest, Qazvin-Rasht.

The central Alborz Cenozoic volcanic rocks, specifically volcano-sedimentary rocks, are much considerable in the Southern slopes and are known as "Karaj Formation". The type section of the Karaj Formation is exposed in the Karaj Valley, Chalus Road (Dedual 1967). The Karaj Formation is called with several other names as:

The Green Beds (Schroder 1945; Dellenbach 1964); the Green Beds and volcanic lava (Gansser and Huber 1962); the Alborz Green tuffite (Darvishzadeh 1991).

The total thickness of type section, is about 3300 m, and has been divided into 5 sections, from top to bottom, as follows (Dedual 1967):

1. Kandovan Shales
2. The Upper Tuff,
3. Asara Shales,
4. The Middle Tuff and
5. The lower Shales

The lithological characteristics of the members:

1. The lower part (1055 m) is composed of the dark gray calcareous and siliceous shale which occur with the gray ash and vitric tuff in the tuffite part and 20 m augite bearing porphyry lava occurs near the base.
2. In the Middle Tuff member (1177 m) thick-bedded vitric layers of green to blue-light green tuffs with some layers of limestone are occurred. Sometimes, the white vitric tuffs rich in the potassic feldspar (K-feldspar) are exposed in the upper part.
3. The Asara shales member (167 m) is mainly composed of calcareous shale with minor layers of tuff and tuffaceous shale.
4. The Upper Tuff member (917 m) which is mostly composed of green tuffs with some shale intercalations, tuffaceous sandstone and calcareous shale.
5. The Kandovan shales relocated in a tight syncline at the northern part of Kandovan defile.

There is no consensus about the thickness of the formation. Lorenz (1964) has estimated a thickness about 1700–2200 m in west of the Karaj valley. Whereas, Moatamed (1987) and Yasaghi (1989) have estimated 5000 and 4000 m, respectively. This ambiguity is due to the different outcrops and faults (Darvishzadeh 1991).

Characteristics of the Karaj Formation

1. Regarding to the studies carried on the preserved volcanic fragments within tuffites at northern Lashkarak and identifying plagioclase types of tuffites found in Karaj Valley show that the initial magma was acidic to intermediate (rhyo-dacite and dacite). These rocks signify a variety of explosive volcanism which finally formed Karaj Formation.

2. The widespread and thick sediments clearly represent the continuation of submarine explosive volcanic activities. Also the alternative layers of tuffites along with calcareous deposits, shale and radiolaritic sediments indicate the lack of volcanic activities. The explosive volcanic activities have scattered volcanic fragments in form of ash in the sea water associated with alternated fragments of feldspar, quartz and biotite (being altered).

3. The gradual and regular tuffite layers with respect to the clastic grain sizes represent shallow facies type. The sea floor gradually subsided along the faults, and fractures, so, the viscous acidic and intermediate lavas moved up to the sea floor along the fractures. The explosive characteristic of lava striking sea water intensified in the fractures and scattered the volcanic fragments into the sea water (Darvishzadeh 1991).

According to Annells (1975), the initial phase of the Karaj Formation and development of a playa environment was associated with shallow limestone deposition in Eocene. In the second phase of Eocene, a weak basaltic eruption that occasionally alternates with some dacitic types occurred and during the third phase of Eocene, the Karaj Formation formed along with rhyolitic and dacitic explosive eruptions.

The crustal movements increased in the late Oligocene, consequently created the basaltic and andesitic lavas scattered in some areas.

The major and minor compositions of volcano-sedimentary rocks of the Karaj Formation are crystal bearing vitric tuffs, carbonate vitric tuffs, and tuffites in northern and eastern Tehran, respectively.

Ignimbrites and welded tuffs are occurred sporadically in Karaj Formation as well as the shallow intrusive rocks (diabase) in the lower part of the formation.

According to Iwao and Hushmandzadeh (1971), the Karaj Green Series is slightly metamorphosed, confirmed by the presence of epidote, chlorite and sometimes leumontite in Tochal Mountain.

In general, the age of the Karaj Formation is dated back to the middle Eocene which extends to Late Eocene in some areas. It seems that the Green Series of Karaj Formation are mostly composed of tuffs and partially tuffites, which their lithology does not change in most areas of central Alborz in the middle Eocene. The volcanism in the late Eocene or possibly early Oligocene was as lava. This lava type volcanism occurred mostly in western and central Alborz. Consequently, the Paleogene sequence of western Karaj, Taleqan area, is classified in two main sections:

– Lower section composed of more than 3000 m tuffs, tuffaceous deposits and nummolitic limestone.
– Upper section composed of continental lava flows, likely late Eocene-Oligocene; more than 2500 m in thickness (Annells et al. 1977).

According to Stalder (1971) the pyroclastic units and those rocks form by eruptions of the volcanic series including the Upper and Lower Tuff members of Karaj Formation, possibly are compatible with Middle-Late Eocene and likely to early Oligocene. This sequence of deposits begins with tuffs and ends with lavas.

According to Annells et al. (1977), the transgression of the Eocene Sea has resulted in deposition of Paleogene Series on the uneven surface area at Qazvin-Rasht quadrangle. The deposits are mainly composed of volcanic rocks which formed during three major volcanic phases as follow:

The Eocene Sea transgressed over an uneven surface at the first stage, and the subsequent evidence of such advance is the accumulation of submarine tuffs and marginal conglomerate developed a high topography near the Paleogene highlands. A series of submarine explosive volcanic activities resulted in the formation of mostly andesitic tuffs, trending west and northwest, which are developed at the second stage and show an interfingering structure.

At the same time, the alternative layers of mudstone and limestone were deposited in the deeper but less active local basins at the vicinity of the northern Taleghan River and the northern Qazvin mountains. At this stage, some movements disturbed the depositional environment and developed diastems as well as some dikes in the thick tuffs.

At the last stage, a shallow limestone deposited in a wide area. The limestone is covered by nodular gypsum layers of volcano- sedimentary origin, 100 m, in eastern region. In fact, the submarine volcanic activities, supplied the needed component to form gypsum in the marginal shallow water basin (playa environment).

There are three magmatic phases since early to late Eocene in Taleghan-Alamut region as follows.

1. Acidic and related pyroclastic eruptions that are equivalent to the Karaj Formation are widespread from Taleghan to the northeastern Qazvin. The main components of volcanic activities are the pyroclastic rocks.
2. Alkaline basalt volcanic activities are the second magmatic phase, which occurred in late Middle Eocene to Late Eocene. A series of basaltic lava extends from northeastern to southeastern Taleghan. The thickness of this sequence is about 1000 m, and contains foid minerals.
3. The third phase of magmatism accompanies by small intrusive bodies composed of monzonite-diorite to monzogabbro. They formed as dyke, sill, and stock works. Some of them are Ziaran, Brajan, and Shekarab bodies. The Shekarab extends towards the north and eastern Kordan and is traceable up to the eastern Tehran. Other intrusive bodies include Sereh body in northern Kordan, syenite body in Lavasan, and Abyekbodyin eastern Tehran.

The volcanic rocks of the third phase are formed by the rhyolitic and dacitic explosive eruptions in the western side of the named areas. As a result, a series of tuff, agglomerate, ash and lava flow deposited over the deformed and eroded series of Phase two. Meanwhile, the Paleogene lavas, andesitic and basaltic tuffs were erupted through the continental crust in the southern of previous named areas, along the north of Abyek- Qazvin road. So, they were relatively explosive and formed interfingering structure with the acidic volcanic rocks.

As previously noted, according to Annells (1975), only two phases out of three phases of Paleogene volcanism are traceable in Qazvin-Rasht quadrangle map (Mohammadi 1996).

The Paleogene volcanic rocks are unconformably deposited on the Permian-Cretaceous formations, at the first phase in the Taleghan area. The fossiliferous limestone interlayers indicate their age as middle Eocene-early Oligocene (Stalder 1971). The first phase volcanic rocks are mainly formed of light green to gray-green acidic tuffs and dark gray mudstones, which likely derived from Cretaceous lavas between the Marjan village and the Prajan fault. The layers of dacitic rocks (150 m) occurred as interlayers of acidic tuffs.

The second phase of Paleogene rocks began with alkaline lavas (basanite, basalt and trachyandesite). Although, changing the first phase to the second phase occurred suddenly in some areas such as eastern Orazan which has been a gradually process. In other words, the layers of tuff and lava, 100 m, show a gradual alternation (Stalder 1971). The Kuh Shah accumulated dikes in Alborz has probably acted as a proper feeding system for lava flows. However, there was a central eruptive fissure in north of Amirnan (Annells et al. 1977).

It should be noted that basanitic lava containing large crystals of analcime could be visible only in the Taleqan area of the Qazvin-Rasht Quadrangle Map (Mohammadi 1996).

Geodynamic-magmatism of Central Alborz
Two hypothesis could be suggested for the geodynamic of the Central Alborz such as:

1. To accept that the central Alborz magmatism is parallel to Urmia-Dokhtar belt and is formed in a back arc basin, north of the belt. However, this view is faced two inconsistent: (A) the volume of mostly acidic and pyroclastic volcanic and volcano-sedimentary rocks in Central Alborz is more than those of the Urmia-Dokhtar belt; (B) Based on the age factor, the rocks of the Urmia-Dokhtar belt are approximately formed at the same age of the southern slopes of Alborz volcano-sedimentary rocks. But, at least, the age of latter should be younger than Urmia-Dokhtar rocks.
2. These rocks could be related to a local and undeveloped rift in Central Alborz. According to this view, the main related volcanic and pyroclastic rocks which previously named as "Green Series" on the southern slopes of the Alborz are acidic and pyroclastic rocks and they are assumed to be resulted from partial melting of the lower part of a continental lithosphere.

The middle Eocene Green Series rocks are associated with the late Eocene-early Oligocene basic dikes. The highly dense basic dikes occurred in Bumehen, Roudehen and Imamzadeh Davud in northwestern Tehran (Aghanabati 2004).

Regarding the above-mentioned mechanism, it is assumed that probably a divergent system caused the asthenosphere to move up and gabbroic domes intruded into lower crust at the southern slopes of central Alborz in middle Eocene. As a result of this movement, the gabbroic domes injected in the lithosphere and caused the partial melting of the lower part of it, while the Green series volcanic and pyroclastic rocks formed at the divergent boundary.

The dominated extensional condition converted to the compressional one in the late Eocene-Oligocene which is in coeval with Pyrenean Orogeny phase. At this stage, the gabbroic magma intruded into the underneath of the crust, but it locked up and failed to move up to the surface. Accordingly, only some dikes are developed, which likely represent the existence of large gabbroic batholiths in these areas.

4.5.1.3 Tarom-Hashtjin Magmatic Belt

A range of volcanic rocks such as rhyodacite, dacite to basalt formed as lava, tuffs and occasionally tuffites in the Tarom belt. Based on the previous studies, the volcanic rocks are basalt, andesite-basalt, trachyte, latite, trachyandesite, dacite, rhyodacite, ignimbrites and acidic-intermediate tuffs of which the andesites form the majority of rocks.

According to Moein Vaziri (1999), most of the Eocene volcanic rocks of Tarom (Tarom-Hashjin of the present author) are of potassic, sometimes sodic alkaline and calk alkaline origin; and the andesites are mostly shoshonitic. It should be noted, in normative mineral of the Eocene volcanic rocks, semi feldspathoid minerals are rarely formed. But, significant percents of orthopyroxene are present. The acidic and basic intrusive bodies are genetically related to each other in some areas of the Tarom-Hashtjin Belt (Sarkirinshad 1989; Zarei 1992).

In the Tarom-Hashtjin belt, many intrusions have cut the volcanic rocks coeval to the Karaj formation as well as the other Eocene volcanic rocks. A sequence of rocks, equal to Qom formation covered them by nonconformity, so their age is the late Eocene to Early Oligocene. The intrusions consist of alkaline granite, granite, granodiorite, monzogranite, monzonite, quartz monzodiorite, syenite, alkaline syenite and quartz syenite (Hajalilou 2000).

Based on the studies of several researchers (Moayed 1991; Peyravan 1992; Torkamani 1997), these intrusions are I-type granitoids. Nevertheless, the present author believes regarding to the fieldstudies, petrography and mineralization (iron mineralization), they composed of "I cordillera" type.

Hajalilou (2000) believed that these intrusions formed in 700 to 880 Centigrade, at 1400 to 3000 m depth. The intrusions containing high percent of water and high coefficient of sulfur fugacity caused to develop a wide hydrothermal alteration zone, which is associated with scattered veins, veinlets and disseminated mineralization within the tuffites and the other Eocene volcanic rocks. Lithological evidence shows the Plutonism is associated with a post-collision magmatic arc (Moayed and Hoseinzadeh 2000).

The magmatic series of these intrusions are shoshonite, high -K calk alkaline and "I" type. The shallower porphyry intrusions with general composition of porphyry monzonite, had a significant role in mineralization and development of the hydrothermal alteration. They occurred along with the major intrusions. It should be considered that the structural geology of the area such as the axis of anticlines and synclines, E-W and NW-SE faults had a very crucial role to the emplacement of these intrusions and vast development of hydrothermal alteration.

4.5.1.4   Ahar-Julfa

Geologically, all rock units of this area can be divided into two categories:

– Igneous rocks
– Sedimentary rocks

The magmatic activity of the Ahar-Julfa area has initiated at the Late Cretaceous and discontnuesly extended to Quaternary. Igneous rocks are the most distributed rock units of the area (Fig. 4.24). The mostly Jurassic-Cretaceous sedimentary units are of less distribution in area and are accompanied by minor Cenozoic sedimentary rocks. They are mostly fossiliferous. By investigating the geological relationships of the sedimentary and igneous rocks, the relative ages of igneous units can be estimated. Occasionally the igneous and sedimentary rock units are forming a mélange by which the age of igneous rocks can be logically determined. In addition, several dating methods are used by lots of researchers for determining the absolute age of these rock units. The large-scale maps can be also used.

According to petrographic studies and geochemical analysis as well as following factors, five magmatic phases can be distinguished in the area:

– The relationships of the igneous rocks and sedimentary rocks which are dated by paleontological studies.
– Absolute datings done by several researches on the igneous rocks.

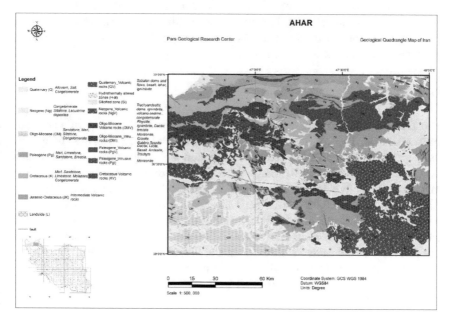

Fig. 4.24 Geological map of Ahar

- The geological relationships of the different igneous units, their similiarities and differences.

Phase I: Cretaceous-Paleocene

In the surroundings of the Mazraeh mine, northern Ahar, and the vicinity of the Kaleybar, both sides of the Aras river, eastern Alian village and western Kaleybar, the Cretaceous basalts are overlaid by the Late Cretaceous recrystallized rocks. The thick volcanic rocks of the area which are overlaid and underplayed by the Late Cretaceous limestones and Early Cretaceous sediments, respectively, can be relatively dated. The Cretaceous andesites and basalts show several important facts.

- The igneous rocks of the Phase I are all volcanics.
- The petrographic composition of the Cretaceous rocks is mostly andesite, andesi-basalt and basalt. Their composition to the east and especially eastern Kaleybar changes to basanite.
- The Cretaceous volcanic rocks can be observed in 1:100,000 geological maps of Ahar, Germi, Kaleybar, Lahroud and Varzeghan (Mobashergermi et al. 2019). In the eastern part of the Ahar-Julfa area, the volcanics are overlaid by Cretaceous limestones while in the Lahroud area they are underlain by Late Cretaceous sandstones. They can be observed at the northwestern part of the Varzeghan 1:100,000 geological map. According to studies performed by Mobashergermi et al. (2019) and Asgharzadeh et al. (2018), they are calc-alkaline in affinity. In the eastern part of the area occasionally the alkaline and shoshonitic rocks can be observed. There is not a comprehensive literature discussing the geodynamics of the Cretaceous volcanic rocks of area (Fig. 4.25).

Phase II: Eocene

The Eocene volcanic rocks are the most exposed volcanic rocks of the Ahar-Arasbaran axis. The volcanic rocks of this phase are of more diversity and are the most distributed units of the area including Iran and neighboring countries such as Armenia, Azerbaijan and Turkey. Although they are mostly andesite, the acidic rocks and especially acidic tuffs are notable (Geological maps of Ahar, Varzeghan and Ahar, 1:250,000, 1:100,000 and 1:500,000 in scale, respectively). In Avan and south western part of Astamal area the Eocene volcanic rocks are underlain by Shale and other rock units of Paleocene-Early Eocene. It is finalized by forming huge calderas (for instance the old calderas of Noqdooz and Sharafabad-Hirjan areas) and dacitic porphyry intrusions (such intrusions can be observed in Astamal, Noqdooz, Avan road and along the Hajilar Chai river) (Fig. 4.26). The petrographic composition of this phase is as below:

Basalt: it is of a limited distribution observable in few local areas.

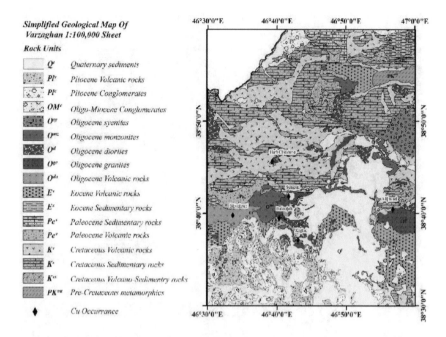

Fig. 4.25   The simplified geological map of Varzeghan, 1:100,000 in scale

Andesite: It is of well distribution in area and what is mentioned in relation to the distribution of volcanic rocks of this horizon is also true for andesitic rocks of the area.

Acidic and trachytic tuffs and rhyolites: These rocks are the most distributed rock units of area. The acidic tuffs occasionally show regional advanced argillic alterations. Both tuffs and andesites are of notable primary pyrite mineralization. All rock units of area are calc-alkaline in affinity (Ghorbani 2004).

Phase III: Late Eocene-Early Oligocene

A major magmatic phase (mostly as intrusions) has affected the area and especially the northwestern part of Iran as well as all parts of the Urmia-Dokhtar magmatic belt during Late Eocene-Early Oligocene. It has resulted in forming several batholites and stocks in Ahar-Julfa axis some of which are as below:

– Shivardagh batholite
– Youseflu batholite
– Ordubad batholite
– Khanbaz-Khankandi and Ahl-e Iman batholites at northern Ahar-Meshkin Shahr road.

Fig. 4.26 The field view of the rocks of Phase II. **a** Alteration, Astamal area. **b** Silicified rhyolites of the Noqdooz area

- Kaleybar nepheline-syenite batholite.
- Mahmoud Abad intrusions.

These intrusions have cut Middle Eocen volcanic rocks (i.e. they are post Middle Eocene intrusions) and include diorite, nepheline-syenite (Kaleybar), granodiorite to granite (Shivardagh), granodiorite, monzonite and granite (Ordubad). Where exposed, these intrusions are overlaid by Qom equivalent rocks (i.e. they are older than Late Oligocene-Early Miocene Qom Fm.). The mentioned intrusions show a northwestern-southeastern trend in Ahar-Arasbaran axis (Aghazadeh 2009).

Phase IV: Early Oligocene-Pliocene

The majority of this phase includes variable volcanic rocks and porphyry small stocks. The main mineralization of the area (especially copper) is related to magmatic rocks of this phase. The Late Oligocene-Late Miocene volcanic rocks of Phase IV can be traced in Ahar-Julfa axis. They are variable in composition and are more scattered in comparison to Phase II magmatic rocks. The intrusion of intrusive rocks has been triggered 25–30 Ma. (Oligocene) (Aghazadeh 2009) and continued to Late Miocene-Pliocene.

The Phase IV includes several phases of different geological times that more studies (especially more dating analytical results) are needed for categorizing them. The volcanic rocks of variable compositions can be observed along with mostly acidic small intrusions. Some of the mentioned intrusions such as Sungun, Masjed Daghi, Haftcheshmeh, Andrian and Kighal are Middle Miocene-Late Miocene in age. Although the volcanic rocks are observed along with intrusive rocks, it seems that during the final stages of this phase the sub-volcanic intrusions were dominated. The Phase IV is terminated by adakitic intrusions such as Kiamaki and its surrounding intrusive rocks (Fig. 4.27a). The Phase IV similar to Phase II is compositionally varied but it is of less distribution compared to the Phase II. They are of more differentiated phenomen. To the west of area, the sub-volcanic and volcanic rocks are of more distribution.

Phase V

The well distributed basalt to andesi-basalt rocks of Phase V are mostly exposed at northern Ahar (southern Mazrae' mine), surroundings of Andrian village and southwestern Kaleybar close to the Peygham village (Figs. 4.27 and 4.28).

4.5.1.5 Torud

The Torud area is located at eastern Central Iran that is confined between Torud (south) and Anjilou (north) faults (Fig. 4.29).

Fig. 4.27 **a** The Kiamaki dome (Phase IV), **b** Young basalts of Phase V and **c** Angular unconformity between younger volcanic rocks of Phase V and Miocene sedimentary rocks

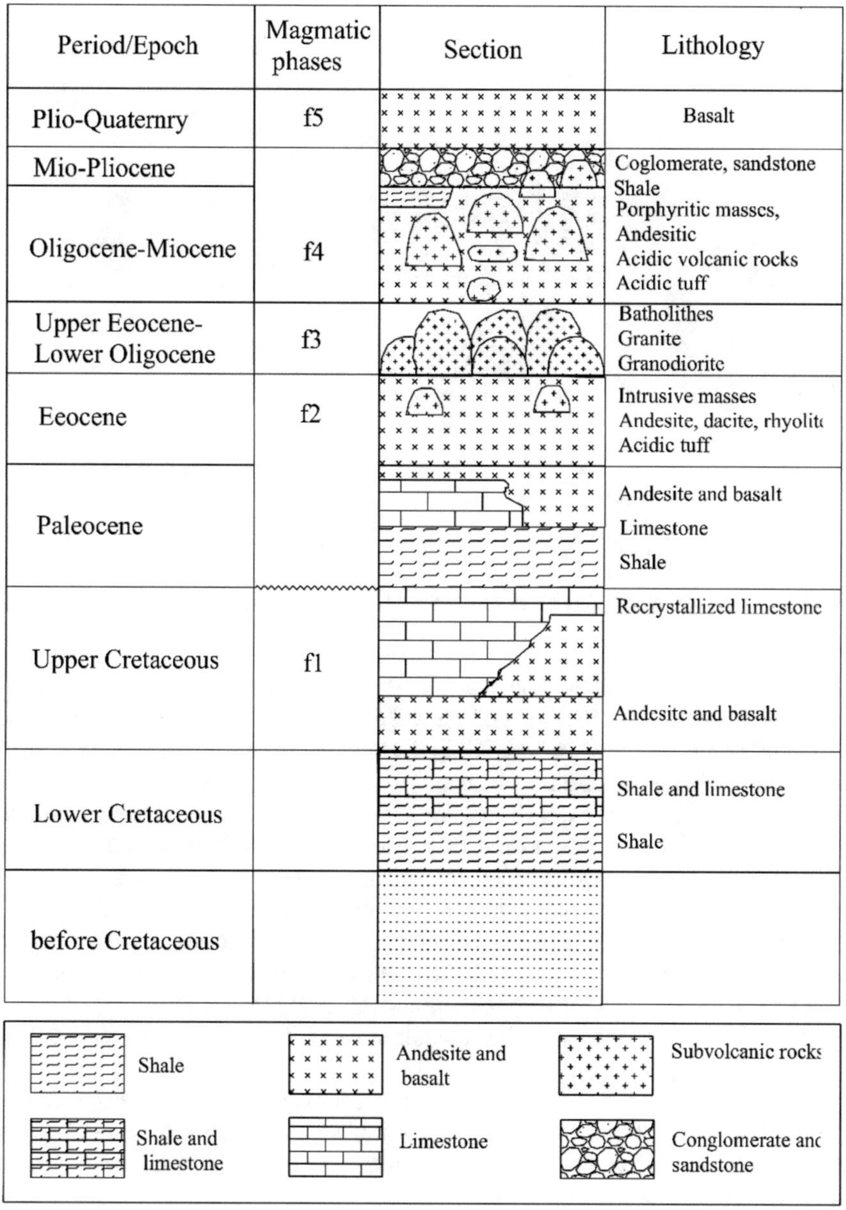

Period/Epoch	Magmatic phases	Section	Lithology
Plio-Quaternry	f5		Basalt
Mio-Pliocene			Coglomerate, sandstone
			Shale
Oligocene-Miocene	f4		Porphyritic masses, Andesitic
			Acidic volcanic rocks
			Acidic tuff
Upper Eeocene-Lower Oligocene	f3		Batholithes
			Granite
			Granodiorite
Eeocene	f2		Intrusive masses
			Andesite, dacite, rhyolite
			Acidic tuff
Paleocene			Andesite and basalt
			Limestone
			Shale
Upper Cretaceous	f1		Recrystallized limestone
			Andesite and basalt
Lower Cretaceous			Shale and limestone
			Shale
before Cretaceous			

Legend:
- Shale
- Shale and limestone
- Andesite and basalt
- Limestone
- Subvolcanic rocks
- Conglomerate and sandstone

Fig. 4.28 The schematic view of Lower Cretaceous-Quaternary rock units of Ahar-Julfa axis

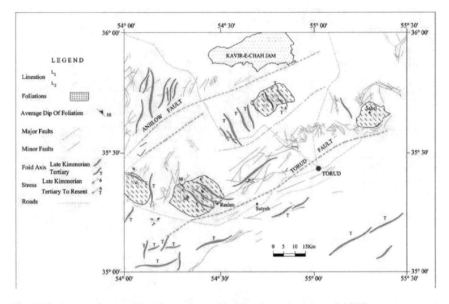

Fig. 4.29 Structural map of Torud area (quoted by Houshmandzadeh et al. 1978)

The area was experiencing magmatic activities since the Palaeozoic. Such activities slowly stared since the first and second eras in response to the tectonics. The climax of magmatic activities was during the third era from which a notable volume of igneous rocks can be observed in area.

The magmatic activities of Torud area affected all geological phenomena and formed magmatic rocks such as tuffs, lavas and various intrusive bodies that entirely covered the area in Tertiary. The Cenozoic magmatic activities began in Lutetian and continued throughout the middle Eocene and finally ended in the Late Eocene. At the end of the late Oligocene, the activities resumes again, but this time the volcanic activities transformed to plutonic activities with lower intensity compared to the first stage of activities and ended very soon (Ghorbani 2005) (Fig. 4.30).

The volcanic rocks include basalt-andesite and dacite in which andesite is dominant (Ghorbani 2005).

The Torud to Chah-e Shirin mining area is confined between two major faults, Anjilou fault in north and Torud fault in South, which is mainly composed of the Eocene volcano-sedimentary rocks, cut by intermediate to acidic intrusive bodies, as follows (Ghorbani 2013).

Eocene volcano-sedimentary units: the volcanic and clastic rocks are integrated in Torud area as a thick succession. According to Houshmandzadeh et al. (1978), the main composition of the volcanic rocks is andesitic type, in Torud-Chah-e Shirin area, which can be divided into four units (Houshmandzadeh et al. 1978).

1. **Basal lava or Satveh andesite**: A sequence of volcanic lava, about 700–800 m, which its upper part does not outcrop is completely exposed in north of Dushakh

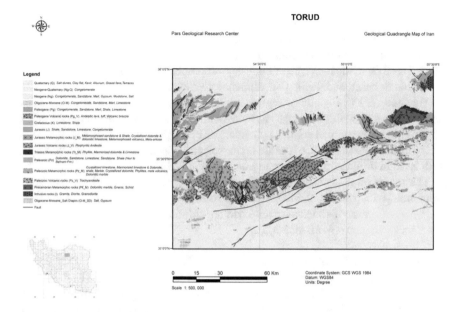

Fig. 4.30 Geological map of Torud area

Mountain. These thick units become thinner towards the west and disappear in a few kilometers farther. As a result of the erosion, the lava debris and fragments supplying the source materials needed to form the basal conglomerates and lower parts of volcanic breccia in the Torud area. Sometimes, the thin layers of andesitic breccia occur within the sequence, but andesitic lavas are dominant. These lavas contain large crystals of pyroxene and hornblende mostly along with biotite.

2. **Lower volcanic breccia**: The sequence begins with clastic rocks like conglomerates and sandstones and continues with alternative layers of tuff breccia, volcanic breccia and lava. In the west of the Baghoo mine, the lower volcanic breccia forms nonconformity on the basal andesite or "Satveh Andesite". The lower volcanic breccia with 800 m thickness is composed of tuffaceous sandstone, dacitic tuff breccia, tuffaceous sandstone, crystalized dacitic tuff, andesitic lava and andesitic dacite tuff breccia from bottom to the top in the area.

3. **Middle volcanic breccia**: The middle unit contains more acidic rocks than the lower and upper units and is composed of fine-grain tuffs and dacite breccia tuff which outcrops in the Kahovan Mount (Sheikhi 2014).

4. **The Upper volcanic breccias**: This section locates at 17 km north of Torud and sits conformably on lower volcanic breccia. It includes conglomerate, sandstone tuff, andesitic lava, tuff breccia, conglomerate, andesitic volcanic breccia and andesitic tuff breccia from bottom to the top.

The volcanic rocks of olivine basalt, differentiated olivine basalt and pyroxene andesite alternate with sedimentary rocks (particularly Nummuliteic limestone and sandstone) in the southern region, south of Torud fault. In the northern region, north of the Torud fault, the rocks are of more acidic composition and form mostly andesite compared to basalt and dacite (Houshmandzadeh et al. 1978).

Volcanic rocks in southern region are lava flows and columns with aphanitic texture. The composition changes to more acidic and intermediate composition and porphyritic texture which contains phenocrysts of plagioclase, amphibole, pyroxene and biotite toward the north of the area. The volcano-clastic rocks exposed alternately with lavas in different parts of the region (Sheikhi 2014).

The major minerals which form the volcanic rocks are plagioclase, pyroxene, amphibole, mica, olivine, alkali feldspars and quartz. A small portion of the secondary and accessory minerals such as apatite, zircon, chlorite and opaque minerals exist in some rock samples (Ghorbani 2005).

Intrusions in the Torud area

The acidic-intermediate intrusions have cut the Eocene volcanic rocks in the Torud, Chah- e- Shirin areas. The intrusions emplaced into the volcanic rocks as domes, stocks as well as dikes. They extended up to a few square kilometers and Challoo intrusion with 25 km^2 is the largest one. All the intrusive rocks have micro-granular or porphyritic texture and are considered as sub volcanics, except the Gandi intrusive body (Houshmandzadeh et al. 1978; Ghorbani 2005).

Based on the composition and field studies, the Torud intrusive bodies are divided into three groups:

1. **Monzodiorite, granodiorite and granite**: The oldest intrusive bodies in the area formed as stocks within the Cretaceous carbonate rocks in southeastern Robaie and northern Challoo intrusive body or Baghoo, Gandi and Challoo Eocene volcanic rocks. Since the age of the host rocks is the middle Eocene, it is believed that they were formed during the late Eocene-Oligocene (Houshmandzadeh et al. 1978; Jafarian 2000). The textural, mineralogical and chemical characteristics of these intrusions indicate that they were emplaced as shallow bodies and have cut the Eocene volcanic and older rocks. Accordingly, they are younger than Eocene. In this region, the most important intrusions are Baghoo, Challoo, Robaie and Gandi. These intrusions are composed of monzodiorite to granodiorite. The intrusions have been cut by intermediate dikes in Baghoo and Challoo. Some of these monzonitic and quartz monzonitic dikes were occurred at the same time of emplacement of the intrusions and have a transitional zone with the host rock. These dikes are exposed in the west Baghoo mine at the bore-holes and mining sites over the highest point of the mine area (Houshmandzadeh et al. 1978; Ghorbani 2005).

2. **Intermediate-basic dikes**: all the dioritic, granodioritic and granitic intrusions in Baghoo, Gandi, Robaie and Challoo are cut by the intermediate to basic

dikes, which mostly have east-west to northeast-southwest trend. The dikes are intruded in the Cretaceous sedimentary and the Eocene volcanic rocks in addition to the intrusions.

3. **Andesite-dacite domes and dikes**: In the southeastern of the Gandi Village, the dacitic domes intruded into the Eocene volcanic rocks and many metallic and nonmetallic deposits such as gold, barite, lead, copper as well as clay occurred. Most of the clay mines are exploited at the present time.

Several light gray acidic to intermediate subvolcanic rocks are exposed within the Eocene volcanism in the north of Torud city. These intrusions are more widespread toward the NW of Torud. Plenty of dark gray microdiorite enclaves are enclosed in the intrusions which both are strongly altered (Houshmandzadeh et al. 1978).

Some small dacitic domes intruded into the plutons and volcanic rocks in the southeastern Baghoo village (Kuh-e Zar gold mine). These intrusions containing Boron (B) bearing solutions caused to form tourmaline and dissolving the feldspars. The Baghoo Turquoise mine is mainly formed within these rock units. In addition, the dacitic dikes have cut the dioritie, granodioritic intrusions. One of these dikes, with east-west trend has 5–10 m thickness and extends up to 700 m. The Baghoo dacitic dikes and intrusions are mostly altered (Ghorbani 2005). The main minerals which forms the intrusions of the area include plagioclase, orthoclase and quartz with the major mafic minerals such as amphibole, biotite and pyroxene.

Based on petrographic and geochemical studies, the intrusions could be classified into two groups of acidic; granite and granodiorite and intermediate, monzonite quartz monzonite types.

The rocks also are of some minor contents of zircon, apatite, titanite, tourmaline, iron and titanium oxides (magnetite and ilmenite).

The EPMA analysis showed that the amphiboles, actinolite, actinolite-hornblende and Mn bearing hornblende can be categorized as calcic group.

In general, for nomenclature of the rocks of the area, different methods have been applied but with same results. Thus, the granitic rocks of the Gandi are mostly granite, granodiorite and quartz monzonite. The Baghoo intrusion ranged from granite, granodiorite to quartz monzodiorite and the monzonite, monzodiorite rocks of Challoo and Robaie areas, vary from monzonite, quartz-monzonite and quartz monzodiorite to monzogabbro.

The magma series that have formed the Eocene volcanic rocks and their equivalent intrusions are mainly high potassic calk-alkaline to alkaline (shoshonite).

4.5.1.6 Cenozoic Magmatism in the Sabzevar Region

The Cenozoic volcanic rocks of the Sabzevar region can be temporally, spatially and geochemically categorized as below:

Andesite rocks: The andesitic rock group is composed of breccia tuffs, agglomerates and Eocene lavas. The rock sequence begins with tuffs formed in a semi marine environment and continues up to a series of pyroclastic rocks formed in a semi continental basin along a small amount of lavas.

Dacite Rocks: This group consists of the Oligocene-Pliocene dacite and andesite containing amphibole (2.7–41 Ma). Unlike the Eocene andesites, dacitic rocks are exposed as separated plugs and dikes. Since in some parts of the Sabzevar zone lava flow, amphibole-bearing andesite in composition, covers dacitic plugs. So, they could be recognizable easily from dacitic intrusions by their petrologic and geologic characteristics. Based on the geochemical studies, the andesitic and dactic groups are calk-alkaline rock type.

Alkaline Rocks: As the point of view of volume, they are the smallest group of volcanic rocks which occur only as individual dikes. Since the dikes cut the Miocene-Pliocene sediments and according to the age determination using K-Ar method, Miocene has been confirmed for them. The alkaline rocks are divided into two rock groups, alkaline basalts and shoshonitic rock.

The K-Ar age determination and field studies indicate a major trend for the region. That is, at the time of closing of the Sabzevar rift, the post collision volcanism has clearly moved from south to north. The displacement trend is traceable by the age dating of the dacite. Because the oldest rocks expose in south of the Sabzevar ophiolitic belt, but the younger occur in the north of belt. Also, the age dating indicates "Pliocene" as the youngest age for the rocks, which only expose to the southern border of the Alborz zone.

Geodynamics of the Sabzevar Region

(A) Primary studies: Based on the lithology, geochemistry, stratigraphy and age dating, the below geodynamic model is recommended for the post-collision volcanism (Spise et al. 1983).

A divergent phase related to an oceanic rift has been active during the late Cretaceous, 70–80 Ma. They believe that the expansion of the oceanic crust terminated at the end of the Cretaceous.

During Paleocene, a subduction zone has been developed and actively continued till development of an andesitic island arc in the foreland basin of Turan Plate (Kopeh-Dagh Zone; Huber 1978a, b, c).

Due to direct pressure trending to the north or northeast (Tapponnier et al. 1981), the micro continental rift between the Central Iran and Turan Plate (Davoudzadeh et al. 1981) was closed in the Eocene. Consequently, the collision of the island arc and the Turan Plate's margin ended in the Eocene.

At that time, some parts of the subduction zone (trench) have broken off and abducted over the other parts of the continental plate. The other descending plate moved downward and begins to be heated and melted. Consequently, the magma rose beside the Sabzevar Ophiolite to build andesitic continental Arc. However, some geologists believe that the conversion activities were ended at the Pliocene, because, during the horizontal movement of Great Kavir fault in Binaloud towards the north or northeast, the vertical and horizontal movements were substituted (Tapponnier et al. 1981). The oceanic lithosphere has a deep slope towards the north, so the andesitic volcanism migrated from south to the north side. As a result, the calc-alkaline volcanic belt between Sabzevar and Ghouchan developed.

The collision of the island arc, oceanic crust and continent, eventually resulted in the accumulation of so much material at the subduction zone causing Moho and magma chamber sink down towards the mantle. This phenomenon leads to the rise of alkaline basalts and development of shoshonitic rocks (Dehghani 1981).

(B) Recent Studies: Shirzadi (1998) believes that the andesite, dacite and rhyolite belong to a post collision complex, where subduction slopes was towards the north and the rocks originated from the same magma.

Khalatbari-Jafari et al. (2008) by study of the Eocene volcanic area in eastern Abbassabad (western Sabzevar) concluded that the abovementioned volcanism has formed in a back arc basin which is consistent to the subduction of Arabian plate beneath the Central Iran. In the process of partial melting of oceanic crust and sedimentary rocks of ultramafic parts of oceanic crust and Supra-subduction zone ophiolites of Paleo Tethys Ocean in northeast Iran, gradually the different units of the Eocene volcanic rocks formed.

4.6   Igneous Rocks Distribution in Iran

Except the Kopet Dagh and Zagros zones, the igneous rocks can be seen in most parts of Iran formed during several geological periods of times. From this point of view, their distribution can be defined as below (Fig. 4.31):

– The distribution of all igneous rocks of Proterozoic-Quaternary.
– It's normal that in some especial parts of time, the igneous rocks are of more temporal-spatial distribution that it can be illustrated as the following:
– The volcanic-intrusive rocks of Late Proterozoic: They are notable but most of them are overlain by younger sediments. They can be found everywhere the Pre-Cambrian rocks are exposed. So, it can be concluded that they are very well distributed in Iran while the exposures were only during the Pre-Cambrian.
– Paleozoic: They were not notable during all Palaeozoic era. According to their temporal-spatial distribution they can be categorized into two groups as below:

1. Northern Central Iran and Alborz (Ordovician-Devonian): they are mostly volcanic rocks of basaltic composition related to the Paleo-Tethys rifting.
2. Late Paleozoic-Early Triassic: these volcanic rocks are mostly observed at western parts of Iran showing a NW-SE trend accompanied by small intrusions. They are mostly related to the extensional environments related to the Neo-Tethys rifting. Albeit, the intrusive bodies of Late Paleozoic and especially the Early-Middle Triassic can be observed in northern parts of country which are related to the Paleo-Tethys collision.

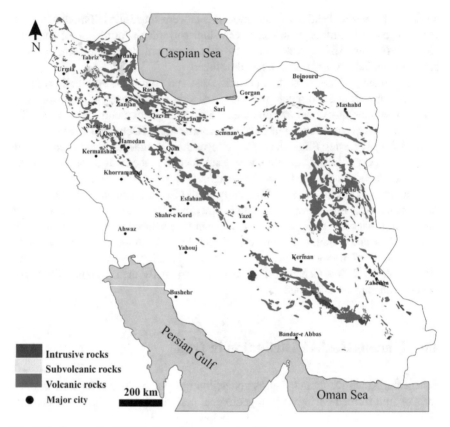

Fig. 4.31 Geographical distribution of igneous rocks of Iran

4.7 Intrusions of Iran

Due to their importance, one chapter of the present book is discussing the Intrusions of Iran.

Pyroclastics: The pyroclastics are mostly found close to the volcanic rocks of Iran and in some cases they are of more volume and distribution compared to volcanic rocks (Fig. 4.32).

The pyroclastic rocks are observed in all geological times in which the volcanic activities are reported. It worths mentioning that their volume in Cenozoic and Late Proterozoic is notable (Fig. 4.33).

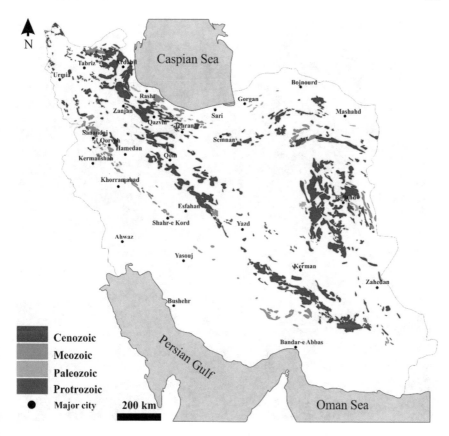

Fig. 4.32 Geographical distribution of volcanic rocks of Iran, based on the age

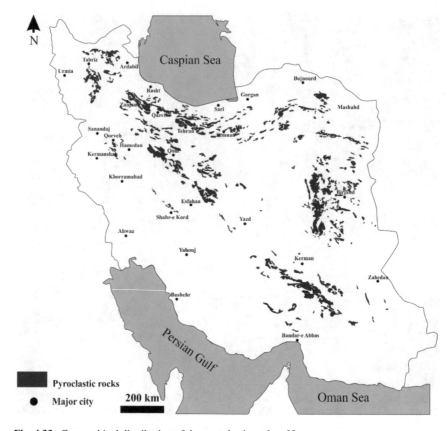

Fig. 4.33 Geographical distribution of the pyroclastic rocks of Iran

References

Aale-Taha B (2003) Petrography and petrology of igneous rocks and related Cu mineralization at the Southeast of Bam (Jebal-e Barez). PhD thesis, Azad University, Tehran, 388p

Abdi Gh (1996) Petrological study of northern Qorveh (Kurdistan) volcanic rocks. MSc thesis, Shahid Beheshti University

Abdolnshad Mamaghani H (2007) Petrology of the internal mass and its emplacement in the southern part of Kashan rocks. MSc thesis, Shahid Beheshti University

Aghanabati A (1998) The Jurassic stratigraphy of Iran. Gel Surv Iran

Aghanabati A (2004) The Geology of Iran. Geological Survey of Iran, Tehran, p 586

Aghanabati SA, Shahrabi M (1987) Geological Quadrangle Map No.K.4 Mashad-Geological Survey of Iran, Tehran

Aghazadeh M (2009) Petrology and Geochemistry of Anzan-Khankandi and Shaivar Dagh granitoids (North and East of Ahar, Eastern Azerbaijan) with references to associated mineralization. PhD thesis, Tarbiat Modares University

Ahmadi Khalaj A, Esmaeily D, Valizadeh MV, Rahimpour-Bonab H (2007) Petrology and geochemistry of the granitoid complex of Boroujerd, SSz, Western Iran. J Asian Earth Sci 29:859–877

Ahmadi Pourfarsangi H (2004) Dynamism of the Mozahem volcano, Babak City. MSc thesis, Shahid Bahonar University, 366p

Ahmadzadeh-Heravi M, Houshmandzadeh MA, Nabavi MA (1990) New concept of Hormuz Formation's stratigraphy and the problem of salt diapirism in south Iran. Proc Symp Diapirism Spec Ref Iran 1:1–21

Alavi M (1980) Tectonostratigraphic evolution of the Zagrosides of Iran. Geology 8(3):144–149

Alavi M (1991) Sedimentary and structural charactristic of paleotethys remanants in northeastern Iran. Geol Soc Amer Bull 103:983–992

Alavi M (2004) Regional Stratigraphy of the Zagros Fold-Thrust Belt of Iran and Its Proforeland Evolution. Am J Sci 304:1–20

Alizadeh H, Arian M, Lotfi M, Ghorashi M, Ghorbani M (2015) Determination of Porphyry Copper Deposit Locations Using Photo Lineament Factor in Northern Parts of the Dehaj-Sardouieyeh Belt. Geosci J 24:247–252

Allenbach P (1966) Geologie und Petrographie des Damavand und seiner Umgebung (Zentral-Elburz), Iran. Abhandlung zur Erlangung der Wurde eines Doktors der Naturwissenschaften der Eidgenossischen Technischen Hochschule Zurich, 1–145

Alric G, Virlogeux D (1977) Pétrographie et géochimie des roches métamorphiques et magmatiques de la région de Deh-Bid. Bawanat: Chaîne de Sanandaj-Sirjan. Iran, Universite scientifique et medicale de Grenoble

Amidi SM (1977) Etude géologique de la région du Natanz– Surk (Iran Central) stratigraphie et pétrographie. PhD thesis, Geological Survey of Iran, Rep. No.42, 316p

Amidi SM, Emami MH, Michel R (1984) Alkalin character of Eocene volcanism in the middle part of central Iran and its geodynamic situation. Geol Randsch 3(3):917–932

Annells RN (1975) Explanatory text of the Qazvin and Rasht quadrangles map. Geological survey of Iran, 3

Annells RN, Arturton RS, Bazley RAB, Davis RG, Hamedi MAR, Rahimzadeh F (1977) Geological map of shakram, Scale 1: 100000, Geol. Surv. Iran

Asadpour M, Heuss S (2018) Evidence for pan-african basement in Ghalghachi Leucogranite (west the Urmia Lake) with using U-Pb zircon dating and whole-rock Sm-Nd and Rb-Sr isotopic analysis. Sci Q J Geosci 28(109):211–220

Asgharzadeh H, Fazel E, Mehrabi B, Masoudi F (2018) Geochemical-metallogenic evolution of agh-daragh igneous rocks (north of ahar) links to cu-au ± w occurrences. J Petrol 8(32):21–44

Assereto R (1963) The Paleozoic formations in Central Elburz (Iran) -(Preliminary note). Riv Ital Paleont Strat 69(4):503–543, 11 figs., 2 pls

Babakhani A, Ghalamghash J (1995) Geological map of Takht-e Soleyman (1/100000). Geological Survey of Iran

Berahmand L (2017) Petrology of porphyry intrusive bodies associated with copper mineralization and felsic dams in Meiduk-Dehajarea. PhD thesis, Shahid Beheshti university

Berberian F, Berberian M (1981) Tectono-plutonic episodes in Iran. In: Gupta HK, Delany FM (eds) Zagros Hindukush, Himalaya Geodynamic Evolution. American Geophysical Union, Washington, DC, pp 5–32

Berberian M, King GCP (1981) Towards a paleogeography and tectonic evolution of Iran. Can J Earth Sci 18(2):210–265

Clark GC, Davies R, Hamzehpour B, Jones CR, Ghorashizadeh M, Hamidi B, Navai N (1975) Bandar-e-Pahlavi, Geol. Quadrangle map of Iran; 1:250,000. Geological Survey of Iran, No. D3

Crawford AR (1972) Iran, continental drift and plate tectonics. In: 24th International Geological Congress, Montreal, Section 3, pp 106–112

Crawford AR (1977) A summary of isotopic age data for Iran, Pakistan and India. Mem Hors Ser Geol Fri 8:251–260

Dakhili MT (1995) Petrology of magmatic and metamorphic rocks in the northwestern of Fariman. MSc Shahid Beheshti University

Darvishzadeh A (1991) Geology of Iran. Neda Publication, Tehran, p 901

Davoudzadeh M, Soffel H, Schmidt K (1981) On the rotation of the Central-East Iran microplate. N Jb Geol Palâont Abh 3:180–192

Dedual E (1967) Zur Geologie des mittlern and unteren Karaj-Tales, Zntral-Elborz (Iran). Eidgen Tech Hochsch, Univ Zurich, Mitt Geol Inst, Zurich, 123p

Dehghani GA (1981) Schwerefeld und Krustenaufban im Iran Hamburger Geophys Einzelschr Reihe A 54:74

Dellenbach J (1964) Contribution a l'etude geologique de la region situee a l'est de Teheran (Iran). These presentee a la faculte des sciences de l'Universite de Strasbourg pour obtenir le grade de docteur es sciences naturelles, 1–123

Dimitrijevic MD (1973) Geology of Kerman Region Geological Survey of Iran, Report Number Yu/52, pp 334

Doroozi R, Vaccaro C, Masoudi F (2017) Mesozoic alkaline plutonism: Evidence for extensional phase in Alpine-Himalayan orogenic belt in Central Alborz, north Iran. Solid Earth Sci, 1–18

Doroozi R, Vaccaro C, Masoudi F, Petrini R (2015) Cretaceous alkaline volcanism in south Marzanabad, Northern Central Alborz, Iran: Geochemistry and petrogenesis, Geoscience Frontiers, pp 1–15

Eftekharnezhad J (1975a) Brief history and structural development of Azarbaijan. Geological Survey of Iran, Internal Rep. N.8, p 9

Eftekharnezhad J (1975b) Brief description of tectonic history and structural development of Azarbaijan, Field excursion guide, No. 2. Note A Sym. Geodynamic of Southeast Asia, Tehran, pp 469–478

Eftekharnezhad J (1980) Tectonic divisions of different parts of Iran regarding sedimentary basins. J Oil Soc 82:19–28

Emami H (1981) Geologie de la reigon de Qom- Aran (Iran) Contribution a l'etude dynamique et geochimique du volcanisme Tertiaire de l'Iran Central. (These)

Emami MH (2000) Magmatism in Iran. Geological Survey of Iran, Tehran, p 608

Emami MH, Sadeghi MM, Omrani SJ (1993) Magmatic map of Iran. Scale 1:1,000,000, Geological Survey of Iran

Fardaei M (2009) Petrography and Petrology of the Cretaceous Volcanic rocks near Majid Abad, Ahar (northwest Iran). MSc thesis, Shahid Beheshti University

Farrokhmanesh D (1998) Petrological and Geochemical Survey of Takab Granitoid Masses. MSc thesis, Shahid Beheshti University

Fürsich FT, Wilmsen M, Seyed-Emami K, Majidifard MR (2009) The Mid-Cimmerian tectonic event (Bajocian) in the Alborz Mountains, Northern Iran: evidence of the break-up unconformity of the South Caspian Basin. In: Brunet, M-F, Wilmsen, M, Granath JW (eds) South Caspian to Central Iran Basins. Geological Society, London, Special Publications 312, pp 189–203

Gansser A, Huber H (1962) Geological Observations in the Central Elburz. Iran: Schweizerische Mineralogische und Petrographische Mitteilungen 42(2):583–630

Ghafari S (2010) Petrology and petrography of igneous rocks of salt domes (east and south Laur). MSc thesis, Shahid Beheshti University

Ghalamghash J, Babakhani AR (1996) Geological map of Kahak area, scale 1: 100,000. Geological Survey of Iran, Tehran

Ghorbani M (1999) Petrological investigations of Tertiary-Quaternary magmatic rocks and their metallogeny in Takab area. PhD thesis, Shahid Beheshti University

Ghorbani M (2004) Petrologic characteristics of magmatic rocks in alteration zones and their association with mineralization within Ahar–Jolfa axis, Research project at Shahid Beheshti University

Ghorbani Gh (2005) Petrology and Petrogenesis of igneous rocks in south of Damghan. PhD thesis, Faculty of Earth Sciences, Shahid Beheshti University

Ghorbani M (2007) Economic geology of natural and mineral resources of Iran. Pars Geological Research Center (arianzamin), 492p

Ghorbani M (2010) Evaluation of porphyry Copper potential in Jebal-e Barez Sub zone, volume II. Pars Geological Research Center

Ghorbani M (2013) The Economic Geology of Iran: Mineral Deposits and Natural Resources. Springer Geology

Ghorbani M (2015) Magmatism-Metamorphism of Iran. Arian Zamin, 325p

Ghorbani M (2019) Lithostratigraphy of Iran. Springer

Hadizadeh Shirazi M (2010) Petrography and petrogenesis of metamorphic-magmatic rocks in the Kulyakesh Sourian region and their links with Mineralogy. MSc thesis, Shahid Beheshti University

Haghipour A (1974) Etude géologique de la région de Biabanak-Bafgh (Iran central), Pétrologie et tectonique du socle precambrien et de sa couverture, France. PhD thesis Sci Nat, Gronoble University, p 403

Haghnazar S (2009) Petrology and geochemistry of mafic rocks in Javaher Dasht area, East Guilan Province, North of Iran. PhD thesis, Shahid Beheshti University

Hajalilou B (2000) Metallogeny of Cenozoic rocks of western Alborz and Azerbaijan (Mianeh-Siahrud), with special insight into Hashtjin. PhD thesis, Shahid Beheshti University

Hassanzadeh J (1993) Metallogenic and tectonomagmatic events in the SE sector of the Cenozoic active continental margin of central Iran (Shahr-e Babak area, Kerman Province). PhD thesis, Los Angeles, University of California, 204p

Hosseini B (2011) Identification of two phases of metamorphosed granitoid masses in the Tutak anticline based on the U-Pb dating. Earth Sci J 84:57–66

Houshmandzadeh A, Nabavi MH (1986) Metamorphic Map of Iran Scale 1: 2,500,000. Tehran: Geological Survey Of Iran

Houshmandzadeh AR, Alavi MN, Haghipour AA (1978) Evolution of Geological Phenomenon in Toroud Area (Precambrian to Recent). Tehran: Geological Survey of Iran, Report H5

Houshmandzadeh A, Nabavi M, Hamdi B (1988) Precambrian-Lower Cambrian gemstones in Iran, Proceedings of the Seminar on Mining Resources and Mineral Power in Yazd Province

Huber H (1978a) Geological Map of Iran, 1 sheet, Geological Survey of Iran

Huber H (1978b) Tectonic Map of Iran, 1 sheet, Geological Survey of Iran

Huber H (1978c) Geological Map of Iran, 1:1,000,000; with explanatory note. North-East Iran, Exploration and Production Affairs, Tehran, 1 sheet

Huckriede R, Kürsten M, Venzlaff H (1962) Geotektonische Kartenskizze des Gebietes zwischen Kerman und Sagand, (Iran), Bundesanstalt für Bodenforschung

Iwao S, Hushmandzadeh A (1971) Stratigraphy and Petrology of the low-grade regionally metamorphosed rocks of Eocene formation in the Alborz range. North of Tehran, Iran

Jafarian MB (2000) Geological map of Kalateh-Reshm (scale 1:100,000)sheet No 6860. Geological Survey of Iran Tehran Iran

Jahangiri A (2007) Post-collisional Miocene adakitic volcanism in NW Iran: geochemical and geodynamic implications. J Asian Earth Sci 30:433–447

Jamei S, Ghorbani M, Williams IS, Moayyed M (2020) Tethyan oceans reconstructions with emphasis on the Early Carboniferous Pir-Eshagh A-type rhyolite and the Late Palaeozoic magmatism in Iran. Inter Geol Rev. https://doi.org/10.1080/00206814.2020.1768443

Jenny JG (1977) Géologie et stratigraphie de l'Elbourz oriental entre Aliabad et Sharud, Iran: thèse Imprimerie nationale

Kani A, Ghorbani M (2016) Late Precambrian-early Paleozoic Stratigraphy of northern Gondwana Region with Special emphasis on Iran, 1st congress on the Paleozoic of northern Gondwana, Prugian, Italy

Kazmin VG, Sbortshikov IM, Ricou LE, Zonenshain LP, Boulin J, Knipper AL (1986) Volcanic belts as markers of the Mesozoic-Cenozoic active margin of Eurasia. Tectonophysics 123(1–4):123–152

Khalatbari-Jafari M, Mobasher K, Davarpanah A, Babaie H, La Tour T (2008) A backarc basin origin for the Eocene volcanic rocks North Abbas Abad, East of Shahrud, Northeast Iran. American Geophysical Union, Fall Meeting

Lorenz C (1964) Die Geologie des oberen Karadj-Tales (Zentral-Elburz), Iran. Inaugural-Dissertation zur Erlangung der philosophischen Doktorwurde vorgelegt der Philosophischen Fakultat II der Universitat Zurich, 113

Mehrpartou M, Mirzaei M, Alaei S (1997) Geological map of the Siahrood 1:100,000 scale. Geological Survey of Iran

Moafpourian Gh (2010) Petrological study of the Kometaetic rocks in the northern Mashhad and Fariman regions. PhD thesis, Shahid Beheshti University

Moatamed A (1987) Sedimentology 1,2. Tehran university publication

Moayed M (1991) Study of petrography and petrochemicals of volcanic and plutonic rocks of Zakir (Southwest Tarom) in relation to copper genesis. MSc thesis, Tabriz University

Moayed M, Hoseinzadeh Gh (2000) Investigating metasomatic and iron mineralization of Alteration, Manganese in Skarn Tikme-Dash, East Azerbaijan. 4th Conference of the Geological Society of Iran, Tabriz University

Mobashergermi M, Zarei Sahamieh R, Aghazade M, Ahmadikalaji A, Ahmadzadeh Gh, Le Roux P (2019) Petrological and isotopic study of basaltic rocks of Barzand area and comparison of them with Poshtasar basalts in South of Germi. Petrology 10(1, 37):234–252

Moein Vaziri H (1999) An Introduction to Iran Magmatism. Tarbiat Moallem University, Iran (in Persian)

Mohammadi B (1996) Petrography, Petrology and economic potential of the extrusive rocks in the Taleghan area. MSc thesis, Shahid Beheshti University

Moradian A (2005) Petrography and Geochemistry of minerals in Feldspathoid—bearing rocks, located in north Shahr—babak, west Meiduk village, Vol. 13, No. 2, 1384/2005 Fall & Winter

Mousavi S (1994) Study of petrology and mineralogy of the southern Lahijan intrusive Massif. MSc thesis, University of Tehran

Mousavi Makavi SA (1998) Narigan granite petrology. MSc thesis, Shahid Beheshti University

Nabavi MH (1976) An introduction to the geology of Iran. Geological survey of Iran, 109

Nedimovic R (1973) Exploration for Ore Deposits in Kerman Region. Report No. 53, Geological Survey of Iran Publication, Tehran

Niroomand S, Lentz DR, Sepidbar F, Tajeddin HA, Hassanzadeh J, Mirnejad H (2020) Geochemical characteristics of igneous rocks associated with Baghu gold deposit in the Neotethyan Torud-Chah Shirin segment, Northern Iran. Geol J 55(1):299–316

Nokhbeh-alfoghahaii A (2016) The study of economic geology of Dehkuyeh, Karmostaj and Paskhand salt domes in Lar-Bastak, southern Iran. PhD thesis, Azad University

Nouraei A (1997) Pathogenesis and geochemistry of granitoids in south Iran. Shahid Jahan, MSc thesis, Shahid Beheshti University

Omrani J, Agard Ph, Whitechurch H, Benoit M, Prouteau G, Jolivet L (2007) Arc-magmatism and subduction history beneath the Zagros Mountains. Iran: A new report of adakites and geodynamic consequences

Parvinpour F (2007) Porphyry copper deposits within Abdar-Dahaj area. MSc thesis, Islamic Azad University

Peyravan HM (1992) Petrography, petrology and geochemistry of northern Abhar igneous rocks and their mineralogical relations with plutonism. MSc thesis, Tehran University

Rasouli J, Ghorbani M, Ahadnejad V, Poli G (2016) Calc-alkaline magmatism of Jebal-e Barez plutonic complex, SE Iran: implication for subduction-related magmatic arc. Arab J Geos 9(4):287

Reyer D, Mohafez S (1972) A first contribution to the NIOC-ERAP agreements to the knowledge of Iranian geology. Editions Technip, Paris, pp 1–58

Ruttner AW (1984) The Pre-Liassic Basement of the Eastern Kopet Dagh Range. N Jb Geol Paläont Abh 168(2–3):256–268

Ruttner A, Nabavi M, Hajian J (1968) Geology of theShirgesht area (Tabas area, East Iran). Tehran: Geological Surveyof Iran. Report 4, 133p

Sabzehei M (1974) Les mélanges ophiolotiques de la région d'Esfandagheh (Iran meridional), étude pétrographique et structurale, thése Doct Etat, Univ Grenoble, 306p

Sabzehei M, Majidi B, Alavi-Tehrani N, Etminan H (1970) Preliminary report, geology and petrography of the metamorphic and igneous complex of the central part of Neyriz Quardangle (compiled by Watters, W.A., Sabzehei, M.). Geological Survey of Iran. Internal Rep., p 60

Sarkirinshad Kh (1989) Petrology and Geology of Ophiolite Neyriz-Hashtjin. 8th Earth Science Symposium, Geological Survey of Iran, pp 35–39

Schroder JW (1945) Quelques aspects de lageologic de 1 Iran: Geneve Bulletin Vereingung Schwizerissher Petroloumgeologen un Petroleuming enieure, band 12, pp 11–20

Shahabpour J (2007) Island-arc affinity of the Central Iranian volcanic belt. J Asian Earth Sci 30(5):652–665

Shahbazi H (1999) Petrological, geochemical and petrofabric study of older rhyolitic and granitic of Mahabad-Bukan. MSc thesis (in Farsi), University of Tehran, Iran

Shahbazia H, Siebelb W, Pourmoafeea M, Ghorbania M, Sepahic AA, Shangb CK, Vousoughi Abedini M (2010) Geochemistry and U-Pb zircon geochronology of the Alvand plutonic complex in Sanandaj-Sirjan Zone (Iran): new evidence for Jurassic magmatism. J Asian Earth Sci 39(2010):668–683

Sharkovski M, Susov M, Krivyakin B (1984) Geology of the Anarak area (Central Iran), Explanatory text of the Anarak quadrangle map, 1: 250,000. V/O Technoexport Report TE/No. 19. Geological Survey of Iran, Tehran

Sheikhi Gheshlaghi R (2014) Volcanism of South and South-West Troud and its Relationship with Formation of Semiprecious Rocks. MSc thesis, Faculty of Earth sciences, Shahid Beheshti University

Shirmohammadi M (2014) Petrology and Geochemistry of intraformation magmatic rocks of Geirud Formation (Central Alborz). MSc thesis, Shahid Beheshti University

Shirzadi AS (1998) Petrography, petrology and geochemistry of aggregate phyllite and ophiolitic and post-ophiolite of the strips north of Sabzevar (northern Froumad village). MSc thesis, Islamic Azad University

Spies O, Lensh G, Miha A (1983) Geochemistry of the post–ophiolithic Tertiary volcanics between Sabzevar and Quchan /NE Iran. Geol Surv Iran Rep 51:247–265

Stalder P (1971) Magmatism Tertiaire et subrecent enter Taleghan et Alamout Alborz, Central, Bullsuiss. Min Petrol 51:1–138

Stocklin J (1968) Structural history and tectonics of Iran; a review. Am Assoc Petrol Geol Bull 52(7):1229–1258

Stocklin J (1974) Evolution of the continental margins bounding a former Southern Tethys. In: The Geology of Continental Margins. Springer, pp 873–887, BIBL 2P, 5Illus UN Geol Surv Inst

Stocklin J, Setudehnia A (1971) Lexique Stratigraphique International, V. III, Asie 9B, Iran, 1. Iran central septentrional et oriental, 2. Iran du Sud-ouest, CNRS (Paris) 3(9):376

Süssli PE (1976) The Geology of the lower Haraz valley area, central Alborz, Iran (No. 38). Geological Survey of Iran

Tapponnier P, Mercier JL, Proust F, Andrieux J, Armijo R, Bassoullet JP, Brunel M, Burg JP, Colchen M, Dupre B, Girardeau J (1981) The Tibetan side of the India-Eurasia collision. Nature 294(5840):405–410

Tarkhani MS (2010) Petrology and Geochemistry of Mesozoic volcanic rocks, around saqez-Piranshahr. PhD thesis, Islamic Azad University

Torkamani E (1997) Petrologic study of intrusive rocks north of Abhar-Khorramdarreh. MSc thesis, Shahid Beheshti University

Valizadeh MV, Esmaili D (1996) Petrography and petrogenesis of Doran's granite. J Sci Univ Tehran 22(1):12–36

Yasaghi AS (1989) Structural Analysis and auilding development of southwest Alborz (Northern Tehran). MSc thesis, Tarbiat Modarres University

Zahedi M (1976) Explanatory text of Esfahan, Geological QuadrangleMap1:250000, No. F8 Geological Survey of Iran, Tehran

Zarei Sahamieh R (1992) Petrography and geochemistry of volcanic rocks north of Abhar. MSc thesis, Tarbiyat Moaalem University

Zohur Ghorbani GhR (1995) Petrological and geochemical study of Lower Paleozoic magmatic rocks (Khosh Yeilagh area). MSc thesis, Shahid Beheshti University

Chapter 5
Intrusive Rocks of Iran

Abstract Intrusive bodies of Iran play a significant role in geology and the knowledge of Iran's geological phenomena, as well as they have strong relations with mining areas. In other words, most of the epigenetic and orthomagmatic metal deposits are dependent on intrusive bodies especially copper and iron; so, they have been studied separately in this chapter.In this chapter, 254 intrusive bodies so far recognized in Iran have been studied and evaluated. All these intrusive bodies are arranged in a table for the ease of judgment and comparison. The following information is given in this table for each intrusive body: Geographical location and 1.25000 and 1.100000 geological map in which the intrusive bodies are located. Characteristics of petrography and magmatic series, geochemistry, absolute age and geological age of the massif, their tectonic environments and the results of many studies are summarized and categorized in a Table. All intrusive masses in this table have been categorized based on their age according to the structural zone in which they occur. Finally, the distribution map of intrusive masses is given.

Keywords Intrusive bodies · Granite · Grano-diorite · Diorite · Gabbro · Porphyry massive · Adakite · Absolute age and geological age

5.1 Iranian Intrusive Bodies

Intrusive bodies of Iran play a significant role in geology and knowledge of Iran's geological phenomena, as well as they have strong relations with mining areas. In other words, most of the epigenetic and orthomagmatic metal deposits are dependent on intrusive bodies especially copper and iron. On the other hand, their intrusive masses are associated with important geological events, showing the extensional environments and the occurrence of a rift. Most of the Precambrian masses in Iran, or the Late Devonian masses.

The early Triassic in the Sanandaj–Sirjan zone and the central part of Central Iran are part of this group. Some of them are signs of collision and orogeny such as intrusive masses with the late Paleozoic–Triassic age in the north, northeastern

Iran, and intrusive masses with Jurassic–Paleocene age in the Sanandaj–Sirjan zone.

In addition, the Eocene–Pliocene intrusive bodies in the Urmia-Dokhtar magmatic belt represent a subduction environment and a continental margin magmatic arc. In this chapter, all of Iran's intrusive bodies except for ophiolite-related ones are classified in terms of age and structural zone and all the characteristics of petrography and geochemistry and their tectonic environment are expressed which is the result and abstract of many studies summarized and categorized in a Table 5.1.

In terms of petrography, intrusive bodies can be divided into two groups:

Mafic to medium bodies including gabbros, diorites, monzonites, and syenite. It seems that wherever one of the bodies are found, there is an evidence of the influence of gabbro to diorite batholith of the mantle, which the rest of the bodies derive from magma differentiation.

Acidic bodies include tonalites, granodiorites, granites, and alkali granites. These intrusive bodies can be of type I, S, A and even M, which are mostly referred in Table 5.1.

Structural Classification of Intrusive Bodies of Iran

Due to geological features such as tectonic, sedimentary basin, magmatic phase and metamorphism, orogenic activities and trends, Iranian lands are divided into several zones (structural map of Iran, Fig. 5.1), all Iranian intrusive masses in the following structural zones have been categorized:

– Central Iran, the main geological zone of Iran, is divided into several sub-zones.
– Azerbaijan, which is west of central Iran Zone, but due to its importance of intrusive and magmatic bodies, it has been referred to as the Azerbaijan, west of Central Iran.
– Sanandaj–Sirjan intrusive bodies, the most important magmatic-metamorphic zone of Iran.
– Intrusive bodies of Urmia-Dokhtar which is a Cenozoic magmatic belt in central western Iran with a NW–SE trend.
– Intrusive bodies of the eastern parts of Iran including the Lut Block and the flysch Zone
– Intrusive bodies of Zagros Zone.

Zagros Zone is a sedimentary basin with igneous rocks found in its basement that is often volcanic which were previously referred.

Table 5.1 Intrusive rocks of Iran based on structural zones and age in each of structural zones

Sr. No.	Name	Geological Map 1:100,000 1:250,000	Geographical coordinates	Geographical location	Lithological composition	Geological age and age determinations	Magmatic series	Tectono-magmatic setting and type of magma	Note (comments and references)
Central Iran									
Precambrian									
1	Taknar-Sar Borj (Bornavard)	Bardaskan Kashmar	57 45 42 35 21 30	28 Northwest of Bardaskan	Granite, granodiorite, tonalite and gabbrodiorite	Granodiorite: 552.69 ± 10.89 Ma. Granite: 538.22 – 1.82 + 4.28 Ma. (Late Neoproterozoic time)	Calc-alkaline	I type post-orogenic	Sepahi Garo (1992) et al. (2004) Karimpour et al. (2011) Malekzadeh et al. (2004)
2	Shotor kuh (Kuh-e Kaftary)	Torud Rezveh	55 27 37 35 42 47	Northeast of Torud	Gabbro, granodiorite, tonalite	Precambrian 526–547 Ma.	Alkaline to Calc-alkaline	Extentionl within plate	Shekari et al. (2018)
					Biotite granite	566 Ma. U–Pb			Hassanzadeh et al (2008)
3	Saqand Yazd	Rizab Ardakan	55 29 19 32 20 30	Saqand Douzakh Darreh	Diabase and volcanics (Granite is recognized by drilling and geophysics)	Precambrian	Alkaline	I Within plate	–
4	Morad	Bagheyn Rafsanjan	56 46 40 30 23 38	Northwest of Kerman (Ab Morad)	Granite	Precambrian	Alkaline	I Extentionl Rift-related	Equivalent with Narigan and Zarigan granite
5	Behabad	Behabad Ravar	56 00 14 31 45 40	10 km south west of Behabad	Granitoid	Late Precambrian-Early Cambrian 525–547 Ma.	Alkaline	I Rift-related	Based on the studies Bafg

(continued)

Table 5.1 (continued)

Sr. No.	Name	Geological Map 1:100,000 1:250,000	Geographical coordinates	Geographical location	Lithological composition	Geological age and age determinations	Magmatic series	Tectono-magmatic setting and type of magma	Note (comments and references)
6	Kuh-e Sefid Donbeh	Saqand Ardakan	55 25 30 32 58 03	North of Sefid Donbeh Mount	Granite and granodiorite	Late Precambrian 554 Ma.	Alkaline-Calc-alkaline	S and I	Hassanzadeh et al. (2008)
7	Esfordi	Esfordi Ravar	55 33 00 31 47 00	The west of Esfordi mine, Around Bafgh	Syenite, Syenogranite	Precambrian-Cambrian	Alkaline-Calc-alkaline	Within plate I	Sharifi (1997)
8	Chapedoni	Saqand Ardakan	55 15 01 32 49 57	Ardakan area	Anatectic granite and Diorite	Precambrian	Alkaline and Calc-alkaline	I and S	–
					Granite and granodiorite	Late Precambrian			Hassanzadeh et al. (2008)
9	Zarigan	Saqand Ardakan	55 09 03 32 32 34	Bafgh area-Chadormalu	Granite and granodiorite	Precambrian-Early Cambrian	Alkaline	I Rift-related	Ramezani and Tucker (2003)
10	Narigan	Esfordi Ravar	55 30 36 31 51 27	Bafgh district	Granite, granodiorite, tonalite, diorite to gabbro	Precambrian-early Cambrian 527 ± 7 Ma. U–Pb zircon 530–520 Ma	Calc-alkaline to Alkaline	I Rift-related	Mousavi Makui (1998) Ramezani (1997)
11	Khar Turan	Semnan	56 02 16 35 51 01	15 km SSW Biarjmand	Granite	Neoproterozoic-early Cambrian (552–534 Ma.)	Calc-alkaline	Alkaline to calc-alkaline	Extentionl within plate
12	Hamijan	Behabad Esfordi	55 57 40 31 46 38	12 km southwest of Behabad	Granite		Calc-alkaline to shoshonite	S to I type Rift-related Post collosion	Mohammadi et al. (2016)

(continued)

Table 5.1 (continued)

Sr. No.	Name	Geological Map 1:100,000 1:250,000	Geographical coordinates	Geographical location	Lithological composition	Geological age and age determinations	Magmatic series	Tectono-magmatic setting and type of magma	Note (comments and references)
13	Kuh-e Khoshoumi	Zarrin Ardakan	54 38 54 32 39 20	Saqand-Posht-e Badam area, west of Kuh-e Khoshoumi	Granite, granodiorite and tonalite	In older reports: Precambrian and In recent reports: 44.3 ± 1.1 Ma. U–Pb zircon	Calc-alkaline-alkalin	Continental arc	Ramezani and Tucker (2003)
14	Shotor Kuh	Torud	56 02 16 35 51 01	Northeast Torud	Gabbro and monzonite	Granite orthogenesis (598–509 Ma.) Felsic (41.3 ± 0.5 Ma.)	Calc-alkaline	S type	Hassanzadeh et al. (2008)
15	Robatkhan (Kuh-e Lakharbakhshi)	Tabas Robat Khan	56 14 01 33 25 27	Robatkhan-e Tabas	Granite to granodiorite	Late Precambrian Pre-Ordovisian	Calc alkaline to Alkaine	I type Rift-related?	–
16	Chevarzagh	Central Soltanieh mountains	56 38 30 30 06 59	The south of Kerman, in Biduiyeh area	Granite	Neoproterozoic	Calc-alkaline	A and S-type granite (syn-collision)	Hasanzadeh et al. (2008)
17	Delbar	Bastam Gorgan	55 25 09 36 17 26	150 km southeast of Shahrud	leucogranite	534 Ma. U–Pb	Alkaline	Extentional	Hassanzadeh et al. (2008)

Central Iran

Mesozoic

Sr. No.	Name	Geological Map 1:100,000 1:250,000	Geographical coordinates	Geographical location	Lithological composition	Geological age and age determinations	Magmatic series	Tectono-magmatic setting and type of magma	Note (comments and references)
18	Mashhad	Mashhad	59 37 40 36 08 37	South and southwest of Mashhad	Granitoid (porphyritic granite, granodiorite, Tonalite)	Permian to Late Triassic 211–215 Ma. and Hercynian phases and Early Cimmerian	Calc-alkaline	S and I Continental collision	Mirnejad (2000)

(continued)

Table 5.1 (continued)

Sr. No.	Name	Geological Map 1:100,000 1:250,000	Geographical coordinates	Geographical location	Lithological composition	Geological age and age determinations	Magmatic series	Tectono-magmatic setting and type of magma	Note (comments and references)
19	Khajeh Morad	Mashhad	59 32 00 36 07 07	South and southeast of Mashhad	Granodiorite and Granite	Late Paleozoic–Triassic 205 Ma. U–Pb	Calc-alkaline	S and I Continental collision	Karimpour et al. (2011)
					Granodiorite	Miocene 17 Ma.			
20	Vakilabad	Torqabeh Mashhad	59 23 03 36 21 02	North of Mashhad	Tonalite, granodiorite	Late Paleozoic–Triassic	Calc-alkaline	S and I Continental collision	CI
21	Esmailabad	Saghand, Zarrin and Ardakan	55 26 36 32 58 52 54 38 30 32 39 28	50 km northwest of Saghand-Posht-e Badam	Biotite-granite	After Permian–Before Cretaceous 218 ± 3 Ma. U–Pb zircon	Calc-alkaline	Continental arc I	Ramezani and Tucker (2003)
22	Kuh-e Darreh Anjir	Anarak Anarak	53 31 52 33 26 17	Kuh-e Darreh Anjir is next to the chah-e Derakhtak (Central Iran) in Anarak area	Plagiogranite, quartz diorite and tonalite, ultramafic rocks, gabbro, diabase	Triassic	Calc-alkaline	S, I	Ramezani and Tucker (2003) Sharkovski et al. (1984)
23	Arusan Mohammadabad	Arusan Jandagh	55 05 49 34 07 38	Anarak area, Jandaq district	Granodiorite, granite, Diorite and quartz diorite, sometimes gabbro	Jurassic-Eocene	Calc-alkaline	–	–
24	Zarrin	Zarrin Ardakan	54 38 42 32 42 59	80 km north of Ardakan, Zarrin area	Granite	After Jurassic-before Neogene	Alkaline	S Within plate post-orogenic	Omrani (1992)

(continued)

Table 5.1 (continued)

Sr. No.	Name	Geological Map 1:100,000 1:250,000	Geographical coordinates	Geographical location	Lithological composition	Geological age and age determinations	Magmatic series	Tectono-magmatic setting and type of magma	Note (comments and references)
25	Airacan	Posht-e Badam Tabas	55 31 34 33 03 33	The north of Khur and Biabanak	Granite	Middle Jurassic to early Cretaceous Rb–Sr, K–A	Calc-alkaline	Within plate S	Aistove et al. (1984)
26	Shir Kuh	Yazd	54 06 39 31 34 19	Southwest of Yazd	granite Granodiorite	Mid. Jurassic 166 ± 1.58 Ma.	Calc-alkaline	I and S Continental arc	Biotite and Garnet Chiu et al. (2013)
27	Toot-Anjiravand	Mehdi abad Ardakan	54 20 41 32 30 36	30 km northeast of Ardakan	Monzogranite	Jurassic	Calc-alkaline	continental arc I	Kanaanian (2001)
28	Khur	Bayazeh Khur	55 28 34 33 01 50	Khur area in Isfahan province	Granite and Syenite	Late Jurassic	Alkaline	S-type granite A-type Syenite	Sharkovski et al. (1984)
29	Torbat-e Heydarieh	Torbat-e Heydarieh	59 13 02 35 25 13	15 km north of Torbat-e Heydarieh	Ultramafic (Dunite, Harzburgite and …) to mafic (Gabbro, Diabas) and Quartz diorite rock	Late cretaceous	Gabbro and Diabas (*Toleitic series*), Quartz diorite (Calc-alkaline)	Related to ophiolite complex	In Associated with ophiolites Torbat-e Heydarieh
30	Sabzevar	Sabzevar Sabzevar	57 47 14 36 19 02	Sabzevar	Mafic rocks to Intermediate-rocks (Gabbro and diabases) and acidic rocks (quartzdiorite and Granophyre)	Late cretaceous	Microgabbro (Tholeitic) Other Stones (Calc-alkaline)	–	in associated with Sabzevar colored melange

(continued)

Table 5.1 (continued)

Sr. No.	Name	Geological Map 1:100,000 1:250,000	Geographical coordinates	Geographical location	Lithological composition	Geological age and age determinations	Magmatic series	Tectono-magmatic setting and type of magma	Note (comments and references)
Central Iran									
Cenozoic									
31	Ghohroud (Vash)	Natanz Kashan	51 36 01 33 39 38	South of Kashan	Granite and granodiorite	Early Oligocene to Middle Miocene 21.2 ± 0.3 Ma.	Calc-alkaline	I active continental margin	Chiu et al. (2013) Abdul Nejad Mamghani (2007)
					Microgabbro	Late Paleocene			
32	Khur	Kabudan Khur	54 13 12 33 24 46	Khur Area, south of Dasht-e Kavir	Granite and granodiorite	Eocene	Calc-alkaline	I active continental margin	–
33	Kuh-e Latif	Latif Kashan	52 10 43 33 58 40	Around of Kashan	Gabbro and diabase	Eocene	Calc-alkaline	Active continental margin	–
34	Toveireh	Anarak Nakhlak	53 52–54 00 33 53– 54 00	Southwest of Jandagh	Grandiorite and granite	Eocene	Calc-alkaline	I Syn-collision	Sargazi And.Torabi (2017)
35	Khonj	Khur Ordib	54 57 06 33 17 03	In Khonj village, east of Bayazeh	Grandiorite and granite	Eocene	Calc-alkaline	Continental margin arc	–
36	Baghou	Kaboudar Ahang Kaboudar Ahang	48 39 17 35 27 14 48 39 31 35 27 22	Northwest of Kaboudar Ahang	Microgranodiorite	Middle Eocene U–Pb zircon 43.4 ± 1.3 Ma. 40Ar/39Ar 43.4 ± 1.3 Ma.	calc-alkaline to shoshonitic	I-type	Niroomand et al. (2018)
					Diorite	38.0 ± 0.87			
37	Sabzevar	Sabzevar Bashtin	57 08 20 36 15 19	East of Sabsevar with ophiolite mélange	Rhyolite to dacite	48 Ma.	Calc-alkaline	Adakitic Post Ophiolite Dacitic and rhyolitic dom	Jamshidi et al. (2015)

(continued)

Table 5.1 (continued)

Sr. No.	Name	Geological Map 1:100,000 1:250,000	Geographical coordinates	Geographical location	Lithological composition	Geological age and age determinations	Magmatic series	Tectono-magmatic setting and type of magma	Note (comments and references)
38	Panjkuh	Gorgan Damghan	54 26 36 46	50 km Southeast of Damghan city	Pyroxene syenite, Biotite syenite and altered igneous rock (monzonite)	Eocene	Calc-alkaline to Shoshonitic	I-type Extensional?	Sheibi (2013)
39	Gholeh Kaftaran	Troud			10 km northwest of Troud	Diorite to granite	Eocene	Subvolcanic	I-type Extensional?	Ghorbani (2005)
40	Kal-e Kafi	Troud Kalateh	54 15 35 25	60 km northeast of Anarak	gabbro to microgranite	Late Eocene–Oligocene	Calc-alkaline	I type	Ranjbar et al. (2011)
41	Sinaqoun	Neauphle-le-Chateau Qom	50 49 10 34 14 21	Northwest of Fordu-Kermajegan	quartz granodiorite and syenite	After Eocene	Calc-alkaline	Continental margin arc	–
42	Qamsar (Kolah Barfi Mount)	Kashan	51 21 59 33 40 14	Qamsar (Golestaneh-Jowsheqan) Kashan	dolerite	After Eocene	Calc-alkaline	I	Ghorbani (2014)
43	Gojed (Kuh-e Sareh)	Kajan Naein	52 48 12 32 46 25	100 km east of Isfahan and 40 km southwest of Naein	Monzogabbro, monzogranite, granodiorite and granite	After Eocene (Pliocene)	Calc-alkaline	I Magmatic arc	Mansouri Esfahani (1993) Jabbari et al. (2010)
44	Torud Moalleman	Torud Moalleman	54 31 58 35 20 44	11 km north of Moalleman	microgranite, micro granodiorite, micro monzodiorite	Late Eocene–Oligocene	Sub-alkaline, Calc-alkaline	I Within plate	–
45	Qaleh Khargoushi	Sarv-e-bala Naein	53 09 37 32 11 25	Northeast of Batlaq-e-Gavkhuni	Granite	Eocene-Oligomiocene		S	Torabi (1997)

(continued)

Table 5.1 (continued)

Sr. No.	Name	Geological Map 1:100,000 1:250,000	Geographical coordinates	Geographical location	Lithological composition	Geological age and age determinations	Magmatic series	Tectono-magmatic setting and type of magma	Note (comments and references)
46	Nivesht	Saveh	50 08 05 35 08 20	Northwest of Saveh, north of Nivesht	Granodiorite	Oligocene	Calc-alkaline	I Magmatic arc, active continental margin	Helmi (1991)
47	Biduiyeh	Bagheyn Rafsanjan	56 38 30 30 06 59	The south of Kerman, in Biduiyeh area	Gabbro	Oligo-Miocene	Calc-alkaline	Magmatic arc I	
48	Tajareh	Kashan	51 24 19 33 41 25	Qamsar area-Ghohroud-south of Tajareh	Diorite	Oligocene–Miocene	Calc-alkaline	I Active continental margin	–
49	Kalijan	Natanz Kashan	51 37 57 33 39 14	Northwest of Natanz Kuh-e-Hilmand	Granodiorite and Quartz diorite	Oligocene–Miocene	Calc-alkaline	I active continental margin	Amidi (1977)
50	Karkas	Tarq Kashan	51 46 09 33 28 02	13 km southwest of Natanz	Granodiorite and Granite	Oligocene–Miocene	Calc-alkaline	I active continental margin	Amidi (1977)
51	Naraq	Neauphle-le-Chateau Qom	50 53 12 34 00 36	Naraq district	Micro gabbro and gabbro diorite	After Eocene, before middle Miocene	Calc-alkaline	I	–
52	Gavkhouni	Naein	53 01 17 32 39 19	20 km south of Naein	Diorite and Granite / Andesite	Eocene-Miocene / 35 Ma.	Calc-alkaline	I active continental margin	Chiu et al. (2013)
53	Kalout-Chatak	Bayazeh Khur	55 19 12 33 00 14	Khur area	Granite and granodiorite	Late Proterozoic	Alkaline	I and S	–

Table 5.1 (continued)

Sr. No.	Name	Geological Map 1:100,000 1:250,000	Geographical coordinates	Geographical location	Lithological composition	Geological age and age determinations	Magmatic series	Tectono-magmatic setting and type of magma	Note (comments and references)
Sanandaj–Sirjan zone									
Precambrian									
54	Pichaqchi (Qareh Zaj)	Shahin Dezh Takab	46 51 53 36 49 15	30 km east of Shahin Dezh	Granodiorite	Late Precambrian?, Cretaceous, Paleocene 70 Ma.	Alkaline, calc-alkaline	I and S-type granite post-collision	Kholghi (1991)
55	Bubaktan	Sonqor Kermanshah		31 km west of Sonqor and in the Bubaktan county	Biotite granite	544 Ma. U–Pb	Calc-alkaline	–	Hassanzadeh et al.(2008)
56	Chah Khatoon	Delijan Golpayegan	50 41 14 33 42 07	11 km northwest of Muteh	Biotite leucogranite leucogranite	596 Ma. U–Pb Pre-Permian (?)	–	–	Hassanzadeh et al. (2008) Abdollahi et al. (2009)
57	Muteh	Delijan Golpayegan	50 41 12 33 42 09	10 km of Muteh village	granite leucogranite	Precambrian (547–605, Pb–U Zircon) Pre-Permian (?)	Alkaline	A post-collision	Hassanzadeh et al. (2008) Abdollahi et al. (2009)
58	Darreh-Ashki and Senjedeh	Delijan Golpayegan	50 41 12 33 42 09	In the Muteh area	leucogranite	Pre-Permian (?)	Alkaline	A post-collision	Abdollahi et al. (2009)
59	Golpayegan	Mahallat Golpayegan	50 16 43 33 33 29	15 km Golpayegan-Khomeyn road	Diorite and pegmatite granite	Precambrian?-Jurassic?	Alkaline	–	–

(continued)

Table 5.1 (continued)

Sr. No.	Name	Geological Map 1:100,000 1:250,000	Geographical coordinates	Geographical location	Lithological composition	Geological age and age determinations	Magmatic series	Tectono-magmatic setting and type of magma	Note (comments and references)
Sanandaj–Sirjan zone									
Late Paleozoic–early Triassic									
60	Hasanrobat	Dehaq Golpayegan	50 51 13 33 25 11	15 km north of Golpayegan	biotite-bearing syenogranite, alkali-feldspar granite and minor amount of monzogranite	Late Precambrian	alkali-calcic to calc-alkalic	Extensional environment	Mansouri Esfahani (2000)
						Lower .Permian 289 ± 5 Ma.		A type, continental break-up	Alirezaie and Hassanzadeh (2011)
61	Hasansalaran	Mahabad Saqqez	46 14–46 29 36 00– 36 07	20 km southeast of Saqqez	alkali-feldspar granite, syenogranite and alkali-feldspar quartz syenite	U–Pb (360.3– 360.7) Late Devonian 103 ± 3.6 Ma. Cretaceous	ferroan alkali-calcic, peralkaline	A type/within plate extensional	Athari et al. (2006) Kazemi et al. (2016)
					Monzogranite, granodiorite, tonalite		Calc-alkaline metaluminous	I type/volcanic arc environment	
62	Baba Ali	Hamadan Tuyserkan	48 10–48 15 34 40– 35 00	39 km to the northwest of Hamadan, east of Almogholagh batholite	quartz-syenite	Late paleozoic	Alkaline	Extentional	Zamanian and Radmard (2016)
63	Ghareh Bagh	Jolfa Tabrize	45-45 12 38–38 09	60 km northeast of Urmia	Diorite	300.7 ± 1.5 Ma.	Sub-alkaline to tholeiitic	Active continental margin	Asadpour et al. (2013)
					Gabbro	300.0 ± 1.3 Ma.			
					Leucogranite	558.6 ± 3.8 Ma.			

(continued)

Table 5.1 (continued)

Sr. No.	Name	Geological Map 1:100,000 1:250,000	Geographical coordinates	Geographical location	Lithological composition	Geological age and age determinations	Magmatic series	Tectono-magmatic setting and type of magma	Note (comments and references)
Sanandaj–Sirjan zone									
Mesozoic									
64	Sikhvoran	Shamil Bandar Abbas	56 34 10 27 53 40	South Baft country in Kerman province	Gabbro, Ultramafic to granite	Triassic 100 ± 1.5 Ma.	Tholeiitic	Oceanic crust and continental rifting	*Sabzehie (1974) assigned them to a Triassic age*
65	Abadeh	Khezrabad Abadeh	53 54 22 31 47 32	Abadeh	Basic intrusive rocks	Triassic	Calc-alkaline	Active continental margin	–
66	Bamak	Khezrabad Abadeh	53 55 10 31 52 01	70 km west of Yazd	Monzogranite to granodiorite	After Cretaceous K–Ar	High-K calc-alkaline	Active continental margin I	–
67	Saman	Shahr-e Kord Chadegan	32 35–32 45 50 40–51 00	North of Saman	Granite, granodiorite, quartzmonzonite	182 ± 4 Ma. Late jurassic	High-K calc-alkaline	I type Subduction and active continental margin	Hosseini and Ahmadi (2016)
68	Band-e Now (Dehbid)	Dehbid Eghlid	53 24 34 30 39 11 53 12 54 30 46 53	20 km northeast of Dehbid in Fars Province	Granite, Diorite	Middle Triassic-Jurassic	Calc-alkaline	I active continental margin	Based on geological evidence –
69	Siah kuh	Neyriz	54 14 59 29 18 54	Southwest Neyriz	Granitoid with Ophiolitic melange	Triassic-Cretaceous 199 ± 30 Ma.	Sub alkalin to Calc-alkaline	A-type granite, continental magmatic arc, Neo-Tethys subduction (?Late Triassic)	Arvin et al. (2007)

(continued)

Table 5.1 (continued)

Sr. No.	Name	Geological Map 1:100,000 1:250,000	Geographical coordinates	Geographical location	Lithological composition	Geological age and age determinations	Magmatic series	Tectono-magmatic setting and type of magma	Note (comments and references)
70	Aligudarz (Mollataleb) (Khorheh)	Aligudarz Golpayegan	33 23–33 34 49 35–49 47	North of Aligudarz (Khorheh) Northwest of Aligudarz (Mollataleb)	Granite and granodiorite quartz–diorites, granodiorites	Jurassic (Liassic –Dogger) Zircon U–Pb data on the granites a crystallization age of 165 Ma. (middle Jurassic) Inherited grains g in age from 180 Ma-190 Ma.	Calc-alkaline and alkaline high calc alkaline series	Subduction/active continental margin	Thiele et al. (1968) knows that these mass granitoides are the same age of Boroujerd and Hamedan, means after liassic, before Cretaceous.-Esna-Ashari et al. (2012)
71	Mollataleb (Darre Bagh)	Aligudarz Golpayegan	49 37 56 33 37 20	10–16 km northwest of Aligudarz	Granite, granodiorite and Monzogranite	Post-Cretaceous	Calc-alkaline	S	Hashemi (2016) Bagherian et al. (2006)
72	Aligudarz (Dehe no)	Khoramabad Durod	33 24–33 27 49 20–49 25	20 km northeast of Aligoudarz	Granodiorite, tonalite	Mesozoic	cCalc alkaline	I & S	Mohammadi et al. (2011)
73	Chah Dezdan Chah Bazargani	Gur-e-sefid Neyrize	54 42 07 29 43 51	36 km southwest of Shahr-e-Babak	Gabbro to Granite, Batolite	Jurassic 165.3 ± 5.3 Ma. 164–173 Ma. 40 K-40Ar muscovite	Calc-alkaline	Active continental margin	Sheikholeslami et al. (2003)
74	Almogholagh	Tuyserkan Hamedan	48 08 59 34 54 01	The northeast of Sonqor, 30 km west of Hamedan, 12 km north of Asad abad	Gabbro, Diorite, Quartzsyenite (1) Felsic rocks: Granite	Jurassic 140 Ma. 148–143 Ma.	Calc-alkaline high-k calc alkaline	Mantle wedge melting, Continental arc I type and some A type	Shahbazi (2011) Jamshidibadr et al. (2018)

(continued)

Table 5.1 (continued)

Sr. No.	Name	Geological Map 1:100,000 1:250,000	Geographical coordinates	Geographical location	Lithological composition	Geological age and age determinations	Magmatic series	Tectono-magmatic setting and type of magma	Note (comments and references)
					(2) Mafic rocks: Quartz diorite and diorite gabbro			Island arc of subduction	Jamshidibadr et al. (2018)
75	Kolah Ghazi	Shahreza Isfahan	51 52 46 32 20 57	50 km southeast of Isfahan	Granodiorite, monzogranite, syenogranite	Jurassic 164.6 ± 2.1 Ma. by U–Pb method on zircon	Calc-alkaline	Post-orogenic/S	Chiu et al. (2013)
76	Borujerd	Shazand Khorramabad	49 06 40 33 41 04 49 01 12 33 48 23	Between Malayer and Borujerd The east and southeast Borujerd and along the road of Borujerd-Arak	Granite, granodiorite, quartz-diorite and monzodiorite	169.6 ± 0.3 Ma. And 169.0 ± 1 Ma. Often middle Jurassic 170 Ma.	Calc-alkaline	Active continental margin/I	Goudarzi (1995) PourJahromi (1994) Mahmoudi (2011) Ahmadi Khalaji (2006)
77	Tutak	Surian Eghlid	53 53 23 30 17 12	East of Surian (Fars province) 250 km northeast of Shiraz	Granite, granodiorite and tonalite	170 Ma.	Calc-alkaline	I	Ahmadi (2004) Hosseini (2011)
78	Kuh-e Darvazeh	Qorveh Sanandaj	47 44 46 35 02 49 47 53 34 35 06 30	South of Qaleh village in Qorveh distict	Gabbronorite to quartz monzonite Olivine gabbronorite to Hornblende gabbro, granite and leucocratic granitoid	171–145 Ma. Jurassic -Cretaceous 149–157 Ma. (gabbro), 163 Ma. (granite) and 154 Ma. (leucocratic granitoid)	Calc-alkaline	I Subduction-related settings	Sang-Qaleh (1995) Sheikh Zakariaee (2008)

(continued)

Table 5.1 (continued)

Sr. No.	Name	Geological Map 1:100,000 1:250,000	Geographical coordinates	Geographical location	Lithological composition	Geological age and age determinations	Magmatic series	Tectono-magmatic setting and type of magma	Note (comments and references)
79	Cheshmeh Qassaban	Tuyserkan Hamedan	48 25 04 34 51 07	Hamedan area	Gabbro	Jurassic	Alkaline and Calc-alkaline	-	It is a part of Alvand Complex
80	Ghuri	Ghuri Neyriz	54 06 28 29 35 13	The northwest of Neyriz	Volcanic complex	147 ± 1	-	-	Fazlnia et al. (2007)
81	Kuhban-Siri	Sonqor Kermanshah	47 52 24 34 45 02	Kermanshah province, The area between Asadabad and Sonqor	Granodiorite	Middle Jurassic	Calc-alkaline	Active continental margin/I	-
82	Yones (Samen)	Malayer Hamedan	48 41 33 34 09 27 48 35 45 34 16 57	15 and 20 km southwest of Malayer	Granite, granodiorite	Middle Jurassic	Calc-alkaline	Active continental margin/I and S	Goudarzi (1995)
83	Sargaz	Esfandaqeh Sabzevaran	57 21 03 28 41 03	Sargaz, Esfandeqeh	Coarse-grained granite	Late Jurassic 175 ± 8.1 Ma. Mid. Jurassic?	Calc-alkaline	I	Chiu et al. (2013) Ahmadipour and Rostamizadeh (2012) (Leucocratic)
84	Astaneh	Shazand Khorram Abad	49 18 41 33 49 51	Shazand area, Astaneh Village	Granite, granodiorite and slightly Quartzdiorite	Jurassic-Late Cretaceous	Calc-alkaline	Active continental margin/I,S	Afshooni (2007)
85	Nezam abad	Khorramabad Shazand	49 10 33 40	Southwest of Arak	Quartzdiorite, granodiorite	Jurassic-Cretaceous	Calc-alkaline	Subduction and active continental margin/I and S	Jafari et al. (2018) Motori (2010)

(continued)

Table 5.1 (continued)

Sr. No.	Name	Geological Map 1:100,000 1:250,000	Geographical coordinates	Geographical location	Lithological composition	Geological age and age determinations	Magmatic series	Tectono-magmatic setting and type of magma	Note (comments and references)
86	Moradbeig	Tuyserkan Hamedan	48 28 12 34 42 04	Moradbeig area, south of Hamedan	Granite	Jurassic-Cretaceous 170 Ma. 70–90	Sub alkaline to Calc-alkaline	Continental arc/I	Baharifar et al. (2004) Shahbazi (2011)
87	Shahin Dezh	Chapan Takab	46 35 18 36 25 30	South of Shahin Dezh	Granite and granodiorite	After Cambrian-Before Cretaceous	Alkaline	SandI	Kholghi Khasraghi (1991)
88	Eghlid	Dehbid Eghlid	53 24 34 30 39 16	Eghlid area	Granite to gabbro	Late Cretaceous	Calc-alkaline	Active continental margin/I	–
89	Takab	Chapan Takab	46 45 33 36 26 02	West of Takab, the south of Takab–Shahin Dej road	Granodiorite-monzogranite	Postcretaceous-Pre-Oligo-Miocene	Calc-alkaline	I	Farrokhmanesh (1998)
90	Balestan	Oshnavieh Urmia	45 23 37 37 09 32	The West of Lake Urmia	Alkali granite, granodiorite, diorite and quartz-monzonite	Early Cretaceous to earlier than Miocene (It was introduced ax Precambrian)	Calc-alkaline	within plate A	Houshmand manavi (2003)
91	Qamishlu	Lenjan Isfahan	51 29 10 32 02 28	70 km southwest of Isfahan	Diorite	94.8 Ma.			Ghalamghash et al. (2009)
92	Zeid Kandi	Chapan Takab	46 47 12 36 26 03	30 km southeast of Shahin Dezh	Alkali granite to syenogranite	Cretaceous	Calc-alkaline	S Continental collision or late orogenic phase	Noraei (1998)

(continued)

Table 5.1 (continued)

Sr. No.	Name	Geological Map 1:100,000 1:250,000	Geographical coordinates	Geographical location	Lithological composition	Geological age and age determinations	Magmatic series	Tectono-magmatic setting and type of magma	Note (comments and references)
93	Serow	Gajgin Serow	44 44 26 37 48 25	North and northeast of Gajgin	Granite, granodiorite and diorite	Cretaceous-Paleocene	Alkalin-Calc-alkaline	I	this masse can be seen in other parts of this map
94	Chah Harigan	Gur-e-sefid Neyrize	54 34 01 29 31 43	43 km northeast of Neyrize	Igneous and metamorphic complex	159–204 Ma. 40 K-40Ar	–	–	Sheikholeslami et al. (2003)
					Amphibolites	103 Ma		–	
95	Alvand	Hamedan Tuyserkan	48 26 33 34 45 24	The southeast and southwest of Hamedan	Based on past information: granite, granodiorite and diorite (Late Cretaceous-Early Paleocene) 64 Ma. U–Pb zircon		Calc-alkaline	Hybrid continental margin arc/I and S	Shahbazi et al. (2010) Mahmoudi (2011)
					Gabbro	166.5 ± 1.8 Ma.			
					Granite	163.9 ± 0.9 161.7 ± 0.6 Ma.			
					Leucocratic granitoid	154.4 ± 1.3 153.3 ± 2.7 Ma.			
96	Maraq-Cheshmeh sefid	Mahallat Golpayegan	50 17 10 33 33 36	North of Golpayegan-southeast of Khomeyn	Monzogranite	Mesozoic	Calc-alkaline-Alkaline	Post-collision/I	Ebrahimi (1991)
97	Naqadeh	Mahabad	45 23 57 36 32 27	5 km Naqadeh county and 30 km southeast of Piranshahr	Olivine gabbro, gabbro, diorite, anorthite, syenite	After Cretaceous 96 ± 2.3 Ma.	Tholeiitic and Alkaline (Syenite)	I (Acidic rocks)	In some parts, tends to change into magmatic amalgamation, Ghalamghash (2002) Mazhari et al. (2009)

(continued)

Table 5.1   (continued)

Sr. No.	Name	Geological Map 1:100,000 1:250,000	Geographical coordinates	Geographical location	Lithological composition	Geological age and age determinations	Magmatic series	Tectono-magmatic setting and type of magma	Note (comments and references)
98	Azna	Golpayegan Golpayegan	49 31 00 33 28 44	7 km North of Azna	Granodiorite, monzogranite, granite and diorite	After Cretaceous-before Oligocene	Calc-alkaline	I type	Moazzen et al. (2004)
99	Piranshahr	Oshnavieh Mahabad	45 22 19 36 48 45	Shahrestan area, Piranshahr (West Azerbaijan)	Gabbro and monzonite	Ultramafic (40.7 ± 0.2 Ma.) Felsic (41.3 ± 0.5 Ma.)	Calc-alkaline	I	Mazhari et al (2009)
100	Robat Chah Ghand	Rabat Anar	54 37 47 30 02 34	Southwest of Shahr-e Babak, East of Harat (Yazd province), 30 km south of Robat	Wehrlite, Olivin-gabbro, monzogabbro, quartz monzonite	Upper Triassic -Jurassic 159.0 ± 40 Ar biotite	Alkaline	Mantel-derived continental rift	Chah Ghand intrusive is located in 50 km southeast of Rabat, Sheikholeslami et al. (2003)

Sananadaj–Sirjan zone

Cenozoic

101	Buin-Miandasht	Golpayegan	50 10 44 33 07 46	The southeast of Aligudarz, 5 km north of Buin-Miandasht	Syenogranite, alkali feldspar granite	Late Cretaceous-Early Paleocene	Calc-alkaline	Active continental margin/I	–
102	Saqqez	Saqqez Mahabad	46 20 46 36 03 15	20 km southeast of Saqqez, In Saqqez-Marivan road	Granite and granodiorite	Cretaceous-Paleocene	Calc-alkaline	Active continental margin/I	Amani (2000)

(continued)

Table 5.1 (continued)

Sr. No.	Name	Geological Map 1:100,000 1:250,000	Geographical coordinates	Geographical location	Lithological composition	Geological age and age determinations	Magmatic series	Tectono-magmatic setting and type of magma	Note (comments and references)
103	Deh-Zalou	Varcheh Golpayegan	49 58 02 33 49 58	Varcheh near Khomeyn	Biotite granite gneiss	Jurassic-Cretaceous-Paleocene	Calc-alkaline	Post-collision	This is a biotite bearing granite which is of a complete gneiss view
104	Baneh-Salouk	Baneh Marivan	45 45 00 35 57	Southwest of Baneh	Granite	Paleocene-Eocene	Calc-alkaline	I	Nogole Sadat and Houshmandzadeh (1974)
105	Penjwen	Marivan Marivan-Banehe	46 01 18 35 31 48	Kurdistan province, near the Iraqi border	Gabbro	Eocene–Oligocene	Calc-alkaline	Active continental margin	–
					Gabbro, granodiorite, monzogranite and syenogranite	37.9 ± 1.9 Ma -granodierite (Zr) 39.3 to 35.6 Ma. (U-Th-Pb)	High k calc-alkaline	S type, Syn-collision	Ranin, et al. (2010) Sepahi et al.(2014)
106	Kharsareh (Kharzahre)	Harsin Kermanshah	47 51 34 34 21 32	80 km southwest of Hamedan	Gabbro	Eocene–Oligocene	Calc-alkaline	Active continental margin	–
107	Miyan Rahan	Harsin Kermanshah	47 44 35 34 24 19	South of Sonqor	Gabbrodiorite and syenite	Eocene–Oligocene	Calc-alkaline	–	Orogenic
108	Kamyaran	Kamyaran Kermanshah	46 47 43 34 58 60	North of Kamyaran in Kurdistan-Kermanshah border	Gabbro to diorite and basalt	54.6 ± 1.8 Ma. U–Pb zircon	Calc-alkaline	I active continental margin	Azizi et al. (2011)
109	Kharsareh (Kharzahre)	Harsin Kermanshah	47 51 34 34 21 32	80 km southwest of Hamedan	Gabbro	Eocene–Oligocene	Calc-alkaline	Active continental margin	

(continued)

Table 5.1 (continued)

Sr. No.	Name	Geological Map 1:100,000 1:250,000	Geographical coordinates	Geographical location	Lithological composition	Geological age and age determinations	Magmatic series	Tectono-magmatic setting and type of magma	Note (comments and references)
110	Takab	Chapan Takab	46 36 16 36 25 04	South west of Takab	Syenogranite	Infracambrian?	Calc-alkaline	I Within plate	–
					Diorite	Oligocene 29.9 ± 0.5 Ma.			
Urmia-Dokhtar									
Cenozoic									
111	Bostanabad	Bostanabad Mianeh	46 52 02 37 47 48	About 45 km south of Tabriz	Quartz monzonite	Eocene–Oligocene	–	Active continental margin I	The magma in some extent indicates a magmatic amalgamation
112	Qarah Aghaj	Gangachin Serow	44 48 31 37 46 51	West Azerbaijan Province, northwest of Urmia	Wehrlite, gabbro to gabbrodiorite	After Cretaceous	Alkaline	I subduction	– –
113	Noosha	Jebal-e Barez Jiroft	58 26 43 28 31 14	West of Alam-Küh	Coarse grained Granodiorite	After Cretaceous-Eocene U–Pb zircon	Calc-alkaline	I Continental	Chiu et al. (2013)
					Gabbro	Late Cretaceous-Eocene 76.6 ± 0.9 Ma.			
114	Hajiabad	Khiaraj Saveh	49 57 54 35 37 16	70 km southwest Buin Zahra	Synogranite, monzogranite, granodiorite, quartz monzonite	39.9 Ma. K–Ar	Calc-alkaline	–	Karimpour et al. (2011)
115	Dudhak	Neauphle-le-Château Qom	50 34 06 34 05 10	South of Qom, Northeast of Mahallat	Granite	Late Eocene–Oligocene	Calc-alkaline to Tholeiitic	Volcanic arc I, active continental margin	–

(continued)

Table 5.1 (continued)

Sr. No.	Name	Geological Map 1:100,000 1:250,000	Geographical coordinates	Geographical location	Lithological composition	Geological age and age determinations	Magmatic series	Tectono-magmatic setting and type of magma	Note (comments and references)
116	Sang-e Sayyad	Bardsir Sirjan	56 46 24 29 40 27	30 km southeast of Bardsir (Kerman province)	Granite, granodiorite, monzogranite and tonalite	Eocene–Oligocene	Calc-alkaline	I continental arc	Sajjadi Nasab (2002)
117	Avaj	Kabudrahang	48 54 06 35 22 42	The west of Avaj, Around shaneh, Gerehak and Qozlu Villages	Gabbro	Eocene–Oligocene	Alkaline	Continental arc/I	Mohammadi 2005
118	Ardestan	Ardestan Kashan	52 08 10 33 14 50	Around east and west of Ardestan County, (Dovarjin)–Taqiabad	Quartz-gabbros, diorite, granodiorite and granite	Late Eocene-Early Oligocene	Calc-alkaline	Active continental margin I	Acidic magma by mafic magma has a dual mode
119	Saveh (Zarand)	Saveh Saveh	50 08 20 35 09 41	Northwest of Saveh Zarand of Saveh area (Mamuniye)	Granodiorite monzogranite, granite and monzonite	Early Oligocene 39.2 ± 3.2 Ma.	Calc-alkaline	I active continental margin	–
120	Shahr-e Babak	Rabat Anar	54 49 57 30 19 41	Around of Shahr-e Babak (Chah-e Bagh)	Granite and gabbro	Eocene–Oligocene	–	I active continental margin	–
121	Qoroghchi	Shahin Dezh Takab	46 33 23 36 59 30	35 km east of Miandoab	Monzogabbro or syenogabbro and syenogabbro	Earlier than Eocene-Upper Oligocene-Lower Miocene	Calc-alkaline	I active continental margin	–
122	Eshtehard (Hajiabad, Dehbala, Razak and Darband)	Hashtgerd Saveh	50 43 19 35 41 40	Eshtehard and Saveh areas	Granite and diorite	Early Oligocene	Calc-alkaline	Active continental margin	Safarzadeh (2006)

(continued)

Table 5.1 (continued)

Sr. No.	Name	Geological Map 1:100,000 1:250,000	Geographical coordinates	Geographical location	Lithological composition	Geological age and age determinations	Magmatic series	Tectono-magmatic setting and type of magma	Note (comments and references)
123	Darreh Zereshk	Khezrabad Abadeh	53 52 11 31 36 02	20 km northeast of Dehshir and south of Aliabad	Granite and granodiorite	Oligo-Miocene	Calc-alkaline	I Continental arc margin	–
124	Khezrabad	Khezrabad	53 38 02 31 38 25	Abadeh (Kuh-e Bandazan)	Granite, granodiorite to diorite	Olig-Miocene	Calc-alkaline	I Continental arc margin	The northeast of Quadrangle Abadeh
125	Tajareh	Kashan	51 24 19 33 41 25	Qamsar area-Ghabrud-south of Tajareh	Diorite	Oligocene–Miocene	Calc-alkaline	I Active continental margin	–
126	Qamsar	Kashan	51 24 42 33 41 44	South of Qamsar	Tonalite, micro pegmatite	Oligo-Miocene	Calc-alkaline	I Magmatic arc	–
127	Darrehzar	Pariz Sirjan	55 51 40 29 57 04	14 km south of Sarcheshmeh mine	Granite porphyry	Miocene	Calc-alkaline	Continental arc I	Based on Studies of Darrehzar age
128	Sarcheshmeh	Pariz Sirjan	55 51 30 29 57 06	20 km south of Rfsanjan	Granodioritic porphyry stock, granitic late fine-grand porphyry and andesite to dacitic hornblende	(Miocene) 12.97 ± 0.23 12.37 ± 0.1 12.16 ± 0.8 Ma. Respectively	Calc-alkaline	I active continental margin and magmatic arc	Chiu et al. (2013) Aghazadeh (2015)
					Granodiorite	Oligocene–Miocene			
					Porphyry massive andesite	Late Eocene 37.6 ± 0.5 Ma.			

(continued)

Table 5.1 (continued)

Sr. No.	Name	Geological Map 1:100,000 1:250,000	Geographical coordinates	Geographical location	Lithological composition	Geological age and age determinations	Magmatic series	Tectono-magmatic setting and type of magma	Note (comments and references)
129	Amir Momeomenin	Sirjan Parize	55 54 21 29 58 09	Northeast of Pariz	Dacitic porphyry to rhyolite	Mio-pliocene	Calc-alkaline	Active continental margin and magmatic arc	–
130	Merdvar	Anar Shahr-e Babak	55 07 28 30 15 02	North of Shahr-e Babak	Dacitic porphyry	Pliocene	Calc-alkaline	I active continental margin and magmatic arc/ adackitic magmatism	–
131	Abdar	Shahr-e Babak Anar	55 19 00 30 18 25	30 km northeast of Shahr-e babak	Microdiorite	Miocene-Pliocene 0/1 Ma. ± 7/5	Calc-alkaline	Magmatic arc I	Chiu et al. (2013)
132	Rigan	Khaneh Khatun Bam	57 57 58 29 00 00	Southeast of Bam (Narmashir)	Granodiorite Basalt and rhyolite	Late Eocene–Oligocene 29.3 ± 0.2 Ma.	Calc-alkaline	I Magmatic arc	Aletaha (1993)
						Late Miocene 18.20 Ma.			Chiu et al. (2013)
133	Venarj	Neauphle-le-Chateau Qom	50 41 29 34 26 02	19 km southwest of Qom next to the Venarch Manganese Mine	Quartzdiorite, monzonite, monzodiorite and syenite	Eocene–Oligocene	Calc-alkaline	I Magmatic arc, active continental margin	Eslamizadeh (1993) UM
					Granitoid	Early Miocene 17.4 ± 0.2			

(continued)

Table 5.1 (continued)

Sr. No.	Name	Geological Map 1:100,000 1:250,000	Geographical coordinates	Geographical location	Lithological composition	Geological age and age determinations	Magmatic series	Tectono-magmatic setting and type of magma	Note (comments and references)
134	Iju	Koshkuieh Anar	54 56 48 30 30 26	Shahr-e Babak	Diorite	Miocene-Pliocene 9.2 ± 0.1 Ma. 9.27	Calc-alkaline	Magmatic arc I	Chiu et al. (2013) Mohammaddoost et al. (2017)
135	Katehkurha	Shahr-e Babak Anar	55 19 05 30 18 24	Shahr-e Babak	Porphyritic diorite	Miocene	Calc-alkaline	I active continental margin	Amidi (1977)
136	Meiduk	Shahr-e Babak Anar	55 10 14 30 25 19	Shahr-e Babak	Porphyritic granodiorite	Miocene 12.5 ± 0.1 Ma.	Calc-alkaline	I	Chiu et al. (2013) Alirezaie et al. (2012) Mohammaddoost et al. (2017)
137	Parkam	Anar Shahr-e Babak	55 10 14 30 25 28	3 Meiduk porphyry copper deposit	Dacitic porphyry to microdiorite and quartz dioritic porphyry	Late Miocene	high K calc-alkaline, metaluminous to peraluminous	I type subduction continental arc	Mohammadi Laghab et al. (2012)
	Sara	Anar Shahr-e Babak	55 08 10 30 26 27	4 km north of Meiduk	Porphyritic dacite				
138	Javazm, Dehaj and khabr	Anar Dehaj	54 45– 55 00 30 30–30 45	Northwest of Shahre Babak	Porphyritic dacite masses	Mio-Peliocene	adakitic composition alkaline magmatism	I type Active continental margin and belong to volcanic arc granitoides	Ghadimi (2016)

(continued)

Table 5.1 (continued)

Sr. No.	Name	Geological Map 1:100,000 1:250,000	Geographical coordinates	Geographical location	Lithological composition	Geological age and age determinations	Magmatic series	Tectono-magmatic setting and type of magma	Note (comments and references)
139	Narkuh	Anar Shahr-e Babak	55 00 50 30 26 21	Thare are 3 masses The coordinate is for the biggest one	Diorite, quartzdiorite, granodiorite	Pliocene	Calc-Alkaline and Shoshonitic	Subduction, I type Adakitic magmatism	Barahmand et al. (2018)
	Ayob Ansar	Anar Dehaj	54 51 33 30 37 35	North of Shahr-e Babak					
140	Dalli	Qom-Arak Salafchegan	50 32 34 27	60 km northeast of Arak	Diorite to granodiorite	40Ar/39Ar for porphyritic diorite: 21 Ma.	Calc-alkaline	Continental margin arc	Ayati et al. (2012) Ayati et al. (2008)
141	Aj Pain	Anar Dehaj	54 50 58 30 50 11	Northwest of Dehaj	Porphyritic dacite	Miocene	Calc-alkaline	Subduction, I type Adaitic magmatism	–
	Aj Bala		54 47 16 30 45 35						
142	Razi Abad (Madin)	Sabzevaran	57 41 24 28 58 30	20 km north of Jiroft	Porphyritic tonalite to granitic porphyry	Eocene–Oligocene	Calc-alkaline	Subduction	–
	Razi Abad (Tenarouye)	Sabzevaran	57 40 29 28 57 30						
143	Saridoun	Sirjan Pariz	55 52 27 29 58 26	Northeast of Pariz	Granodioritic porphyry stock, granitic late fine-graind porphyry and andesite to dacitic hornblende	(Miocene) 12.97 ± 0.23 12.37 ± 0.1 12.16 ± 0.8 Ma. Respectively	Calc-alkaline	I active continental margin and magmatic arc	–

(continued)

Table 5.1 (continued)

Sr. No.	Name	Geological Map 1:100,000 1:250,000	Geographical coordinates	Geographical location	Lithological composition	Geological age and age determinations	Magmatic series	Tectono-magmatic setting and type of magma	Note (comments and references)
144	Kerver	Jiroft Sabzevaran	57 45- 58 00 28 30- 29 00	Southeast of Jiroft	Porphyritic tonalite to granitic porphyry		Calc-alkaline	Subduction	Rasouli and Ghorbani (2012)
145	God-e Kelvari	Anar Dehaj	54 59 39 30 35 33	Shahr-e Babak	Diorite	Pb-U 16.36	Calc-alkaline	Subduction I	Mohammaddoost et al. (2017)
146	Keder	Anar Dehaj	54 45 05 30 36 59	Shahr-e Babak	Diorite to granodiorite	Pb-U 7.54 Ma.	Calc-alkaline	Subduction I	Mohammaddoost et al. (2017)
147	Sereno	Anar Robat	54 55 49 30 28 30	Shahr-e Babak	Diorite to granodiorite	Pb-U 10.2	Calc-alkaline	Subduction I	Mohammaddoost et al. (2017)
148	Kuh-e Kolah Sar	Tarq Kashan	51 46 09 33 28 02	10 km northwest of Morvarid defile	Gabbro	Cenozoic	Calc-alkaline	I active continental margin	–
149	Jebal-e barez-Delfard	Sabzevaran Sabzevaran	57 37 29 28 53 38	30 km northwest of Jiroft	Granite, granodiorite and tonalite	Late Eocene-Miocene	Calc-alkaline	I Active continental margin	The age of granite in 1:25,000 map of the area is mentioned Jurassic
150	Kuh-e Tezerj	Anar Anar	55 09 08 30 33 30	50 km north of Shahr-e Babak	Quartz monzonite	Neogene	Calc-alkaline	I Magmatic arc	–
Azerbaijan									
Precambrian									
151	Sarv-e Jahan	Soltaniyeh Zanjan	48 58 34 36 14 35	West of Soltaniyeh, in Sarv-e Jahan village	Alkali granite to syenogranite / Leucogranite	U–Pb zircon / 559 Ma.	Alkaline-Calc-alkaline	Rift-related	Hassanzadeh et al. (2008), Ghorbani (1999)

(continued)

Table 5.1 (continued)

Sr. No.	Name	Geological Map 1:100,000 1:250,000	Geographical coordinates	Geographical location	Lithological composition	Geological age and age determinations	Magmatic series	Tectono-magmatic setting and type of magma	Note (comments and references)
					Sheeted Biotite granite accompanied by chloritization alteration	554 Ma.			
152	Khalaj	Mahneshan Takab	47 34 08 36 43 26	North of Mahneshan	High-Na alkaline granite	Late Precambrian	Alkaline	Rift-related	
153	Mahneshan	Mahneshan Takab	47 32 48 36 45 05	Pari village	Holo leucocratic granite	Late Precambrian	Alkaline	Rift-related	Salehi (2019)
154	Doran	Zanjan	48 23 09 36 34 23	12 km south of Zanjan in Doran village	High-Na alkaline granite	Late Precambrian 567 Ma. U–Pb zircon	Alkaline and Calc-alkaline	A and S-type granite (Post-collision)	Valizadeh and Esmaili (1993) Hassanzadeh et al. (2008), Ghorbani (1999)
155	Salim Khan Yeylaghi	Shahin Dezh Takab	46 51 38 36 47 02	Salim Khan mountain, 25 northeast of Shahin Dezh	High-Na alkaline granite	Late Precambrian	Alkaline	Rift-related I and A	Ghorbani (1999)
156	Shah Bolagh-e Bala-Moghanlu	Zanjan	48 00 30 36 38 22	35 km southwest of Zanjan	High-Na alkaline granite	Precambrian 548 ± 27 Ma.	Alkaline	S and A-type granite Tensil	Ghorbani (1999)
157	Mahmudabad	Qujur Takab	47 42 09 36 25 44	Southeast of Dandi-East of Takab county	Granite to monzodiorite	Late Precambrian	Alkaline and Calc-alkaline	Extensional environment/I	Ahmadi Rohani (1999)

(continued)

Table 5.1 (continued)

Sr. No.	Name	Geological Map 1:100,000 1:250,000	Geographical coordinates	Geographical location	Lithological composition	Geological age and age determinations	Magmatic series	Tectono-magmatic setting and type of magma	Note (comments and references)
158	Inche	Mahneshan Takab	47 33 47 36 40 30	In Takab area, the north of Ghazi Village, Inche and northeast of Zeid Kandi	Anatectic granite	Late Precambrian?	Alkaline?	S, A	This kind of granite is also seen around Inche and the northeast of Zeid Kandi. Probably, it is a kind of Doran granite Jamshidibadr (2002), Ghorbani (1999)
159	Alam-Kandi	Takht-e Soleymān and Mah Neshan	47 29 58 36 42 12	In Mah Neshan area, Alam kandi road, Mah Neshan-Pari-Dandi-Aq Kandi	Na-alkali granite	Precambrian	Alkaline	A and S Tensil	–
160	Zarrinabad Mahmudabad	Qujur Takab	47 42 28 36 27 01	Southeast of Dandi	Gabbro, Gabbrodiorite and Granite	Late Precambrian? - Cretaceous?	Tholeiitic	I	–
161	Aq Darreh Bala	Shahin Dezh Takab	46 59 51 36 40 07	30 km north west of Takab	Granite, granite porphyry	Precambrian	Alkaline	A-type granite	Kholghi (1991)
Azerbaijan									
Late Devonian-Early Triassic									
162	Moro	Tabriz Tabriz-Poldasht	46 08 54 38 14 06	25 km northwest of Tabriz	Diorite and gabbro monzogranite and gabbro	Devonian-Carboniferous	Alkaline	A continental rifting	Alkaline magma, changing to calc-alkaline due to magma amalgamation (Jamei et al. 2020) (continued)

Table 5.1 (continued)

Sr. No.	Name	Geological Map 1:100,000 1:250,000	Geographical coordinates	Geographical location	Lithological composition	Geological age and age determinations	Magmatic series	Tectono-magmatic setting and type of magma	Note (comments and references)
	Mishu				Leucogabbro	Early carboniferous U–Pb (356.7)		Within plate extensional	Scanni et al. (2013)
163	Ghoshchi (Ghoshchi batholith)	Urmia	45 00 40 37 47 27	50 km north of Urmia	Gabbro-diorite, biotite granite, alkali granite, syenites	Late Carboniferous	Alkalin to Sub alkalin (Toleitic)	Within plate A-type granites	Advay et al. (2010)
					Granite gabbronorite	Granite: 317.3–318.4–322 gabbronorite: U[Pb: 316.8–319.9			Shafaii Moghadam et al. (2013)
164	Jolfa (Pir-Eshagh)	Tabriz Julfa	45 31 53 38 43 53	Northwest of Pir-Eshagh	Syenite	Carboniferous 339.7 ± 2.7 Ma	Alkalin	Extensional	Jamei et al. (2020)
Azerbaijan									
Cenozoic									
165	Kaleybar	Kaleybar Ahar	47 00 01 38 48 41	North of Ahar, South of Kaleybar	Nepheline syenite nepheline monzodiorite, nepheline diorite	Eocene	Alkaline-hyperalkaline	Active continental margin/I	–
					Nepheline monzodiorite, nepheline diorite	Oligocene			Hajialioghli et al. (2011)
166	Arasbaran (Mahmudabad and MarzrudVillage)	Kaleybar Ahar	47 03 52 38 38 18	The northeast of Varzaghan (East Azerbaijan)	Diorite, monzonite, granite and gabbro	Eocene–Oligocene	Calc-alkaline	Active continental margin/I	–

(continued)

Table 5.1 (continued)

Sr. No.	Name	Geological Map 1:100,000 1:250,000	Geographical coordinates	Geographical location	Lithological composition	Geological age and age determinations	Magmatic series	Tectono-magmatic setting and type of magma	Note (comments and references)
167	Youseflou (Anzan)	Ahar	47 18 24 38 23 38	20 km Ahar-Meshginshahr	Granite, granodiorite	Eocene–Oligocene U-Ph (40 Ma.)	Calc-alkaline	Active continental margin	Namnabat (2019)
168	Khan baz	Ahar	47 26 43 38 24 53	East of Ahar	Granite, monzonite, quartz monzonite, monzondiorite and quartz monzondiorite	Eocene–Oligocene	Calc-alkaline	Active continental margin	Zamani and Vasigh (2010)
169	Khalifian	Ahar	47 25 36 38 26 21	East of Ahar	Granite, granodiorite	Eocene–Oligocene	Calc-alkaline	Active continental margin	Ghorbani (2004)
170	Sharaf Khan	Ahar	47 17 36 38 20 16	Southeast of Ahar	Granite, granodiorite	Eocene–Oligocene	Calc-alkaline	Active continental margin	Ghorbani (2003)
171	Gholan (Ordubad)	Siahrud Tabriz	46 22 34 38 53 24	180 km south of Tabriz, 15 km southwest of Kharvana, 20 km east of siah rood	Diorite, quartzdiorite, quartz monzodiorite, quartzmonzonite, granodiorite, monzogranite	Eocene-Oligocene	Calc-alkalan	I type	Mokhtari et al. (2005)
172	Bozqush	Qareh Chaman Mianeh	47 28 17 37 44 43	15 km south of Sarab	Nephelin Syenite	Eocene-Oligo-miocene	Alkaline-hyperalkaline	Active continental margin/I	–
173	Razgah	Ahar	47 20 38 38 04 10	45 km of southwest of Meshkinshahr	Syenite and nepheline syenite	Eocene–Oligocene	Alkaline	Active continental margin/I	–

(continued)

Table 5.1 (continued)

Sr. No.	Name	Geological Map 1:100,000 1:250,000	Geographical coordinates	Geographical location	Lithological composition	Geological age and age determinations	Magmatic series	Tectono-magmatic setting and type of magma	Note (comments and references)
174	Tarom Sofla	Abhar Zanjan	48 46 16 36 39 32	East of Zanjan, Tarom Mountains	Quartz monzonite, monzodiorite, diorite	Late Eocene-Early Oligocene	High-K Calc-alkaline, Shoshonite	I	Dehghani (2011)
175	Khorasanlu	Abhar Zanjan	49 09 14 36 20 07	North of Khorramdarreh In Zanjan province	Quartz monzonite, quartzsyenite	Eocene–Oligocene	Calc-alkaline	Active continental margin/I	–
176	Tarom (Chodarchay)	Zanjan Rudbar	49 07 03 36 37 18	East of Zanjan	Granodiorite, granite	Late Eocene-Early Oligocene	Calc-alkaline	Continental margin/I	Ghorbani (2015)
177	Tarom (Khalifehlo)	Zanjan Tarom	48 59 23 36 30 11	Southeast of khorramdarreh	Granodiorite, monzogranite	Late Eocene-Early Oligocene	Calc-alkaline	Continental margin/I	Ghorbani (2015)
178	Tarom (Kuhian)	Zanjan Tarom	48 49 22 36 48 05	Northeast of Zanjan	Granodiorite, granite	Late Eocene-Early Oligocene	Calc-alkaline	Continental margin/I	Ghorbani (2015)
179	Maragheh-Hashtrud	Qarah Aghaj Mianeh	46 39 31 37 02 42	58 km southwest of Hashtrud	Alkali syenite, monzonite and quartzmonzonite	Eocene–Oligocene	Sub alkaline to Calc-alkaline	Active continental margin/I	Ilghami (2005)
180	Heris	Ahar	47 15 48 38 13 27	11 km east of Heris county	Diorite, quartz diorite, granodiorite, monzodiorite	Eocene–Oligocene	Calc-alkaline	Active continental margin or post-collision/I	Ghasemi-Asl (2003)
					Granite	Granite, U–Pb: 306 Ma. Late carboniferous	–	Within plate extensional	Advay and Ghalamghash (2011)
181	Tarom mountains	Tarom Zangan	48 47 49 36 48 56	Eastern heights of Abhar-Zanjan	Granodiorite, gabbrodiorite, tonalite	After middle Eocene–Oligocene	–	Active continental margin/I	Zaraei-Sahamieh (2002)

(continued)

Table 5.1 (continued)

Sr. No.	Name	Geological Map 1:100,000 1:250,000	Geographical coordinates	Geographical location	Lithological composition	Geological age and age determinations	Magmatic series	Tectono-magmatic setting and type of magma	Note (comments and references)
182	Khorram Daraq	Soltaniyeh Zanjan	48 33 49 36 25 37	West of Soltaniyeh, Around Khorram Daraq village	Granite to granodiorite	Early Eocene 53.4 ± 0/4 Ma. U–Pb	Calc-alkaline	Active continental margin/I	Hassanzadeh et al (2008)
183	Golestaneh	Tarom Zanjan	48 41 44 36 41 24	10 km south of Golestaneh	Gabbro, diorite	Eocene–Oligocene	Calc-alkaline	Active continental margin	–
184	Hashtrud	Qarah Aghaj Mianeh	46 39 30 37 02 41	East Azerbaijan Province-north of Hashtrud	Granite, granodiorite and tonalite	Oligocene	Calc-alkaline	Cordillera granite type I (Continental collision) and hybrid granites (HSS)	Ilghami (2005)
185	Masjed Daghi	Tabriz Siah Rud	45 56 25 38 52 40	30 km southeast of Jolfa city	Diorite porphyry	Oligocene	–	–	Atalou et al. (2017) Akbarpour (2005) Hajalilou and Aghazadeh (2016)
186	Khankandi	Ahar Ahar	47 28 18 38 25 56 / 47 28 18 38 25 56	150 km NE of Tabriz city, East of Ahar-The road of Ahar-Meshgin Shahr	Gabbro, monzonite / Granitoied	28.9 Ma. / Oligocene	Shoshonitic / Calc-alkaline	Subduction / Active margins	Aghazadeh et al. (2010a, b)
187	Mirzahasankandi (Qaresou)	Lahrud Ahar	47 45 17 38 48 28	South of Mugan plain, Salavat mountain range and 55 km north of Sabalan mount	Quartz micromonzonite	Late Oligocene	Calc-alkaline	Active continental margin/I	–

(continued)

Table 5.1 (continued)

Sr. No.	Name	Geological Map 1:100,000 1:250,000	Geographical coordinates	Geographical location	Lithological composition	Geological age and age determinations	Magmatic series	Tectono-magmatic setting and type of magma	Note (comments and references)
188	Sheivar Dagh Ahar batholith	Ahar Kaleybar	47 01 48 38 49 10	Mazraeh village and 35 km north of Ahar town	Quartz-diorite to granodiorite	Oligo-Miocene Ghorbani: Eocene–Oligocene 30.8 ± 2.1 Ma.	high-K calc-alkaline and adakitic	I-type	Mollai et al. (2009) Aghazadeh et al. (2010a, b)
189	Oghlandagh	Meshkinshahr Ahar	47 47 44 38 25 08	The west of Sabalan mount	Leuco-monzodiorite	Oligo-Miocene	Calc-alkaline	Active continental margin/I	–
190	Ali Javad	Ahar Ahar	47 01 59 38 39 15	20 km north of Ahar city	Quartz monzonite, quartz monzodiorite and granodioritic composition	Eearly Miocene(22–20)	Shoshonitic and high-K calc-alkaline	Volcanic arc granites field and post-collision granites	Hajalilou and Aghazadeh (2016)
191	Sonajil	Ahar Ahar	47 14 26 38 14 24	Southeat of Ahar	Diorite to granite, tonalite	Miocene	Calc-alkaline	–	–
192	Andaryan (Miveh Rud)	Ahar Khajeh	46 36 16 38 41 16	West of Varzaghan In the Middle of Andaryan and Astarghan villages	Monzodiorite, monzonite to quartzmonzonite, granitic dike	Late Miocene	–	I type	
193	Haft-cheshmeh	Ahar Varzaghan	46 38 52 38 45 09	37 km northwest of Varzaghan	Granite, gabbro to granodiorite	Late Miocene	Calc-alkaline	Volcanic arc granite (subduction)	Hassanpour et al. (2010)
194	Sungun	Azerbaijan Sharghi	46 42 17 38 41 47	75 km north west of Ahar	An early monzonite/quartzmonzonite and a later diorite /granodiorite	Late Miocene	high-K andesite or diorite	Continental arc I-type	Hezarkhani (2005)

(continued)

Table 5.1 (continued)

Sr. No.	Name	Geological Map 1:100,000 1:250,000	Geographical coordinates	Geographical location	Lithological composition	Geological age and age determinations	Magmatic series	Tectono-magmatic setting and type of magma	Note (comments and references)
195	Jolfa	Tabriz Jolfa	45 40 30 38 55 20	SE of Jolfa City and Aras River	30 subvolcanic porphyritic dacitic in	based on stratigraphical studies, these domes are Middle to Late Miocene in age	Calc-alkaline	Partial melting of either subducted oceanic crust	Jahangiri (2007) This stratigraphical age is consistent with dating of rhyolitic domes by Moharrami (2015), which yield ages of 10–17 Ma. based on whole rock Rb/Sr and K/Ar dating methods for early stage post-collision rhyolitic domes in the Armenia Highland
	Marand	Tabriz Marand	45 45 30 38 25 00	To the west and east of Marand City	North Tabriz, North Misho and Darediz dextral faults (Adakitic composition)				
	Nahand	Tabriz	46 27 56 38 15 11	Nahand region	Hornblende-andesite, dacite to rhyodacite and rhyolite, dacite porphyry				
	Kiamaki	Tabriz Jolfa	45 51 21 38 46 57	East of Jolfa	(Adakitic composition) Dacite porphyry				Jahangiri (2007) Moharrami et al. (2014) Miocene 10 Ma.
196	Sahand	Mineh Bostanabad	46 30–46 45 37 35–37 42	southern part of Tabriz Fault	Dacite porphyry and rhyodacite (adakitic composition)	Middle to late miocene	High K calc-alkaline	Partial melting of subduction zone in an active continental margin	Pirmohammadi Alishah (2016)
197	Siah Cheshmeh	Siah Cheshmeh Maku	44 24 23 39 05 02	Southeast of Siah Cheshmeh in Maku area	Granite	Mesozoic-Cenozoic	Calc-alkaline	I	(Alavi-Naeini and Blourchi 1973)

(continued)

Table 5.1 (continued)

Sr. No.	Name	Geological Map 1:100,000 1:250,000	Geographical coordinates	Geographical location	Lithological composition	Geological age and age determinations	Magmatic series	Tectono-magmatic setting and type of magma	Note (comments and references)
Eastern Iran									
Precambrian									
198	Torbat-e Jam	Kahrizno Torbat-e Jam	60 20 58 35 14 57	25 km west of Torbat-e Jam	Granite	Late Precambrian	Calc-alkaline	I	–
199	Shahrakht	Ahangaran Shahrakht	60 03 47 33 23 11	180 km of east Qaen	Granodiorite to monzonite	?Late Precambrian, Paleozoic and Mesozoic	–	–	Geometal BRGM (1981)
Eastern Iran									
Mesozoic									
200	Torbat-e Jam	Torbat-e Jam	60 50 52 35 27 46	25 km of northeast of Torbat-e Jam	Diorite, quartz monzodiorite, quartz monzonite	Upper Triassic 153 ± 5 Ma.	–	–	Jamal Ghavi 2018 (gabbro to quartz diorite: ircon U–Pb age of 215.5 $\pm$ 0.9 Ma.)(granodiorite to monzonite: zircon U–Pb age: 217 ± 2 Ma.), Itype, Alkali-ealcic, Late Triassic)
201	Chahar Farsakh	Chahar Farsakh ChahVak	59 47 35 31 44 11	165 km south of Birjand, eastern part of Lut block	Granite to granodiorite	Mid. Jurassic Early cretaceous	Calc-alkaline	Active continental margin/I	–
202	Shahkuh	Deh-Salm Chahvak (Deh-Salm)	59 19 02 31 02 29	Lut block, Deh-Salm area	Biotite granite, granodiorite / Monzogranite	Mid. Jurassic K-Ar / 162.9 Ma.	Calc-alkaline	Active continental margin	Karimpour et al. (2011)

(continued)

Table 5.1 (continued)

Sr. No.	Name	Geological Map 1:100,000 1:250,000	Geographical coordinates	Geographical location	Lithological composition	Geological age and age determinations	Magmatic series	Tectono-magmatic setting and type of magma	Note (comments and references)
203	Qaen (Zal)	Qaen	59 06 51 33 34 45	18 km south-southwest Qaen	Syenogranite; Granite, leucogranite and biotite granite	161.6 Ma.; ?Late Precambrian-Late Paleozoic-Jurassic	Alkaline	I type	ate Paleozoic is assigned by B.R.G.M report, Keshtkar (2010)
204	Mokhtarabad and Rameshk	Rameshk Fanuj	58 47 31 26 46 16	East of Iran in Fanuj	Gabbro, granitoid rocks mostly trondhjemite	Cretaceous-Paleocene	Tholeiitic	MORB	In associated with ophiolite complexes
205	Mahrud (Mahirud)	Mahrud Gazik	60 48 17 32 13 37	East of Iran, 150 km south of Birjand	Hornblende-biotite tonalite, gabbro and diabase	Upper Cretaceous, lower Paleozoic	Calc-alkaline and tholeiitic	Mixed with crustal melts/M	–
206	Bajestan	Ferdows and Bajestan	58 15 40 34 30 23	40 km north west of Gonabad	Hornblende granite; Granite; Porphyritic granite	77 Ma..; 76.6 Ma.; 76.6 Ma.	High-K calc-alkaline	Active continenta margin/I	Karimpour et al. (2011)
207	Qaen	Qaen	59 47 56 33 29 45	60 km south of Qaen	Gabbro and tonalite	Cretaceous	Alkaline-calc-alkaline	–	in associated with Qaen ophiolite
208	Basiran	Basiran Chahvak	59 07 30 31 58 21	The south of Birjand and northeast of Basiran	Gabbro intrusion, Gabbro and diabase dikes	Late Cretaceous to Late Paleocene/70 Ma.	Tholeiitic	MORB	In Associated with Birjand ophiolites
209	Kuh-e Sefid	Rum Qaen	59 15 30 33 03 14	24 km northeast of Birjand	Granite and monzonite	Upper Cretaceous	Calc-alkaline	Post-collision/I	–

(continued)

Table 5.1 (continued)

Sr. No.	Name	Geological Map 1:100,000 1:250,000	Geographical coordinates	Geographical location	Lithological composition	Geological age and age determinations	Magmatic series	Tectono-magmatic setting and type of magma	Note (comments and references)
210	Gazu	Ozbak kuh Ferdows	57 15 37 34 39 40	78 km northwest of Bajestan	Granodiorite	75.2 Ma. K–Ar	–	–	Karimpour et al. (2011)
211	Khash	Bazman Iranshahr	60 13 48 27 50 27	The northeast of Jaz Murian	Alkali-granite and Hornblende granite granodiorite to diorite	Late Cretaceous 74.2 ± 0.2 Ma.	Calc-alkaline	Active continental margin	the Bazman batholith is of circular form with a central granite and a marginal diorite and granite Berberian et al. (1982) Karimpour et al. (2011)
212		Gazu Narreh Now	61 44 04 28 16 12	60–90 km east and southeast of Khash	Two-mica granites	After Late cretaceous	Calc-alkaline	S Within plate	–
213	Gazik	Gazik	60 14 50 32 59 46	Khorasan-e Jonoubi, Gazik distict	Harzburgite, lherzolite, gabbro and dolerite	Cretaceous	Almost alkaline and slightly Calc-alkaline	–	in associated with ophiolite complexes
214	Ghale Rig	Chahvak Chah Dashi	31 03–31 10 59 47–59 59	Southwest of Nehbandan	Granite, granodiorite	Jurassic	Calc-alkaline	S type Syn-collision	Toulabi Nejad et al. (2016)
Eastern Iran									
Cenozoic									
215	Zahedan (Kuh-e Loochan)	Zahedan	60 41 44 29 21 50	10 km south of Zahedan-west of Khash-Zahedan road	Granite, quartz granodiorite, monzonite and granite	Oligocene	Calc-alkaline	Continental Arc granitoids (CAG)/I	Mirabad granite is located within south of Zahedan granitoid (Hosseini 2002)

(continued)

Table 5.1 (continued)

Sr. No.	Name	Geological Map 1:100,000 1:250,000	Geographical coordinates	Geographical location	Lithological composition	Geological age and age determinations	Magmatic series	Tectono-magmatic setting and type of magma	Note (comments and references)
216	Kuh-e Abdullahi	Deh-Salm Chahvak (Deh-Salm)	59 00 40 31 28 34	East of Kuh-e Abdullahi, south of Nehbandan	Quartz granodiorite, monzonite, granite	Eocene–Oligocene 32 Ma. After Paleocene	Sub alkaline	Intracontinental/I and S	
217	Givshad	Birjand	59 04 21 32 39 04	28 km southwest of Birjand	Pyroxene diorite Quartz monzodiorite to monzonite	Eocene	Calc-alkaline	–	
218	Robat-e Shur	Robat-e Khoshab Boshruyeh	57 56 00 33 53 16	27 km southwest of Ferdows	Quartz syenite	Eocene 42 Ma. K–Ar	Alkaline	–	Karimpour et al. (2011)
219	Ghaleh Zari	Basiran Chahvak	31 42 29-31 44 13, 59 05 35- 59 09 12	196 km south of Birjand and 16 km southeast of Ghaleh Zari	Diorite, diorite porphyry, monzodiorite, monzodiorite porphyry, granodiorite and granodiorite porphyry	Eocene–Oligocene	Calc-alkaline	I type	Nakhaie et al. (2011)
220	Sar now Sar	Taybad Taybad	60 30 46 34 31 26	280 km southeast of Mashhad, near Sangan Iron ore	Granite and granodiorite	Late Eocene–Early Oligocene	Calc-alkaline	Continental arc/I	Magamtic differentiation can be observed in this mass Ghavami-Riabi (1992)
221	Kuh-e Ridge	Birjand	59 02 12 32 45 46	Ridge area-south of Birjand	Diorite	Eocene–Oligocene	Calc-alkaline	Post-collision/I	–

(continued)

Table 5.1 (continued)

Sr. No.	Name	Geological Map 1:100,000 1:250,000	Geographical coordinates	Geographical location	Lithological composition	Geological age and age determinations	Magmatic series	Tectono-magmatic setting and type of magma	Note (comments and references)
222	Dokuheh/ Dodarreh	Khusf Birjand	58 58 50 32 42 22	Southeast of Birjand, north of Dodarreh village	Granite	Eocene–Oligocene	Calc-alkaline	Post collision/I	–
223	Azghand	Feyzabad Torbat-e Heydarieh	58 48 50 35 18 32	30 km southwest of Kashmar	Quartz monzonite, granodiorite, monzogranite and tonalite	Eocene–Oligocene	Calc-alkaline	Active continental margin/I	–
224	Kuh-e Givshad	Birjand	59 04 39 32 39 17	30 km south of Birjand	Quartz monzodiorite to monzonite	Eocene–Oligocene	Calc-alkaline	–	–
225	Khounik	Mukhtaran Birjand	59 09 52 32 23 51	60 km southwest of Birjand	Hornblende monzonite	38.2 Ma. K–Ar	Calc-alkaline	–	Karimpour et al. (2011)
226	Lakhshak	Taleh siyah Zahedan	60 44 23 29 34 29	15 km northwest of Zahedan, along the Zahedan-Kerman road	Granodiorite, monzodiorite	Oligocene 32 Ma.	Calc-alkaline	Active continental margin/I	Rezaei-Kahkhani (2006)
227	Zahedan (Kuh-e Loochan) lucho	Zahedan	60 41 44 29 21 50	10 km south of Zahedan-west of Khash-Zahedan road	Granite, quartz granodiorite, monzonite and granite	Oligocene	Calc-alkaline	Continental Arc Granitoids (CAG)/I	Mirabad granite is located within south of Zahedan granitoid (Hosseini 2002)
					Quartz granodiorite, monzonite, granite	Eocene–Oligocene 32 Ma.	Sub alkaline	Intracontinental/I and S	

(continued)

Table 5.1 (continued)

Sr. No.	Name	Geological Map 1:100,000 1:250,000	Geographical coordinates	Geographical location	Lithological composition	Geological age and age determinations	Magmatic series	Tectono-magmatic setting and type of magma	Note (comments and references)
228	Graghu, Narrehtu and Sarvan mountains	Nukabad Khash	60 48 19 28 44 15	South and southeast of Zahedan	Granite, granodiorite, diorite and quartz diorite	Oligomiocene 32 Ma.	Calc-alkaline	–	Sadeghian (1994)
229	Saravan	Kushkok Saravan	62 06 17 27 57 05	Bam	Granodiorite	Mio-Pliocene	–	–	–
230	Chah Anjirdan (Cheshmeh Bid)	Mirjaveh Zahedan	61 04 55 29 07 23	Zahedan area	Metamorphic complex, metamorphosed igneous rocks, granite to granodiorite	40 K–40Ar	Calc-alkaline	I and S	Sheikholeslami et al. (2003)
231	Shadan	Mokhtaran Birjand	59 00 22–59 08 54 32 18– 32 24 48	Southwest of Birjand	Porphyritic tonalite	Tertiary	Calc-alkaline	Continental arc	Sadri (2011)
		Sarchahshour Birjand	58 58 46 32 21 25		Porphyritic diorite	37.26 ± 0.26 Ma.	–	Contractional Deformation events related to the subduction	Richards et al. (2012) Siahcheshm et al. (2014)

Alborz zone

Mesozoic

Sr. No.	Name	Geological Map 1:100,000 1:250,000	Geographical coordinates	Geographical location	Lithological composition	Geological age and age determinations	Magmatic series	Tectono-magmatic setting and type of magma	Note (comments and references)
232	Lahijan	Rasht Qazvin-Rasht	49 56 27 37 07 12	South of Lahijan, Green road	Gabbro, dolerite and granite	Late Paleozoic to upper Triassic	Tholeiitic (gabbrodiorite) Calc-alkaline (granitoid)	–	Granitoid type I

(continued)

Table 5.1 (continued)

Sr. No.	Name	Geological Map 1:100,000 1:250,000	Geographical coordinates	Geographical location	Lithological composition	Geological age and age determinations	Magmatic series	Tectono-magmatic setting and type of magma	Note (comments and references)
233	Lahijan	Langarud Qazvin-Rasht	50 07 34 37 09 28	Southeast of Lahijan	Granodiorite, gabbro, dolerite	Triassic	Calc-alkaline tholeiitic	I	This granite is resulted from the continental collision between Iranian and Turan plates
234	Masuleh	Bandar-e Anzali Bandar-e Anzali	49 05 36 37 06 01	5 km southeast of Masuleh	Tourmaline granite With diorite and gabbro	Triassic 180 ± 5 Ma.	Calc-alkaline	Collision /I?,S?	It is associated with the dioritic and gabbro intrusive
235	Gasht-Rudkhan	Bandar-e Anzali Bandar-e Anzali	49 15 13 37 10 24	North of Gasht-Rudkhan, Talesh Mountains	Granite and migmatite	Triassic-Jurassic	Calc-alkaline	S	–
236	Javaherdasht	Javaherdasht	36 55–36 56 50 53–50 56	East of Gilan	Non-layered and layered gabbros, and layered gabbros	Cretacouse	Calc-alkaline to alkaline	MORB	Haghnazar et al. (2011)
237	Lisar (Soobatan)	Khalkhal, Astara	48 47 39 37 58 25	30 km of Hashtpar (Talesh)	Granite and granodiorite	Triassic To Middle Jurassic 180 Ma.	Sub Alkaline	Collisionl post orogenic	Due to the high resistance and pink color, it can be economically significant Asghari-Nezhad (1996)
238	Baladeh	Amol Baladeh & Marzanabad	51 23 50 36 17 00	South of Kamarbon village	Gabbro	Permian–Triassic	Alkaline	Extentional	Doroozi and Masoudi (2012)

(continued)

Table 5.1 (continued)

Sr. No.	Name	Geological Map 1:100,000 1:250,000	Geographical coordinates	Geographical location	Lithological composition	Geological age and age determinations	Magmatic series	Tectono-magmatic setting and type of magma	Note (comments and references)
Alborz zone									
Cenozoic									
239	Tehran north intrusive rocks	Fasham Tehran	51 45 56 35 36 34	Darabad. Kan. Egoul and 20 km southwest of Roudehen	A lot of monzogabbro bodies	Eocene–Oligocene	Alkaline and Calc-alkaline	Active continental margin/I	
240	Mobarakabad	Fasham Tehran	51 57 42 35 46 27	6 km northeast of Roudehen (Ardineh and Mobarakabad)	Melagabbro, diorite	After Eocene	Calc-alkaline	I	Saadat (2015)
241	Lashkarak	Fasham Tehran	51 35 09 35 48 51	East of Tehran, western elevations of Litian and Zarrin-Kuh dome	Quartzdiorite	After Eocene	Calc-alkaline	I continental	–
242	Chenasak	Qazvin Qazvin-Rasht	50 24 33 36 16 53	25 km east of Qazvin, Northwest of Abyek	Syenite, monzonite, melagabbro	Oligocene–Miocene	Shoshonite to alkaline	I	Bakhtiari (1996)
243	Alam-Kuh	Baladeh Amol	51 42 49 36 01 51	Near Alam-kuh mount, In Central Alborz	Alkali granite Granodiorite	Late Eocene-Oligo-miocene	alkaline -Calc-alkaline	Active continental margin/I	–
244	Varkash-Valian-Ziyaran	Shokran Qazvin-Rasht	50 13 34 36 09 26	North of Ziyaran	Granodiorite to monzonite	Late Eocene–Oligocene	Alkaline, Calc-alkaline	Active continental margin/I	–
245	Talesh	Khalkhal Bandar-e Anzali	48 47 30 37 58 33	18 km northeast of Talesh (Soobatan)	Granite to diorite	Eocene–Oligocene	Calc-alkaline	Active continental margin/I	–

(continued)

Table 5.1 (continued)

Sr. No.	Name	Geological Map 1:100,000 1:250,000	Geographical coordinates	Geographical location	Lithological composition	Geological age and age determinations	Magmatic series	Tectono-magmatic setting and type of magma	Note (comments and references)
246	Qasr-e Firuzeh	Fasham Tehran	51 33 57 35 39 53	Northeast of Tehran	Granodiorite and granite	Eocene–Oligocene	Calc-alkaline	Active continental margin/I	–
247	Kordan-Soreh	Tehran	51 05 23 35 57 56	Karaj Dam's abutment	Gabbro and monzonite	Eocene–Oligocene	Calc-alkaline	Active continental margin/I	in fact his mass is a sill with monzonite composition
248	Lavasan	Fasham Tehran	51 48 44 35 50 48	North of Lavasan	Syenite	Late Eocene–Oligocene	Alkaline	Active continental margin/I	–
249	Mendejin	Hashatjin Bandar-e Anzali	48 12 25 37 15 06	15 km south of Mendejin	Tonalite, diorite, aplite	Eocene–Oligocene	Calc-alkaline	I	–
250	Ramsar	Ramsar Qazvin-Rasht	50 45 10 36 38 40	Around of Ramsar, northwest of Baraseh	Gabbro, picrite, syenite	Eocene–Oligocene	–	I	–
251	Akapol	Marzanabad Amol	51 04 17 36 25 00	5 km northeast of Alam-Kuh, Kelardasht	Granodiorite, quartz monzonite	Late Paleocene-Early Eocene 54 ± 4 Ma. to 56 ± 3	Alkaline (Calc-alkaline)	Active continental margin/I	Khalaj (2006)
252	Talesh	Masouleh-Hashtjin, Bandar-e Anzali	48 35 55 37 01 15 48 17 12 37 10 55	Talesh mountains (north of Gasht-Rudkhan, near Shavalem)	Granite, pegmatite, peridotite body and olivine gabbro	Paleogene	–	–	–

(continued)

Table 5.1 (continued)

Sr. No.	Name	Geological Map 1:100,000 1:250,000	Geographical coordinates	Geographical location	Lithological composition	Geological age and age determinations	Magmatic series	Tectono-magmatic setting and type of magma	Note (comments and references)
253	Fuman	Bandar-e Anzali	49 06 09 37 08 30	20 km southwest of Fuman	Diabase, gabbro, dolerite and diorite	Cretaceous -Oligocene	High-K alkaline -Calc-alkaline	I	–
Zagros									
Precambrian									
254	Gachin	Kohurestan Bandar Abbas	55 56 05 27 05 52	35 km southwest of Bandar Abbas	Granite and gabbro	Precambrian-lower cambrian	Alkalin-Calc-alkaline	–	Ghorbani 2014

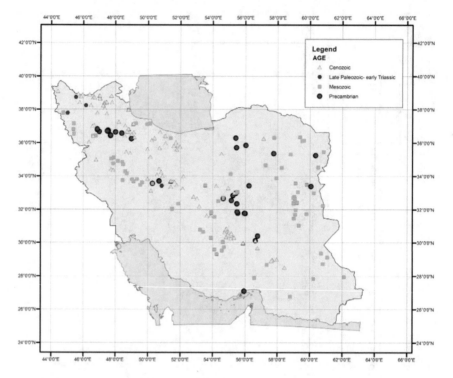

Fig. 5.1 Spatial distribution of intrusions in Iran

5.1.1   Age Classification of Intrusions

The intrusive bodies in Iran can be categorized as follows:

Late Precambrian
Late Precambrian intrusive bodies were associated with the Caledonian orogeny phase. At this time, a lot of intrusive bodies have been intruded in Alborz, central Iran and the Sanandaj–Sirjan zone, such as Doran in Azerbaijan and Narigan -Zarigan in central Iran.

Late Paleozoic–Triassic
There are many intrusive bodies in the north, northeast, and Azerbaijan, which are formed in the direction of the Paleo-Tethys closure (Tethys II). Like intrusive around Mashhad and Lisar which exhibit collisional properties. There are also intrusive bodies of this age in the Sanandaj–Sirjan and southern Azerbaijan, which are associated with the opening of the Neo-Tethys (Tethys III) that exhibit extensional characters.

Mesozoic

The Mesozoic bodies of Iran, which are mainly concentrated in the Sanandaj–Sirjan and Central Iran zones, have been generally formed in the middle Jurassic, although at middle Jurassic-late Cretaceous several intrusive pulses can be studied, but the bodies around the middle Jurassic are more evident in the Sanandaj–Sirjan zone.

Cenozoic

As previously mentioned, the Cenozoic magmatic activity of Iran, in terms of volcanism and intrusive, has been extremely intense in various regions of Iran such as Central Iran, Urmia-Dokhtar, Azerbaijan, and eastern Iran.

The bodies are mainly related to the subduction of Neo-Tethys Ocean beneath Iran and collision of the Arabic plate to central Iran plate, and ultimately the shrinking of the Iranian plateau. Although the intrusive masses at this interval indicate widespread spatial and temporal distribution, however, in some time interval they reach their peak of activity as follows:

Late Eocene-Early Oligocene

At this time, the batholithic intrusive are abundant, especially in the Urmia-Dokhtar, Azerbaijan, eastern and central Iran.

Late Oligocene–Miocene

The intrusive masses of this period usually have smaller dimensions compared to those of late Eocene-early Oligocene. Their peak is in the middle to upper Miocene. These masses are often dispersed in Urmia-Azerbaijan and Azerbaijan and are porphyry, and many of the porphyric copper potential in Iran are also of formed in this time.

Upper Miocene–Pliocene

These intrusive are seen in many parts of Iran, including Urmia-Dokhtar, Kerman, Azarbaijan, northern Sabzevar, and eastern Iran. These bodies usually start from about 10 million years ago and their age continues until the Quaternary. These masses are often sub-volcanic and have a dacite composition and are generally Adakitic.

References

Abdollahi MJ, Karimpour MH, Kheradmand A (2009) Petrography and sulphur isotope studies of pyrites in the Muteh gold deposit. Am J Appl Sci 6(6):1086–1092. ISSN: 1546–9239

Abdul Nejad Mamghani H (2007) Petrology of the internal mass of Vash and how it is located in the rocks of southeastern Kashan. MSc thesis, Faculty of Earth science, Shahid Beheshti University

Advay M, Ghalamghash J (2011) Petrogenesis and U–Pb dating zir-con of granites of Heris (NW of Shabestar), eastern Azerbaigan province. Iran J Crystallogr Mineral January 2011, Vol. 18, No. 4, Winter 1389/2011(in Persian)

Advay M, Jahangir A, Mojtahedi M, Ghalamghash J (2010) Petrology and geochemistry of Ghoshchi batholith, NW Iran. J Crystallogr Mineral 17:716–733

Afshooni Z (2007) Alteration and mineralization of gold in the Astana granitoid massif (southwest of Arak). University of Tehran. Faculty of Science. Geology Department. MSc thesis, 159p

Aghazadeh M (2015) Petogenesis and U-Pb age dating of intrusive bodies in the Sar Cheshmeh deposite. Sci Quart J Geosci. 25:97. Autumn 2015 (In Persian)

Aghazadeh M, Castro A, Badrzadeh Z (2010a) Petrogenesis of Ter-tiary alkaline potassic intrusive rocks in the Arasbaran zone, NW Iran. ALKALINE ROCKS conference, Kyiv, Ukraine, 19–21 September, 2010

Aghazadeh M, Castro A, Rashidnejad Omran N, Emami MH, Moinvaziri H, Badrzadeh Z (2010b) The gabbro (shoshonitic)–monzonite–granodiorite association of Khankandi pluton, Alborz Mountains, NW Iran. J Asian Earth Sci 38(2010):199–219

Ahmadi AR (2004) Petrography, petrogenesis and geochemistry o the metamorphic rocks of Tutak complex, MSc thesis, Shahid Beheshti University

Ahmadi Khalaji A (2006) Petrology of the granitoid rocks of the Boroujerd area, PhD thesis, University of Tehran

Ahmadi Rohani R (1999) Study of the metamorphic and igneous rocks of the southeastern Dandi, MSc thesis, Shahid Beheshti University

Ahmadipour H, Rostamizadeh G (2012) Geochemical aspects of Na-metasomatism in Sargaz granitic intrusion (south of Kerman province, Iran). J Sci Islam Repub Iran 23(1):45–58

Aistov L, Melnikov B, Krivyakin B, Morozov L (1984) Geology of the Khur area (Central Iran), Explanatory text of the Khur quadran-gle map 1:250,000

Akbarpour A (2005) Economic geology of Kiamaki area regarding gold and copper mineralization (Masjed Daghi Jolfa) in Eastern Azerbaijan. Unpublished PhD dissertation (in Persian), Science and Research Branch, Islamic Azad University, Tehran, 262

Alavi-Naeini M, Bolourchi MH (1973) Maku quadrangle map, scale.1:250,000, with explanatory text: Geol. Survey Iran, Geol. Quadrangle Al

Aletaha kohbanani, B (1993) Petrology and geochemistry of Igneous rock in east of Zarand, Kerman (Unpublished MSc thesis). Tehran University, Tehran, Iran, pp. 117 (In Persian)

Alirezaei S, Hassanzadeh J (2011) Geochemistry and zircon geochronology of the Permian A-type Hasanrobat granite, Sanandaj–Sirjan elt: A new record of the Gondwana break-up in Iran. Lithos. https://doi.org/10.1016/j.lithos.2011.11.015

Alirezaie A, Aliyani F, Moradian A (2012) The study of petrogra-phy, geochemical characteristics and tectono magmatic setting of Meiduk porphyry copper deposit, Shahrebabak–Kerman, Iranian. J Crystallogr Mineral, Issue 3 (In Persian)

Amani Kh. (2000). Petrology of Saqqez granitoid, MSc thesis, Tehran University

Amidi SM (1977) Etude géologique de la région du Natanz-Surk (Iran Central) stratigraphie et pétrographie, PhD thesis, Gel Surv Iran, Rep. 42, 316 p

Arvin M, Pan Y, Dargahi S, Malekizadeh A, Babaei A (2007) Petrochemistry of the Siah-Kuh grani-toid stock in the southwest of Kerman, Iran: implications for initiation of Neo-Tethys subduction. J Asian Earth Sci 30(3):474–489

Asadpour M, Hovis S, Pourmoafi M (2013) New evidences of magmatic activities of Precambrian and Paleozoice in Ghare Bagh mass, Northwest of Iran. Geosci J. https://doi.org/10.22071/gsj.2013.53593 (In Persian)

Asghari-Nezhad (1996) Petrology and Geochemistry Lisar (Talesh) granitoid mass. MSc Thesis, Science Faculty, Tehran University

Atalou S, Nezafati N, Lotfi M, Aghazadeh M (2017) Fluid inclusion investigations of the Masjed Daghi copper-gold porphyry-epithermal mineralization, East Azerbaijan Province, NW Iran. Open J Geol 07(08):1110–1127

Athari SF, Sepahi AA, Moazzen M (2006) Hasan Salaran granitoids of Saqqez, a complex of two various gran-itoid types in the Sanandaj–Sirjan metamorphic belt Res J Univ Isfahan Sci 25(3):77–98 (In Persian)

Ayati F, Asadi HH, Noghreyan M, Khalili M (2008) Preliminary report of porphyry Cu exploration at central province of Iran. J Earth Resour 34(5):277–299

Ayati F, Kalimi Noghreyan M, Khalili M (2012) Petrographic and mineral-chemistry of the magmatic-alteration zones South of Salafchegan. Petrol J 2(8): 1–20

Azizi H, Chung SL, Tanaka T, Asahara Y (2011) Isotopic dating of the Khoy metamorphic complex (KMC), northwestern Iran: a significant revision of the formation age and magma source. Precambr Res 185(3):87–94

Bagherian S, Darvishi I, Moazzen M, Khakzad A (2006) Investigation on mineralization potential of molataleb granitoid body using geochemical charactristics. Geosci Sci Q J 15(58):158–165

Baharifar AA, Moinevaziri H, Bellon H, Piqué A (2004) The crystalline complexes of Hamadan (Sanandaj–Sirjan zone, western Iran): metasedimentary Meso-zoic sequences affected by Late Cretaceous tec-tono-metamorphic and plutonic events. Comptes Rendus Geosci 336(16):1443–1452

Bakhtiari M (1996) Geology, petrology and petrography of North-west Abyek (Chenasak village) intrusive mass. MSc thesis, Azad University

Barahmand L, Ghorbani M, Pourmoafi M (2018) Review and comparing geochemistry of barren and productive hypabbysal intrusive bodies considering evidence for adakitic rocks in the Dehaj-Meiduk area (North Shahrbabak). Sci Q J Geosci 27(106):193–202

Berberian F, Muir ID, Pankhurst RJ, Berberian M (1982) Late Cretaceous and early Miocene Andean-type plutonic activity in northern Makran and Central Iran. J Geol Soc 139(5):605–614

Chiu HY, Chung SL, Zarrinkoub MH, Mohammadi SS, Khatib MM, Iizuka Y (2013) Zircon U–Pb age constraints from Iran on the magmatic evolution related to Neotethyan subduction and Zagros orogeny. Lithos, 162, pp 70–87

Dehghani S (2011) Petrology and petrography of the intrusive and rocks of the Lesser Tarom, MSc thesis, Shahid Beheshti University

Doroozi R, Masoudi F (2012) Geochemistry, petrogenesis and tectonic setting of Kamarbon Theralitic, Teschenitic gabbroic intrusion (Central Alborz), petrology, 3rd year, No. 12, Winter 2012 (In Persian)

Ebrahimi M (1991) Geology and petrology of magmatic rocks in the north of Golpayegan (Cheshmeh Sefld-Margh area), PhD thesis, University of Tehran

Eslamizadeh E (1993) Petrology of Venarj (Qom) Intrusive mass. MSc thesis, Islamic Azad University

Esna-Ashari A, Tiepolo M, Valizadeh MV, Hassanzadeh J, Sepahi A (2012) Geochemistry and zircon U–Pb geochronology of Aligoodarz granitoid complex. Sanandaj–Sirjan Zone, Iran. J Asian Earth Sci 43(2012):11–22

Farrokhmanesh D (1998) Petrological and Geochemical Survey of Takab Granitoid Masses, MSc thesis, Shahid Beheshti University

Fazlnia A, Moradian A, Rezaei K, Moazzen M, Alipour S (2007) Synchronous activity of anorthositic and S-type granitic magmas in Chahdozdan batholith, Neyriz, Iran: evidence of Zircon SHRIMP and monazite CHIME dating. J Sci Islam Repub Iran 18(3):221–237

Geometal BRGM (1981) Geological map of Ahangaran, geological quadrangle map of Iran, 1:100000, series sheet 8059, Geological and Mining Survey of Iran, Tehran

Ghadimi Gh (2016) Geochemical and petrogenesis of granitoides rocks in south-east of centeral Iranian volcanic belt, north-west of Share-Babak, Kerman province, Iran. J Tethys 4(4):295–311

Ghalamghash J (2002) Petrology of intrusive rocks in Urmia-Oshnoyeh region and their placement mechanism, PhD thesis, Faculty of Earth Sciences, Shahid Beheshti University

Ghalamghash J, Nédélec A., Bellon H, Vousoughi Abedini M, Bouchez JL (2009) The Urumieh plutonic complex (NW Iran): a record of the geodynamic evolution of the Sanandaj–Sirjan zone during Cretaceous times—Part I: Petrogenesis and K/Ar dating. J Asian Earth Sci 35(5):401–415

Ghasemi Asl R (2003) Petrography, petrogenesis and geochemistry of the magmatic rocks in the East Harris (northeastern Tabriz), MSc thesis, Shahid Beheshti University

Ghavami Riabi A (1992) Geochemical and genetic study of igneous and metamorphic rocks and iron ore deposits in Sangan area of Khorasan, MSc thesis, Shahid Beheshti University

Ghavi J, Karimpour MH, Mazaheri SA, Pan Y (2018) Triassic I-type granitoids from the Torbat e Jam area, northeastern Iran: petrogenesis and implications for Paleotethys tectonics. J Asian Earth Sci 164:159–178

Ghorbani Gh. (2005) Petrology and petrogenesis of igneous rock southeast of Damghan, PhD thesis, Shahid Beheshti University (In Persian)

Ghorbani M (1999) Petrological investigations of tertiary–quaternary magmatic rocks and their metallogeny in Takab area. PhD thesis, Shahid Beheshti University

Ghorbani M (2003) Volcanological foundations with an attitude toward volcanoes in Iran, Pars Geological Research Center

Ghorbani M (2004) Petrologic characteristics of magmatic rocks in alteration zones and their association with mineralization within Ahar–Jolfa axis, Research project at Shahid Beheshti University

Ghorbani M (2014) Geology of Iran, magmatism-metamorphism of Iran, Airan Zamin (In Persian)

Ghorbani M (2015) Petrologic studies of intrusive bodies in the Tarom area, a comparison to each other and their mincralogical implication, MSc thesis, Islamic Azad University Ashtian Branch

Goudarzi H (1995) Metamorphism and magmatism in Malayer-Boroujerd region, Master's thesis, Kharazmi University, GSI, No G8

Haghnazar Sh., Malakotian S, Allahyari Kh. (2011) Petrology, investigation of petrological, mineralogical and geochemical properties of Javaherdasht gabbros (east of Guilan province). Iran J Crystallogr Mineral 18(4). Winter 1389/2011 (In Persian)

Hajalilou B, Aghazadeh M (2016) Geological, alteration and mineralization characteristics of Ali Javad porphyry Cu–Au deposit, Arasbaran zone, NW Iran. Open J Geol 6:859–874

Hajialioghli R, Moazzen M, Milke R (2011) Titanian garnet in nepheline syenite from the Kaleybar area, East Azerbaijan province, NW Iran. CentL Eur Geol 54(3):295–311

Hashemi M (2016) Gemstones of the edge of the city of Aligudarz, 3rd National symposium of gemology and crystallography of Iranian society of crystallography and mineralogy, 11–13 November, Tehran, Iran (In Persian)

Hassanpour S, Moayyed M, Rasa I (2010) Geology, alteration and mineralization in the Haft-Cheshmeh Cu–Mo porphyry deposit. Iran J Geol 4(15):15–28. Autumn 2010 (In Persian)

Hassanzadeh J, Stockli DF, Horton BK, Axen GJ, Stockli LD, Grove M, Schmitt AK, Walker JD (2008) U–Pb zircon geochronology of late Neoproterozoic–early Cambrian granitoids in Iran: implications for paleogeography, magmatism, and exhumation history of Iranian basement. Tectonophysics 451:71–96

Helmi F (1991) Petrology and geochemistry of igneous rocks in the northwest of Saveh (north of Nievesht). MSc thesis (in Persian), Tehran University, Iran

Hezarkhani A (2005) Petrology of the intrusive rocks within the Sungun porphyry copper deposit, Azerbaijan, Iran, Department of Mining, Metalogy and Petroleum Enginering Amirkabir University, Asian Earth Sciences

Hosseini B (2011) Identification of two phases of metamorphosed granitoid masses in the Tutak anticline based on the U–Pb dating. Earth Sci J 84:57–66

Hosseini M (2002) Petrology and geochemistry of southwest Zahedan granites, MSc thesis, Science Faculty, Tehran University, 290 p

Hosseini B, Ahmadi AR (2016) Geochemistry and U–Pb dating of North Saman granitoid rocks. Q J Geosci 25(100):109–120. Summer 2016 (In Persian)

Houshmand Manavi S (2003) Petrogenesis of granitoid plutons in Balestan area (NE Oshnavieh), MSc thesis, Faculty of Earth Sciences, Shahid Beheshti University

Ilghami sareskanrud F (2005) Petrography, geochemistry and petrogenesis of magmatic rocks of southwest of Hashtrood (south of Middle Mianeh). MSc thesis, Faculty of Earth Sciences, Shahid Beheshti University

Jabbari A, Ghorbani M, Koepke J, Torabi Gh., Shirdashtzadeh N (2010) Petrography and mineral chemistry of basaltic dykes in the west of Borooni (SW of Ardestan, Iran): evidences of magma mixing. J Petrol 1(2):17–30

Jafari A, Fazlnia A, Jamei S (2018) Geochemistry, petrology and geodynamic setting of the Urumieh plutonic complex, Sanandaj–Sirjan zone, NW Iran: new implication for Arabian and Central Iranian plate collision. J Afr Earth Sci 139(2018):421–439

Jahangiri A (2007) Post-collisional Miocene adakitic volcanism in NW Iran: Geochemical and geodynamic implications. J Asian Earth Sci 30(2007):433–447

Jamei S, Ghorbani M, Williams IS, Moayyed M (2020) Tethyan oceans reconstructions with emphasis on the early Carboniferous Pir-Eshagh A-type rhyolite and the Late Palaeozoic magmatism in Iran. Int Geol Rev. https://doi.org/10.1080/00206814.2020.1768443

Jamshidi K, Ghasemi H, Miao L (2015) U–Pb age dating of Sabzevar Addakite. Petrology 6(23):121–138

Jamshidibadr M (2002) Petrology and petrography of igneous and metamorphic rocks of Incheh bolagh village area, between Shahindezh and Takab. Faculty of Geology, Tabriz University

Jamshidibadr M, Collins AS, Salomão GN, Costa M (2018) U–Pb zircon ages, geochemistry and tectonic setting of felsic and mafic intrusive rocks of Almogholagh complex, NW Iran. Period Di Mineral 87(1):21–53

Kanaanian AS (2001) Petrology and geochemistry of Kahnouj ophiolite collection, MSc thesis, University of Tehran

Karimpour MH, Lang Farmer G, Stern CR, Salati E (2011) U–Pb zircon geochronology and Sr–Nd isotopic characteristic of Late Neoproterozoic Bornaward granitoids (Taknar zone exotic block), Iran, Vol. 19, No. 1, Spring 1390/2011 Pages 1–18 (In Persian)

Kazemi T, Azizi H, Hiroasaha Y (2016) Geochemistry and Sr–Nd isotope behavior of granitoid masses of Hasan Salar, southeast of Saqez, MSc thesis, Kurdistan university (In Persian)

Keshtkar I (2010) Petrography and petrogenesis of granitoid and metamorphic rocks of southeastern Qaen. MSc thesis, Shahid Beheshti university

Khalaj M (2006) Petrology and geochemistry of the Akapel granitoid mass, southwestern Klardasht, MSc thesis, Tehran University

Kholghi Khasraghi MH (1991) Metamorphism plutonism and stratigraphy of east of Shahin Dezh. MSc thesis, Tehran University, 260 pp. (in Persian)

Mahmoudi Sh. (2011) U–Pb dating and emplacement history of granitoid plutons in the northern Sanandaj–Sirjan zone, Iran. J Asian Earth Sci 41:238–249

Malekzadeh A, Karimpour MH, Mazaheri SA (2004) Geology, mineralization and geochemistry of Tak I, Taknar polymetal massive sulfide (Cu–Zn–Au–Ag–Pb) deposit, Khorasan-Bardaskan. Iran J Crystallogr Mineral 12(2):253–272

Mansouri Esfahani M (1993) Geology and petrology of intrusive rocks of Gojed in relation to the surrounding volcanic rocks (Gojed village in the southwest of Nain), MSc thesis, Isfahan University

Mansouri Esfahani M (2000) Petrogenesis of Hasan Robat granitoidic mass. Res J Univ Isfahan Sci 13(1):37–58.

Mazhari SA, Bea F, Amini S, Ghalamghash J, Molina JF, Montero MP, Scarrow J, Williams IS (2009) The Eocene bimodal Piranshahr massif of the Sanandaj–Sirjan Zone, NW Iran: a marker of the end of the collision in the Zagros orogen. J Geol Soc 166:53–69

Mirnejad H (2000) Geochemistry and petrography of Mashhad granites and pegmatites, MSc thesis, Tehran University

Moazzen M, Moayyed M, Modjarrad M, Darvishi E (2004) Azna granitoid as an example of syn-collision S-type granitisation in Sanandaj–Sirjan metamorphic belt, Iran. Stuttgart, 2004(11):489–507

Mohamadi F, Ebrahimi M, Mokhtari MAA (2016) Petrology and geochemistry of Homijan granitoid and associated felsic rocks (SW Behabad, Central Iran). J Geosci Geol Surv Iran 25(98):223–236

Mohammaddoost H, Ghaderi M, Kumar TV, Hassanzadeh J, Alirezaei S, Stein HJ, Babu EVSSK (2017) Zircon U–Pb and molybdenite Re–Os geochronology and sulfur isotope composition of vein materials in the Chah-Firouzeh porphyry copper deposit, Kerman Cenozoic Magmatic Assemblage, southeastern Iran. Ore Geol Rev 88:384–399

Mohammadi R (2005) Petrology of Tertiary igneous rocks, Razan region of Hamadan, PhD thesis, Azad University

Mohammadi Laghab H, Taghipour N, Iranmanesh M (2012). Distribution pattern of Cu, Mo, Pb, Zn and Fe elements in Sara (Parkam) porphyry copper deposit, Shahr–Babak, Kerman province. Iran J Geol 5(20):17–28 (In Persian)

Mohammadi A, Khalili M, Mansouri Esfahani M (2011) The effect of weathering on the geochemistry and mineralogy of the Dehno Granites (northeast of Aligudarz). 18 (4):601–614. https://ijcm.ir/article-1-492-fa.html (In Persian)

Moharrami F (2015) Petrology of igneous rocks of Kimaki region, south east of Julfa, north west of Iran, Faculty of Earth Science, Shahid Beheshti university

Moharrami F, Ghorbani M, Pourmoafee M (2014) Geogemical and tectono-magmatic pattern of the formation of Kimaki Adakitic Dome, southeast of Julfa (northwest of Iran). Iran J Earth Sci 91–102

Mokhtari M, Moein-Vaziri H, Ghorbani MR, Mehr Parto M (2005) Petrography and petrology of Gholan batholith, Earth Sciences Symposium (In Persian)

Mollai, H., Yaghubpur, A.M., Sharifiyan Attar, R. (2009). Geology and geochemistry of skarn deposits in the northern part of Ahar batholith, East Azarbaijan, NW Iran, H. Mollai et al./ Iranian Journal of Earth Sciences 1 (2009)/ 15–34.

Motori (2010) Lithology of granites and metamorphic rocks of Astaneh–Nezam Abad bodies and their relationship with Sn and W mineralization. MSc thesis, Shahid Beheshti university

Mousavi Makui SA (1998) Narigan granite petrology, MSc thesis, Shahid Beheshti University

Nakhaie M, Karimpour MH, Mazaheri SA, Heydarian MH, Zarinkoub MH (2011) Petrology of intrusive rocks associated with iron mineralization in Bisheh area (East Iran, South of Birjand), 2th Symposium of Iranian Society of Economic Geology, 6–7 July 2011

Namnabat E (2019) Magmatism and alteration of Varzaghan–Julfa axis. PhD thesis, Shahid Beheshti University (In Persian)

Niroomand Sh., Hassanzadeh J, Tajeddin HA, Asadi S (2018) Hydrothermal evolution and isotope studies of the Baghu intrusion-related gold deposit, Semnan province, north central Iran. Ore Geol Rev. https://doi.org/10.1016/j.oregeorev.2018.01.015

Nogole Sadat MAA, Houshmandzadeh A (1974) Map of Marivan of scale 1:100000 (In Persian)

Noraei A (1998) Study of petrogenese and geochemistry of granitoids southeast of Shahin-Dezh, MSc thesis, Shahid Beheshti University

Omrani J (1992) Petrology and geochemistry of permeable rocks in Zarrin area (Ardakan Yazd). MSc thesis, Faculty of Science, Geology Department, University of Tehran, 136 pp.

Pirmohammadi Alishah F (2016) Geochemistry and tectonics of the formation of Sahand dacitic dome, southeast of Tabriz (northwest of Iran). J Tethys 3(4):327–339

Pourjahromi MH (1994) Metamorphism and magmatism of the Malayer–Hamedan–Toiserkan, MSc thesis, Kharazmi University

Ramezani J (1997) Regional geology, geochronology and geochemistry of the igneousand metamorphic rock suites of the Saghand Area, central Iran: unpublished PhD thesis, St. Louis, Missouri, Washington University, 416 pp.

Ramezani J, Tucker RD (2003) The Saghand region, central Iran: U–Pb geochronology, petrogenesis and implications for Gondwana tectonics. Am J Sci 303:622–665

Ranin A, Sepahi AA, Moein-Vaziri H, Aliani F (2010) Petrology and geochemistry of the plutonic complexes of the Marivan area, Sanandaj–Sirjan zone (In Farsi with English abstract). J Petrol Isfahan Univ Iran 2:43–60

Ranjbar S, Kalimi-Noghreyan M, Mackizadeh MA (2011) Study of Skarnmineralization at North of Kal-e Kafi and its relation with the Kal-e Kafi intrusive mass (in Persian). Petrology 3(9):107–126

Rasouli J, Ghorbai M (2012) Geochemistry and petrography of late intrusions body of Kerver (East of Jebal-e Barez), 30th Symposium of Geosciences, 2012–02–20 (In Persian)

Rezaei-Kahkhani M (2006) Geology, petrology and mineral potential of ophiolite melange of the northern Torbat Heydarieh (Asadabad Area), MSc thesis, University of Tehran

Richards JP, Spell T, Rameh E, Razique A (2012) High Sr/Y magmas reflect arc maturity, high magmatic water content, and porphyry Cu ± Mo ± Au potential: examples from the tethyan arcs of central and eastern iran and Western Pakistan. Econ Geol 107(2):295–332

Saadat B (2015) Comparison of the intrusive masses of East and West Tehran, PhD thesis, Azad university

Sabzehei M (1974) Les Mélanges ophiolitiques de la région d'Esfandagheh (Iran méridional): étude pétrologique et structurale, interprétation dans le cadre iranien, Université Scientifique et Médicale de Grenoble)

Sadeghian M (1994) Investigating the petrology of igneous rocks and metamorphic rocks of CheshmehGhassaban, Hamedan Region, PhD thesis, University of Tehran

Sadri Y (2011) Petrology of Shadan igneous rocks (southwest of Birjand, east of Iran), MSc thesis (In Persian)

Safarzadeh E (2006) Petrography and petrology of intrusive mass of Haji abad (south of Boin Zahra), MSc thesis, Shahid Beheshti University

Sajjadi Nasab M (2002) Petrography and petrology of granitoid stone Sidadbestser Kerman, MSc thesis, Shahid Beheshti University

Salehi M (2019) Petrological and geochemical analysis of the pyroclastic and volcanic rocks in northwestern Mahneshan (Zanjan province), Iran, PhD thesis, Islamic Azad University, Science and Research Branch, Tehran

Sang-Qaleh R (1995) Petrology of igneous rocks of southern Qaleh (Qorveh area), MSc thesis, Shahid Beheshti University

Sargazi M, Torabi Gh. (2017) Petrography and mineral chemistry of the Eocene granodiorites in the Toveireh area (southwest of Jandaq, Isfahan province). J Econ Geol 10(2): 440–479. Fall 2018– Winter 2019. ISSN 2008–7306 (In Persian)

Scanni E, Azimzadeh Z, Dilek Y, Jahangiri A (2013) Geochronoly and petrology of the early Carboniferous Misho Mafic Complex (NW Iran), and implications for the melt evolution of paleo-tethyan rifting in western cimmeria., Lithos 162–163:264–278

Sepahi Garo A (1992) Petrology of grantoids at Granary, Taknar-Saberj region (northwest of Kashmar), MSc thesis, Isfahan University

Sepahi A, Shahbazi H, Siebel W, Ranin A (2014) Geochronology of plutonic from the Sanandaj– Sirjan zone, Iran zone, Iran and New Zircon and Titanite U–TH–Pb ages for granitoids from the Marivan pluton. Geochronometia 41(3):207–215. https://doi.org/10.2478/s13386-013-0156-z

Shafaii Moghadam H, Ghorbani G, Zaki Khedr M, Fazlnia N, Chiaradia M, Eyuboglu Y, Santosh M, Galindo Francisco C, Lopez Martinez M, Gourgaud A., Arai S (2013) Late Miocene K-rich volcanism in the Eslamieh Peninsula (Saray), NW Iran: implications for geodynamic evolution of the Turkish–Iranian High Plateau, Gondwana Research, GR-01153; No of Pages 23

Shahbazi H (2011) Petrology of igneous rocks and migrations of Alvand complex and Almogholagh infiltration mass and genetic relationship between them. PhD thesis, Shahid Beheshti University

Shahbazi H, Siebel W, Pourmoafee M, Ghorbani M, Sepahi AA, Shang CK, Abedini MV (2010) Geochemistry and U–Pb zircon geochronology of the Alvand plutonic complex in Sanandaj– Sirjan zone (Iran): new evidence for Jurassic magmatism. J Asi Earth Sci 39(6):668–683

Sharifi A (1997) Study of Central Iranian grantoids in the Esfardi-Zarigan Districts, MSc thesis, University of Tehran

Sharkovski M, Susov M, Krivyakin B (1984) Geology of the Anarak area (Central Iran). Explanatory text of the Anarak quadrangle map, 1:250000 (V/O Technoexport Report TE/No. 19). Tehran, Iran: Geological Survey of Iran

Sheibi M (2013) Mineral chemistry and mass changes of elements during alteration of PanjKuh intrusive body (Damghan, Iran). Geopersia 4(1):87–102

Sheikh Zakariaee J (2008) Petrography and petrology of magmatic stones in Qorveh Region, MSc thesis, North Tehran Branch

Sheikholeslami R, Bellon H, Emami H, Sabzehei M, Piqué A (2003) Nouvelles données structurales et datations 40 K–40 Ar sur les roches métamorphiques de la région de Neyriz (zone de Sanandaj–

Sirjan, Iran méridional). Leur intérêt dans le cadre du domaine néo-téthysien du Moyen-Orient. Comptes Rendus Geosci 335(13):981–991

Shekari S, Sadeghian M, Ghasemi H, Zhai M (2018) Mineral chemistry, petrogenesis of metapelitic rocks of metamorphic–igneous Shotor-Kuh complex (SE Shahrood). Iran J Crystallogr Mineral 26(1):184–204

Siahcheshm K, Calagari AA, Abedini A (2014) Hydrothermal evolution in the Maher-Abad porphyry Cu–Au deposit, SW Birjand, Eastern Iran: evidence from fluid inclusions. Ore Geol Rev 58:1–13

Thiele O, Alavi M, Assefi R, Hushmand-Zadeh A, Seyed-Emami K, Zahedi M (1968) Explanatory text of the golpaygan quadrangle map 1:250,000 (No. E7). Tehran: Geological Survey of Iran

Torabi Gh. (1997) Geological, petrological & geochemical studies of the QalehKhargooshi shoshinitic association (Sarve-Bala, west of the Yazd province, Iran). MSc thesis, University of Isfahan

Toulabi Nejad E, Biabangard H, Ahmadi Khalagi A (2016) Petrology and petrogenesis of Ghale-rig granitoid, south west of Nehbandan, east of Iran. New Find Appl Geol J 10(19):80–91

Valizadeh MV, Esmaili D (1993) Petrography and petrogenesis of Doran granite. J Sci 22:1

Zamani R, Vasigh Y (2010) Geochemistry & petrogenesis of Meshkin shahr granitoid (Khanbaz-khankandi), NW IRAN. International Applied Geological Congress, Department of Geology, Islamic Azad University-Mashad Branch, Iran, 26–28 April 2010

Zamanian H, Radmard K (2016) Geochemistry of rare earth elements in the Baba Ali magnetite skarn deposit, western Iran—a key to determine conditions of mineralization, Geologos 22(1):33–47

Zaraei-Sahamieh R (2002) Petrography, petrology and geochemistry of the volcanic rocks in the northern Abhar and relation between volcanism and mineralogy, MSc thesis, Kharazmi university

Chapter 6
Young Volcanoes of Iran

Abstract In this chapter, all Quaternary volcanoes of Iran have been studied and described. Since the volcanic activity of some of them have begun during the Late Miocene and continued until the Quaternary; so they are referred to as Late Miocene-Quaternary volcanoes in this chapter. These volcanoes are described and classified in eight groups according to their geographical location, eruption and activity. For each volcano or any group of volcanoes in Iran, the geographical location, expansion of activity, petrology of volcanic rocks, magmatic series, geological map and the cause of origin have been described and analyzed.

Keywords Volcanoes · Damavand · Sahand · Sabalan · Qorveh · Taftan · Bazman · Quaternary volcanoes

6.1 Young Volcanoes of Iran (Late Miocene-Quaternary Magmatism in Iran)

The Late Miocene-Quaternary magmatism, in fact is the continuation of Cenozoic magmatism in Iran. Most of the volcanic activity has been accompanied by sub-volcanics. A very noticeable fact about the most of the late Miocene-Quaternary volcanic activities and magmatism is that the volcanic activities have begun particularly during the Late Miocene or Mio-Pliocene and continued to Quaternary.

The late Miocene-Quaternary volcanic areas are divided as follows in Iran:

1. Central Alborz, Damavand volcano.
2. Sahand region (south of Tabriz, north of Maragheh, east of Urmia Lake), the Sahand and other volcanoes in the region.
3. Sabalan area, Sabalan volcano.
4. Sistan and Baluchistan region, Taftan-Bazman axis and Shahsavaran Mountains.
5. Qorveh-Bijar-Takab axis.
6. Takab-Qarehaghaj volcanic area.

© The Author(s), under exclusive license to Springer Nature Switzerland AG 2021 277
M. Ghorbani, *The Geology of Iran: Tectonic, Magmatism and Metamorphism*,
Earth and Environmental Sciences Library,
https://doi.org/10.1007/978-3-030-71109-2_6

7.   SE Kerman volcanic areas, volcanic craters around the Hasanalikhan castle at Rayen, Tutak and Haider castle as well as the Mozahem volcano.
8.   Quaternary basalts.

6.1.1   Central Alborz, Damavand Volcano

Geographic location: The Damavand volcanic' cone is located at 60 km east of Tehran, at coordinates of 52° 06′ 24″ E and 35° 57′ 05″ N. Neighboring cities of the Damavand volcano are Rineh (in the southern slope), Polur, Damavand and Firuz-Kuh in the eastern part. Lava flows and pyroclastic materials extend approximately 400 km^2 with a geographical coordinates range from 51° 59′ to 52° 18′ longitude and 36° 04′ 30″ to 35° 48′ 38″ latitude. The altitude of the Damavand volcano is 5610 m above the sea level. There are two routes to the summit. The first way is relatively the easier through the southeastern part and the other is northern route, where climbing is difficult and dangerous. There are several havens for climbers in the way that one of them is located at an altitude of 5000 m. The weather of Damavand is very cold, freezing in winter and mild in summer. In most months of the year, summit of Damavand volcano is covered by snow. The best month to climb is July. Some parts of the white and yellow color observed at the Damavand summit is related to sulfur steam sparkles of the volcano (Fig. 6.1).

Structural position of Damavand: The Damavand volcano is located in the eastern Central Alborz. The eastern and western extensions of Alborz Mountains are of two different trends that meet each other at the Damavand Volcano. As a result, the Alborz Mountains make a ring belt from Aliabad (Gorgan area) to Astara, around the

Fig. 6.1   A view of Damavand volcano (taken by Zadsaleh 2019)

Caspian Sea. According to Stocklin (1974) two major orogenies, Asyntic (Precambrian) and Alpine (Mesozoic and Cenozoic) have developed the Alborz mountains. The last Alborz orogenic movements occurred in the Pliocene-Pleistocene. It caused mild folding, thrusting and increasing the Alborz elevation.

The Damavand volcano occurred during the Quaternary volcanic activities in Central Alborz. All previous tectonic structures such as faults, thrusts and folds are covered lava at the vicinity of Damavand Volcano (Fig. 6.2).

Damavand cone morphology and the spreading out of lava: The Damavand volcano is an asymmetric cone where lava flows expanded widely in the southeastern part (Bout and Derruau 1961).

In the north of the Lenal depression, the crest of Sardovich Mountain is composed of the zigzag shape lava flows. Being 11 km in length and trending west-east, they do not belong to the Damavand lava flows.

The Sardovich lavas are related to the Damavand lavas, where the Nonal valley joins Hareh Mountain, westwardly (Allenbach 1966). They are older than depression's Nonal lava near Haji Dela (Allenbach 1966). According to the geological map of the area the lavas of Karf, Nyneh and lava which are found around Rineh and Ab-e garm are older than those of Damavand (Bout and Derruau 1961).

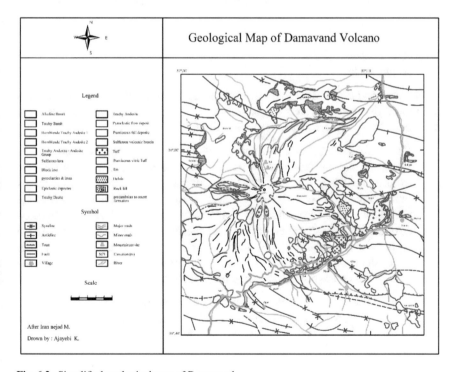

Fig. 6.2 Simplified geological map of Damavand

The Hareh and Sardovich mountains' lava are not quite linear and they extend as flat bodies. There are lavas in the northern Hareh Mountain which are older than Damavand lavas and are likely formed at pre-Quaternary (Irannezhad 1992).

In the NW part of Damavand Volcano, the young lava flows cover more area than the older lavas, so the northeastern Melar lavas certainly are young and could be originated from the Damavand central crater. The west Haji Dela's lavas are compatible with the younger phase of the volcanic activities and they erupted from a secondary crater in 4 km northeast of the central volcano. The southwestern side of the Damavand volcanic cone is much more regular than the northwestern side.

The shape of the Damavand volcanic cone indicates that the activity was not confined to the central crater, but also some side vents at high altitude have a role in the creation of cones at the southwestern and northeastern sides. Obviously the main activity of the Damavand is related to its central crater.

6.1.1.1   Petrographic of the Damavand Volcano

Based on the petrological composition, SiO_2 percentage and mineralogy of Damavand volcanic rocks, three rock groups are signified (Ovcinnikow 1930; Allenbach 1966):

(A)   Basic rocks,
(B)   Intermediate rocks and
(C)   Acidic rocks.

The main part of volcanic rocks is intermediate and the volume of basic rocks is much lower than the other rock types. The chemical composition boundary between intermediate and acidic volcanic rocks is gradual.

(A)   **Basic rocks**: These rocks are located within the Polur (southwestern Damavand volcano) and Rineh in the east and Varkuh Bridge in the far south-east towards the Rineh. The basic rocks are older than other volcanic rocks. Accordingly, they could be the oldest volcanic basic rocks of Damavand which are covered by younger intermediate lavas (trachyandesite) in the Polur area (Irannezhad 1992). The exposures of younger intermediate volcanic lavas have been reported by Allenbach (1966) near Varkuh Bridge. The basic lavas found in lower slopes of the Damavand volcano, sometimes are observed far from the cone, but they are not seen at steep parts of the cone.

The basic lavas are more widely-distributed. Because of their nature and the act that they are covered by younger lava flows. Nevertheless, the younger lava flows have not been able to completely cover the older basic lavas.

The petrology of the area is as follows:

Polur Basic Rocks: These rocks are more basic in origin compared to the other rocks of the area and due to their mineralogical composition, they could be named as alkaline-olivine basalts. These rocks are of porphyric texture and microlitic matrix, but in some cases they show flow texture. The mineralogical composition of one rock sample is as follows:

Phenocrysts (about 30% of rock volume): Olivine (about 10%) is mostly xenomorph, augite (~20%) and apatite (about 1%)

The matrix of rock (~70% of rock volume): The microlite of plagioclase (labradorite; 0.5 mm in length) approximately 35% and augite (~15%), superfine alkali-feldspar within the plagioclase microlite (~10%), opaque minerals and glass (~10%) and apatite less than 1%.

Sometimes the rocks have pores filled by automorphic crystals of calcite.

The basic rocks of Rineh: The basic rocks of Rineh are trachybasalt and are underlain and overlain by the trachyandesite (Irannezhad 1992). Mineralogical composition of a rock sample is as follows:

Phenocrysts: The minerals are mostly automorph augite crystals with maximum length of 1.5 cm and biotite. Some of the biotites are altered to iron oxides. Some of the semi or completely automorphic plagioclase crystals are surrounded by alkali-feldspar. Apatite and analcime are just reported in these rocks (Irannezhad 1992).

Matrix: the matrix consists of the microlites of plagioclase, clino-pyroxene and a minor amount of alkali-feldspar which is xenomorph in shape and lightly fills the space between the other crystals of the matrix.

(B)   **Intermediate rocks**: These rocks are lavas and pyroclastic rocks which form the main part of the Damavand cone and are mainly composed of trachyandesite and trachyte. The rocks that show geochemical characteristics of a rock between trachyandesite and trachyte are varied. There is a gradual change from trachyandesite to trachyte.

Trachyandesite rocks: The matrix and minerals composition of the trachyandesite show some small changes, for instance it is mentioned that the trachyandesite in Rineh and Kraf areas are of hornblende (Allenbach 1966) or found without this mineral (Irannezhad 1992). The analytical results of two rocks' samples are as follows:

Phenocrysts (25% of rock volume): Plagioclase and anorthose (15%), augite, hornblende, biotite and olivine (10%).

Matrix: Plagioclase (65%), anorthose, glass (10%) and opaque minerals.

Trachyandesite-trachyte rocks: In fact, the intermediate rock types are the main component of the Damavand volcanic rocks that are extended in all directions.

Regarding the ferromagnesian minerals found in trachyandesite and trachyte of Damavand, there are sub-rock types as follow (Irannezhad 1992):

- trachyandesite bearing biotite, augite and olivine;
- trachyandesite bearing biotite and augite;
- trachyandesite bearing biotite, augite and hornblende;
- trachyandesite bearing biotite, augite and hypersthene;
- trachyte bearing biotite and augite;
- trachyte bearing biotite, augite and hornblende; and
- trachyte bearing biotite, augite and hypersthene.

(C)   **Acidic Rocks**: These rocks (100 m) are located in the hillside of the northern
      Hareh Mount and cover the Lar limestone. The lava alternating with tuff mate-
      rials that are strongly altered. These condensed, red lavas are of plagioclase
      and hornblende as phenocrysts. Petrographic specification of a sample is as
      follows:

- Phenocrysts (55% of rock volume): plagioclase (35%), biotite (10%),
  hornblende (10%)
- Matrix (45%): Plagioclase (35%), opaque and quartz (10%).

Volcanoclastic Rocks: The Volcanoclastic rocks are more common in the
southern, eastern and western parts of the Damavand volcano and their volume
decreases in the northern parts. The volcanoclastic rocks are less widely-extended
compared to the lavas. Generally, they can be divided into two types:

- Pyroclastic and
- Epyclastic.

  Pyroclastic Rocks
  Damavand's Tuffs

1.   **Vitric tuffs of Haraz Valley**

These rocks having trachytic composition are exposed in the southern and eastern
Damavand along the Haraz-Damavand road (from Ask village to the Baijan).
Mineralogical composition of a rock sample is as follows:

- Phenocrysts (30% of rock volume): alkaline feldspar (Anorthoclase; 10%), biotite
  (5%), augite (2%), apatite (about 10%) and opaque minerals (2%)
- The matrix forms 70% of rock volume.

2.   **Southern Damavand Summit's trachyte tuff**

These rocks are located 400 m above the third climbing haven and one kilometer
west of the site. The area is about one square kilometer and the lava flows are well
outcropped. The rocks' mineral composition is as follows:

- Phenocrysts (40% of rock volume): anorthoclase (10%), plagioclase (15%), biotite
  (5%), augite (5%), orthopyroxene (1%), apatite (1%) and the opaque minerals
  (3%).
- The matrix of the rock (60% of rock volume): Alkaline feldspar and plagioclase,
  the volume of alkaline feldspar is more than the Plagioclase. The opaque minerals
  (2%) and glass (~5%).

3.   **Vitric tuff of the northern Damavand**

These rocks are located at northern hillsides of Damavand volcano, Haji Dela-
Nandel and 8 km southwestern Nandel. They are of a little distribution and show
flow texture.

– Minerals such as plagioclase, augite and apatite as well as opaque minerals appear as phenocrysts.
– The matrix is composed of glass contains tiny crystals of biotite, pyroxene and opaque minerals.

4.　**Vitric pumice tuff of Rineh**

These tuffs are overlain by trachyandesites and are older than them. Microlite feldspars, augite and biotite crystals are severely altered and are dispersed within the vitric matrix. Rock cavities are filled with carbonates.

5.　**Volcanic breccia of the Damavand volcano**

The Damavand volcanic breccias contain fragments of sulfur-bearing trachyte. The Sulfuric breccias are of more exposure in the southern part of Damavand Summit. The area is known as the Sulfur Hill. It seems that all of the sulfur content of the volcano is not originated from the magma but from older gypsiferous sediments.

6.　**Pumice pyroclastic deposits**

The Pumice pyroclastic deposits are distributed in different parts of Damavand volcano. They are most extensive in the south and southeast of Damavand. At some points, these deposits are exploited as pozzolan. Some of these mines are located in northern of the Malar and Fireh villages of southern Damavand and a few kilometers of western Rineh village. Mineralogical composition of a sample of pumice from Fireh mine is as follows:

– Phenocrysts (40% of rock volume): anorthoclase (20%), plagioclases which some of their margins are covered by alkaline feldspar (5%), biotite (5%), augite (5%), orthopyroxene (hypersthene) (2%), opaque minerals (2%) and colorless apatite (1%).
– The matrix of the rock (60% of rock volume): the matrix is made of glass with a large amount of coarse and fine pores. Pumice pyroclastic deposits consist of two types, the pumice fall deposits and the pumice flow deposits.

The pumice deposits in the Damavand area are pumice fall type because first they are well sorted, the same grain size and cluster alongside each other. Second, all topographic levels, relief and recessed areas are covered equally by them (Fig. 6.3).

7.　**Western Damavand pyroclastic flow deposits**

These deposits include various pieces of scoria and lava flows, situated within a context of lava flow with lots of pores. It seems the main crater from which the deposits are erupted, is at an elevation of about 3500 m, west of the Damavand summit.

8.　**Pyroclastic flow deposits of northern Abgarm village**

These deposits consist of fragments of trachyte and trachyandesite rocks that have been severely altered and weathered. These deposits are located directly over the Lar limestone Formation in the northwest and over the Shemshak Formation in northwestern and northeastern parts of Abgarm Village, respectively.

Fig. 6.3 Two views of the Acidic pyroclastic rocks (pumice) in the NE of Damavand

9.   Pyroclastic flow deposits of ash and block

These deposits are well-distributed in different parts of Damavand, particularly along the road. The blocks are homogeneous in composition where both particles and ashes are of the same lithology. The best examples of exposures can be observed along the Abgarm-Polur road.

Epiclastic Deposits
The pyroclastic deposits are more exposed in southern and eastern parts of the Damavand.

6.1.1.2   Formation of the Damavand Volcano

Several theories have been proposed regarding the formation of the Damavand volcano as follows:

– Ovcinnikow (1930) believes that the fault zones of Abgarm and Ask have caused the rise of the lava flow through the channels of faults. Based on his tectonic map the two faults, Abgarm and Ask, join each other just below the Damavand volcano.
– Christa (1940) refers to a bending in the Alborz Mountains arc between Firuz-Kuh and Damavand to the east which has caused the formation of Damavand volcano.
– Allenbach (1966) suggests that the faults that have affected the sedimentary forma-tions have caused lava to raise to the surface. He believes that the structures of the east and west of Damavand volcano are not correlatable, so one cannot definitively say that which fault was responsible in the creation of the Damavand volcano.
   The Baijan fault may have a special role in this regard. The fault valley of Haji Dela-Nandel which extends to the east and southeast could be a caldera. Probably the old lavas of Sardovich, Hareh, around the Karf and Rineh areas have risen through this fault valley to the surface.
   The central Damavand Caldera is located at the area where Ask-Nava and Shahan Dasht thrust faults exist. Most probably by these faults the formation of the Dama-vand volcano made possible. Actually the impact of these thrusts has been caused that the formation and evolution of the Damavand volcano.
– Jang et al. (1975) were dealing with this fact that Arabian and Eurasia (Europe, Asia) plates are collided and as a result the subducted plate has experienced partial melting along the Benioff zone to create magma at depth. They believe that the calc-alkaline volcanic rocks of central Iran are formed by partial melting at high depths of this zone. So, due to being far from the suture zone and the higher depths of partial melting the alkaline volcanic rocks are formed at Damavand volcano.
– According to Brousse et al. (1977), chemical composition of the Damavand volcano indicates that it is a retard developed volcano and far from the Zagros zone. Its formation was initiated by distinctive subduction and melting of the oceanic crust.
– Darvishzadeh (1987) believes that the last compressional movements of Iranian plateau have caused folding, uplifting and shrinking of continental crust. Conse-quently, they have pushed out the point where the Alborz zone changes its trend.

These pressures led to the activation of deep faults and as a result magma found a way to raise to the surface.

- Nogol-e Sadat (1985) believes that the movements of the curved faults created a tensional zone in the region of Damavand volcano. As a result, the volcano has been formed because of these phenomena. The directions of faults are N110-115. If the Vararoud and Ask faults with N130 trend, join each other, the conjunction point is the location of the Damavand volcano.
- Irannezhad (1992) like Allenbach and Nogol-e Sadat (1985) believes that the deep fault zones may have formed some conditions by which the alkaline magma was able to come to the surface.

The recognized faults that continue under the Damavand volcano are as follows:

- Ask fault is traceable up near Damavand.
- Baidjan fault that closely passes the Baidjan village in the vicinity of Karf has been covered by Damavand lava. Westward trend of the fault is visible in the Haji Dela-Nandel valley.
- Nawa fault near the Nawa village. The fault area between the Ask and Gazanak villages is covered by the Damavand lava.
- Sefid-Ab fault in the west side of Damavand volcano is traceable in the Damavand lavas.
- Shahan Dasht fault area between Vaneh and Gazanak is covered by the Damavand tuffs and lavas.
- Vararoud fault in the West of Damavand and upper part of Sefid-Ab fault passes near the junction of Dalichai and Vararoud villages continues up to the Damavand volcano.
- Some geologists, such as Qalamqash, believe that hot spots exist below Damavand (Interview with Asr-e Iran website).

Future Activities of the Damavand Volcano

It seems that the entire Damavand magmatic chamber is crystallized. If such phenomenon is happened, the Damavand Volcano is turned off permanently and cannot be reactivated.

6.1.2   Sahand Volcano

The Sahand volcanic region with more than 5000 km^2 is located in the southern Tabriz city, the northern Maragheh and eastern Urmia Lake.

Volcanic activities of the Sahand area have been intense during the early Miocene. Nevertheless, the geological evidence shows that it was relatively inactive at the beginning of middle Miocene till the late Miocene. From the late Miocene up to the Quaternary it became widely active.

From the viewpoint of petrology, the Sahand volcanos are categorized into the following two groups (Fig. 6.4):

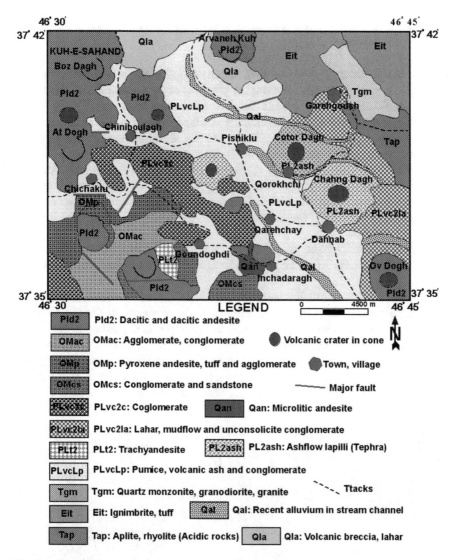

Fig. 6.4 Simplified geological map of Sahand (Behrouzi et al. 1997)

– The main Sahand Volcano and the other surrounding volcanic craters such as southern Alijan village, Khatoonabad volcanic dome, Pakhi Qoludaghy and Qezeldaghy. They are Mio-Pliocene in age and are mostly composed of acidic pyroclastic rocks.

– Plio-Quaternary volcanic rocks of the Islami Island with foid bearing shoshonitic rocks.

6.1.2.1   Geographical and Geological Characteristics of the Sahand Volcanos

The Sahand volcanos with a maximum height of 3710 m are located 40 km south of Tabriz (Ghasemi Asl 2003). The smaller volcanos of the northwestern Urmia Lake and volcanic activities occurred at Ararat Volcano close to the Iran-Turkey border as well as volcanics found in Armenia are assumed to be related to each other. The absolute age determinations show 12–14 Ma. for the samples taken from Sahand volcanic rocks (Moein-Vaziri and Amin Sobhani 1977).

Moein Vaziri (1999) was of the opinion that the Sahand volcanos activities were occurred during several stages. During the process, there were several relatively inactive episodes. Abundant ash with pumice fragments that are dispersed far from the craters, indicates that the Sahand volcanos have experienced strong eruptions (Figs. 6.5 and 6.6).

The highest peak (alternation of breccia, pyroclastic rocks and silicified limestone) is formed in two stages including breccia lavas accompanied by later dacitic volcanic rocks.

The main rocks of the area consist of andesite, dacite, rhyodacite and rhyolite with abundant pyroclastic materials. These rocks are saturated in silica and contain high aluminum contents. The volcanism has occurred in a shallow marine basin due to finding different shallow water fishes in the surrounding area which are mid-Miocene in age. The volcanic activity began as unfavorable conditions for animals' living; consequently, some groups of mammals lost their lives together. The evidence of these animals exists in the sedimentary basins at the vicinity of the area.

Fig. 6.5 Satellite image of Sahand volcano

Fig. 6.6 A view of Sahand volcano

6.1.2.2 Geology and Volcanic Rocks of the Sahand Volcanos

Sahand volcanic body is actually a stratovolcano, including ignimbrites, pyroclastic rocks and lavas erupted through several volcanic chimneys which are scattered over a wide area.

6.1.2.3 Volcano-Sedimentary Rocks of the Sahand Area

Sahand volcanic rocks, 3000 km^2 in area, have covered the Miocene or older sediments. Volcano-sedimentary deposits have extended several kilometers far from craters to the surrounding plains. The Sahand volcanic rocks have better exposures at the northern part, Khalatpoushan area which is now described in details.

6.1.2.4 Sahand Northern Basin

In the western part of Bostanabad, especially near Qurigol Swamp Sahand volcano-sedimentary deposits accompanied by some horizons of ignimbrite are very well exposed. The horizontal layers including an alternation of pumice, lapilli and sinerite observed on either side of the road. The volcano-sedimentary formations at Khalat-poushan area are of well exposure due to floods occurred within the valley resulting in the erosion of basement. As a result, a thickness of more than 40 m of the rocks is clearly exposed.

In Shahgoly, Laleh and Sardroud (southern Tabriz), the volcano-sedimentary rocks are exposed while showing major differences compared to those of Khalatpoushan. Sediments of this region are affected by Bidkhan and Zijenab volcanoes. The volcano-sedimentary rocks are in fact alluvial deposits, formed during floods of numerous rivers. Some of the deposits are eroded and some are remained.

The present rivers flow in valleys with lots of terraces. The sediments of the terraces are composed of sediments carried by floods or those originated from marine or lake deposits. After water retrogression, they have been outcropped and later eroded by the rivers.

6.1.2.5   Khalatpoushan Area

The pumice, lapilli and sinerite layers overly the gypsum and salt formations of upper Tortonian and Sarmatian as an unconformity in the eastern area.

Rieben (1935) named the formations as Alluvium-Tuff deposits. The dip of layers is about 10°. However, the thickness of layers and their grain size in horizontal beds indicate an on shore or deltaic basin. The dip of beds could be related to the dip of older sediments upon which the new sediments are deposited.

The horizontal and diagonal layers of different thicknesses in fore-set and top-set beds signify the geomorphological condition of the depositional basin. There are some very thin veins of black manganese oxide with no economic interest in the lower layers. The presence of manganese oxide veins is another reason to assume on shore and/or deltaic environments as their depositional environment.

In the western part of the Khalatpoushan and Agh-Yoqosh sedimentary horizontal beds of volcano-sedimentary formations are exposed along the fore-set beds. The sediments profile in the region is characterized by the several layers such as ignimbrites, diatomite, sand, fine-grained and coarse-grained conglomerate layers.

Based on the foregoing, we conclude that the activity of Sahand volcanoes started by andesitic magma and followed by later dacitic-ignimbritic and acidic pyroclastic rock. The acidic rocks are dominant compared to andesites.

6.1.3   Sabalan Volcano

The Paleogene volcanic activity was continued through the Neogene -Quaternary in the Ardabil, Meshkinshahr, Ahar and Arasbaran. Their composition is more basic and are of less extent (except for Sabalan Volcano). At this region, only Sabalan Volcano is described (Fig. 6.7).

Fig. 6.7 A view of Sabalan mountain

6.1.3.1 Geographical and Geological Characteristics

The Sabalan volcano with an altitude of 4811 m is located west of Ardabil and is the dividing line of Aras River and Urmia watersheds. The Sabalan is an inactive volcano where extends from the Gharehsou valley in northwestern Ardabil with an east-west trend, about 60 km in length and 48 km in width, continues up to Qoushadagh Mount, south of Ahar. The Sabalan Mount is a strato-volcano where its lava flows cover an area of about 1200 km^2.

6.1.3.2 Geology and Petrology of the Sabalan Volcanic Rocks

The Sabalan is central volcano located on an east-west trending horst system.

Didon and German (1976) suggested a Plio-Quaternary age for the Sabalan Volcano and pre-Miocene age for the initial volcanic activities. Anyway, Babakhani et al. (1980) were of the opinion that in the Meshkinshahr area, the first lava flow has covered the Alvar early Quaternary tuffs and conglomerates. In other words, the volcanic activity started at the early Quaternary and continued to the last glacial period (20–70 kyr.). Didon and German (1976) have divided volcanic activities of the Sabalan volcano into three stages as below:

1. **The old Sabalan lava flow**: the lava flows that have been erupted at this stage form the main part of the Sabalan volcano. The same authors divided this section

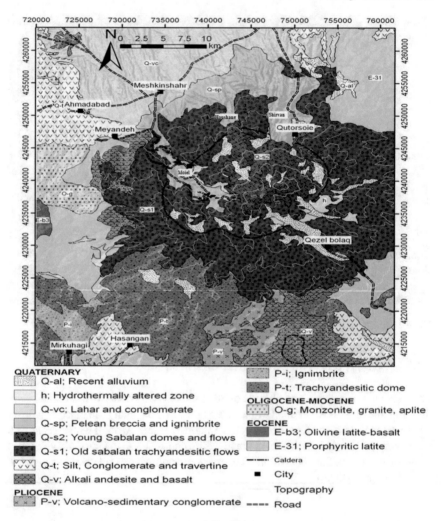

Fig. 6.8 The geological map of Sabalan volcanic (Mousavi et al. 2014)

into five stages, but Babakhani et al. (1980) have summarized them into three steps as follows:

(A)   Underlying andesite outcrops in the north and east;
(B)   Intermediate andesite constitutes the principal part of the Sabalan volcano which is expanded all around the Sabalan Mountain; and
(C)   The last dacitic lava flow completed and ended the old stage of the Sabalan eruption.

2. **Subsidence**: At this stage, the central part of the old caldera was collapsed, resulted in a circular subsidence with a diameter of 20 km. During the subsidence of old caldera, eruptions occurred simultaneously with the explosion and forming of pyroclastic materials. Such a mechanism could be explained by triggering the devastating hot ash avalanches accompanied by the parts of the older rocks going downward to be deposited at the glacial valleys. Later, the low temperature lava flows down as ignimbrites and pumices through Ghotorsui valley and finally a gigantic eruption developed huge volcanic volume of ash in the area.

3. **Domes and lava flows of young Sabalan**: After the caldera collapse, the rhyolitic, ignimbritic and dacitic lava erupted and formed the highest points of the mount forming dacitic domes (Fig. 6.9). At this stage several cones and three lava flows formed within the old caldera. In the present century, the sulfur gases and volcanic vapors are getting out from the surroundings of the volcano. The important point about the Sabalan volcanic rocks is the all rocks are high -Na alkaline in type. The geological map of the volcanic rock units of Sabalan is shown in Fig. 6.8 (Mousavi et al. 2014).

Fig. 6.9 Sabalan Volcano view (Montakhabi 2012)

6.1.4   Sistan and Baluchistan and Northern Jazmurian Volcanoes

6.1.4.1   Geographical and Geological Specification

The E-W trending Taftan and Bazman volcanoes as well as Shahsavaran mount are located at the southeastern part of Urmia-Dokhtar Zone, south of Lut Block, northern Jazmurian (Fig. 6.10). It is about 200 km and 10–60 km in length and width, respectively. The eastern continuation of the volcanic belt is traceable at the Kuh-e Sultan area in Pakistan (Girod and Conrad 1975).

Based on the reports of Girod and Conrad (1975) and Moein-Vaziri and Amin Sobhani (1977), the age of the volcanic rocks is Plio-Quaternary and their lithological composition varies from basalt to rhyodacite, but andesite and dacite are the dominant rock types.

The great Bazman volcano and Shahsavaran Mount are of ellipse shape and an E-W trend. The basaltic young volcanism of Shahsavaran Mount is also traceable eastwards by the basaltic volcanism of Chah-Shahi (northern Iranshahr), Taftan and Kuh-e Sultan (at Pakistan).

The Bazman and Shahsavaran mounts are bounded to the Jazmourian depression in south. Eocene flysch overlying the Upper Cretaceous ophiolite sequence is folded during the Early Miocene and as a result, the Makran Basin has experienced a gradual subsidence (Stocklin 1975; Arshadi and Forster 1983). The above mentioned rock units were southwardly thrusted in Pliocene. So, at the present time, the Shahsavaran volcanic complex is located 150 km north of the greatest thrust zone. The Shahsavaran volcanic complex is westwardly limited by the Nayband dextral fault. It was mentioned that the Nayband fault had a movement of about 120 km. It seems the active Nayband fault detachs the Eocene flysch from the Zagros limestone series in the west (Wellmann 1966). The Chaman Fault movement is estimated to be about 500 km.

The Shahsavaran volcanic activity started since 12 Ma. and lasted until the Quaternary (Girod and Conrad 1975). Its volcanic lavas have disconformably covered the Eocene to Oligocene rocks. Furthermore, some dikes of the same composition have cut the Eocene-Oligocene granodioritic intrusive bodies at Shahsavaran Mount which are covered by the Miocene to Quaternary lavas (Girod and Conrad 1975).

Petrography and petrogenesis: According to Girod and Conrad (1975), Plio-Quaternary lavas of Shahsavaran Mount have experienced magmatic differentiation. The series include peraluminous basalt, andesite, dacite and rhyodacite, calc-alkaline in type.

The older basaltic eruptions form the minority of the rocks. The basaltic flows are widespread as layers separated by acidic volcanic rocks.

The andesitic rocks consist more than 50% of the Shahsavaran volcanic complex. The andesitic lavas are too young and cover the recent alluviums; therefore, the calderas have been well preserved. The andesitic rocks of Bazman volcano are

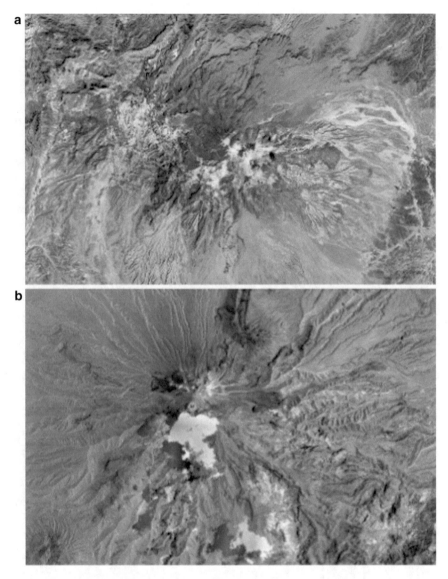

Fig. 6.10 Satellite imagery showing the general structure of Taftan (**a**) and Bazman (**b**) volcanoes

composed of volcanic breccias, pumice tuff and ignimbrites, indicate explosive eruptions. Mineralogical composition of basalts is invariable and they are composed of
phenocrysts including basic plagioclase (85–90% of anorthite as core), augite, olivine
(with a reactional margin composed of orthopyroxene) and a matrix including plagioclase, alkali feldspar and quartz. They are peraluminous and calc-alkaline forming
the basic series of such rocks in the region.

Some of the andesites are hyaloporphyric containing 5–10% of glass. The phenocrysts include plagioclase (30–90% anorthite) and the matrix include oligoclase-andesine, orthopyroxene, quartz (tridymite) $\pm$ augite $\pm$ alkaline feldspar and iron oxide particles.

Dacite and rhyodacite are more abundant at Bazman volcano compared to the Shahsavaran Mount. Therefore, the acidic rocks are of more significance in eastern Shahsavaran Mount. In this region, there are lots of places where the alternation of acidic tuffs, rhyodacite and dacite can be found.

1.   **Taftan volcano**: Taftan is a young Pliocene-Quaternary, semi-active volcano located in Baluchistan, 50 km of Khash City. The volcano has a height of 4050 m above the sea level and 2000 m above the surrounding plains. The basement of the volcano is composed of the Upper Cretaceous and Eocene flysch where Mesozoic ophiolites are exposed. The ophiolites are tectonized during the Upper Cretaceous (Fig. 6.11).

The first eruption of the Taftan was lava and pyroclastic rocks with a dacite and rhyodacite composition triggered at 20 km northwest Taftan summit (Gansser 1966).

The reactivation of Taftan has occurred at Lijvar Crater, about 10 km northwestern of the present crater in the late Pliocene. The other crater, located 2 km of southern present crater, was also active at the same time (Moein-Vaziri 1985).

After an inactive period that led to deposition of agglomerate, a major explosion occurred at a distance of 2 km in southern part of the present crater resulted in a erosional pit. The ash from the explosion created a massive thick layer between the

Fig. 6.11   A view of Taftan volcano

Fig. 6.12 Plio-Quaternary
volcanoes in Qorveh
-Bijar-Takab axis (Ghorbani
2003)

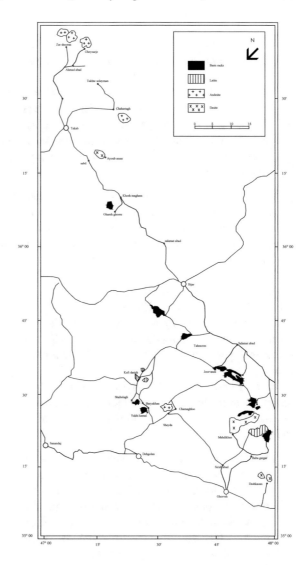

Quaternary andesitic lavas and the older volcanic rocks. The ash layer can be considered as a time indicator between the Pliocene and Quaternary lavas. Afterwards,
andesite lavas flowed away. At the present time, water vapor, sulfur and carbon
dioxide gases are coming out through the relatively high Taftan andesites. Because
of the steep slopes of Taftan and more viscosity of the basic lavas, they have reached
Tamin village at 12 km of northern flank.

   One of the interesting features of Taftan is its inconsistent mineralogy and inverse
transformation of minerals in the Quaternary andesite lavas which are described as
below:

1. Based on petrographic studies of a sample, two plagioclase phenocrysts are observed, one of them was more calcic and the other was more sodic,
2. The marginal part of the plagioclase phenocrysts is richer in anorthite than their cores,
3. Microlites composed of plagioclase in matrix are richer in anorthite compared to the marginal part and/or core of the phenocrysts at the same rock,
4. The margins of orthopyroxene are richer in Mg than the core, and
5. Na_2O and K_2O show negative correlation, whereas these elements have participated in the mineralization of those minerals forming at the late stages of crystallization.

Unlike other young volcanoes in Iran, basic rocks are not found in the Taftan lava flows. The primary activities of the Taftan differ in composition compared to that of final activities which are mostly composed of andesitic-dacitic lava. They consist of acidic pyroclastic rocks similar to pumice and pumicite. Pyroclastic rocks and pumice tuffs cover large parts of eastern and southwestern Taftan Volcano. These rocks are composed mainly of pumice and pumicite.

Some samples of pumice breccia can be found at the southern part of the Sangan village along the Myrjaveh-Khash road. They are white in color and because of their high porosity the density is low. In southwestern Taftan, the pyroclastic rocks are composed mainly of pumicite that are somewhat dense and non-porous. In fact, the pyroclastic particles have been cemented by a matrix as volcanic ash with a same composition.

In the Khash road, along Myrjaveh road, the travertine terraces and calcareous sediments of freshwater cover a wide area. In the area the pumice overlies the travertine and as a result at the same place, clay, travertine and pumice indices can be found.

2. **Bazman volcano**: the volcanic and plutonic rocks of the northern Jazmurian depression are known as magmatic Bazman complex. The magmatic complex is a part of Urmia-Dokhtar Zone. The Bazman intrusive rocks consist of alkaline porphyry granite with potassic coarse-grained feldspar, hornblende granite and granodiorite to quartzdiorite. The age of the rocks is 64–74 Ma. The granitoids cut the older gabbro at their marginal parts where some parts of the later can be found as enclaves in granitoids. The granitoid have intruded into the shale, limestone, sandstone, siltstone and dolomite of Paleozoic Era and caused the development of contact metamorphism resulted in the formation of hornfels and skarns.

The extrusive rocks of this region are dacite, andesite-dacite and rarely rhyolite ignimbrite and vitric tuffs which have cropped out in eastern and southern Bazman Volcano. The Bazman volcanic rocks are mainly andesite, basalt, and minor olivine-basalt. The Bazman stratovolcano has a complex structure and a variety of lava such as andesite, dacite and rhyodacite are highly exposed mostly in eastern side of the volcano. The main crater of Bazman Volcano is composed of an alternation of ignimbrite, breccia, pumice and lava.

6.1.5 Qorveh, Bijar and Takab Volcanic Areas

The volcanic activities were started in the late Miocene-Pliocene at northeastern Qorveh. It began by sub-volcanic magmatism, monzodiorite in composition. Later hornblende andesite and andesitic tuffs were formed. However, the amount of pyroclastics was much more than the lava (Abdi 1996). The color of the most hand specimens is gray and red and they show porphyry texture and porphyritic with a matrix composed of microlites and glass. Phenocrysts are mostly plagioclase and hornblende. The petrographic and geochemical studies indicate magmatic amalgamation and differentiation in the initial magma (Abdi 1996). In the Qorveh area, the volcanism intensity increased again in Quaternary and a variety of different igneous rocks have formed (Abdi 1996). The Quaternary igneous rocks in northeastern Qorveh could be classified as follows:

(A) **Basic and relatively basic igneous rocks**: The rocks in the aforementioned area are formed by lava flows or as small bodies overlying on the Miocene andesites. The basic rocks are characterized by high porosity and black or red color. Their texture is porphyritic with a matrix composed of microlite and glass. The phenocrysts consist of olivine and pyroxene while the matrix is composed of olivine and plagioclase. The basic and ultrabasic rocks are well distributed at northern Qorveh as scoria cones and lava flows (Fig. 6.13). These rocks have xenoliths of peridotite and gneiss. Also, the original magma shows some contamination by the crustal rocks. The basic rocks are rich in CaO, Fe_2O_3, MgO, Na_2O, K_2O, Cr, Ni, Sr, Ba and Co.

(B) **Intermediate igneous rocks**: These rocks are mostly composed of trachyandesite with porphyritic texture and microlite as matrix. There is some evidence of mixing the acidic and basic magmas.

(C) **Acidic Igneous Rocks**: Acidic rocks are relatively younger than the basic rocks. They have formed as high and light colored (white) domes. Their composition varies from rhyodacitic tuff to rhyolite in Baharloo-Dashkesan villages.

The rocks are of calk-alkine character and are the result of partial melting of the crust, due to basic and ultrabasic magma intrusions beneath the continental crust. These rocks are assigned to be associated with a post-subduction rift environment (Abdi 1996).

The volcanic rocks of this axis are often in the form of lava flows. The Miocene volcanoes are composed of high-K trachyandesite, high-K latite and high-K rhyodacite, while in contrast the Pliocene-Quaternary volcanoes are of different compositions. However, in general, they are basalt or rhyodacitic-dacitic lava flows. A few of them are pumice and scoria (Fig. 6.12). The most important volcanos of this axis are described as follows:

1. **Qarehtour Volcano**: This volcano is located between Takab and Bijar. It seems it is located on a fault that separates the Plio-Quaternary's sediments from the metamorphic rocks. The rocks of this volcano are mostly basaltic lava. The

Fig. 6.13 Two views of the Qorveh scoria

geological evidence indicates the volcanic magma has erupted from a small volcanic channel in a few steps in a relatively short time interval.

2. **Nadri Volcano**: This volcano is located 15 km of southwestern Bijar. Its lavas cover an area of about 13 km^2 which are mainly basanitic.

3. **Tahmoures Volcano**: It is located close to the Tahmoures village. The basanitic lava is the main component of the volcano. Its lava erupted from a small volcanic channel in a few steps at a relatively short time interval.

4. **Firouzabad Volcano (Jurvandi)**: Probably the Firouzabad volcanic activity lasted longer than the other basaltic volcanoes in Qorveh and south of Takab area. The volcano was active during Plio-Quaternary. The Pliocene sediments are occasionally mixed with the basalt flows. Hence, the volcanic rocks and the sediments accompanied by them are slightly folded and cut by the river. Also the Ahmadabad lava which is younger than the Firouzabad volcano has poured into the same valley.

5. **Ahmadabad Volcano**: It is a basaltic volcano accompanied by scoria formation and its lava flows can be traced up to the Firouzabad volcano. The Ahmadabad volcano is younger than the Firouzabad volcano. The lava of the volcano contains some enclaves of metamorphosed basement, such as gneiss as well as xenoliths composed of peridotite.

6. **Mahdikhan-Qezeljehkand Volcano**: The basaltic lavas have flowed away in several stages in Mahdikhan-Qezeljehkand area and overlie the Plio-Quaternary carbonates (travertine) and clastic sediments. A travertine layer underlies the basaltic lavas of Mahdikhan and overlies the Pliocene clastic sediments. The basanitic lavas of volcano cover a vast area. The basement of the area which probably belongs to the Precambrian metamorphic rocks is covered by the Plio-Quaternary clastic and carbonate sediments. These metamorphic rocks are exposed by three kilometers of the volcano to the east. The porous gneiss enclaves were found within the volcanic scoria of Qezeljehkand. These enclaves have experienced the partial melting phases.

7. **Gharehbolagh Volcano**: The dacitic lava erupted from these volcanoes mostly developed as domes. The Gharehbolagh dacitic lava flows cover the basanitic rocks. So, the dacitic lava is younger than later.

8. **Sheida and Yakhy Kamal volcanoes**: These volcanoes are dacitic and andesitic domes located on the top of the Miocene sandy marl sediments. One of these eruptions has occurred through a NE-SW fissure (Gheshlagh Khoda-Karam's dike) and is likely responsible for the formation of an ignimbrite horizon around the Zafarabad and Gheshlagh Khoda-Karam.

9. **Volcanos around the Kani Derizh**: The Kani Derizh volcanic domes are composed of high-K dacitic lavas that are apparently underlain by clastic and carbonate sediments, Miocene in age, as scattered bodies (Fig. 6.13).

6.1.6   Volcanoes of the Takab and Qareh-aghaj Areas

There are a number of volcanoes in Takab, Angouran, Mahneshan, Pary and Qareh-aghaj that are located at the Urmia-Dokhtar zone (Fig. 6.12).

Volcanic activities in the area, after the Aquitanian acidic volcanism including acidic tuff, tuffite and tuffaceous marl (lower part of the Qom Formation), reduced. Afterwards, during the late Miocene to Quaternary, a new stage of volcanic activities triggered and continued up to the lower Quaternary. Most of the volcanoes in this region have been formed during this period. The volcanoes are mostly andesite or dacite-rhyolite in composition. The Petrology and field evidence signify that these volcanoes have a great difference in the point of view of geodynamic and petrogenesis compared to those of Qorveh-Bijar axis. A number of volcanic centers in the region are as follows:

1. **Volcanoes around the Takht-e Soleyman area**: The area around Takht-e Soleyman is topographically of a circular shape. There are several scattered exposures of volcanic rocks and volcanoes surrounded the area such as the Zarshuran, Baba-Nazar, Amirabad and Chahartagh. It seems that all the volcanic activities coincide with the upper Miocene to early Quaternary. The rock compositions of the volcanic centers are potassium-rich andesite, dacite and basalt. Hydrothermal springs associated with them have resulted in a large amount of travertine formation in the region that is still continuing.

2. **Volcano of Ayyub Ansar**: This volcano is located 15 km SE of the Takab, close to Arabshahi (Arabcheh) and Sebil villages. The type of rocks is rhyolitic to dacitic lavas associated with tuffs exposed in the east. It seems that the volcanic rocks are situated in the middle and upper part of the Upper Red Formation. Accordingly, it is supposed that the age of the volcano is the late Miocene to early Pliocene.

3. **Kuh-e Baba Volcano**: this volcano is located in the eastern part of the Qareh-aghaj village. Pir Saqqa, Maryam, Ali Kand, Hammam, Abak and Isisu villages are around the volcano. There are hot springs in most parts of these villages which are related to the Kuh-e Baba volcano such as hot springs of Isisu village (Ab-e garm), Hammam and Pir Saqqa geyser.
   Andesitic and basaltic volcanic rocks of the Kuh-e Baba overlie the Upper Red Formation.

4. **Gurgur Volcano**: This volcano is located in the north of Aghadarreh and south of Arabshah, southeast of Arpachay. The Gurgur volcano shows the characteristics that are similar to Kuh-e Baba.

6.1.7 Southeastern Kerman Volcanoes and the Quaternary Volcanic Craters Around Qaleh Hasanali of Rayen, Qaleh Haidar and Tutak

Fourteen Quaternary explosive craters can be observed around the Qaleh Hasanali of Rayen, Qaleh Tutak and Qaleh Haidar, 120 km of southeastern Kerman, which were initially introduced as the meteorite impact (Dimitrijevic 1973; Ahmadi 2004) and later as volcanic craters (Milton 1977). The craters have been developed in the Eocene volcanic or volcano-clastic sediments of Oligo-Miocene or clastic-sediments of the Quaternary. So, their age is Quaternary, and their volcanic evidence suggests that they could be ultravolcano.

Among these 14 craters, the high-K young magmatic rocks with ultrabasic composition have been erupted only from the crater No. 9. Generally, the rock fragments which outpour from the craters are volcano-clastic sediments of Oligo-Miocene and Eocene (Fig. 6.14). However, in some of them, shattered fragments of igneous rocks such as granite, syenite and rarely diorite and gabbro exist with clastic materials.

6.1.7.1 Petrography of the Volcanic Rocks Southeastern Kerman

The volcanic rocks of the area consisting of quartz basalt, andesite, trachyandesite, trachyte, rhyodacite, rhyolite and pyroclastic rocks (tuff and breccia) are scattered around the craters of Qaleh Hasanalli of Rayen or even at the inner part of some of the craters. The crushed volcanic rocks with an Eocene age during the blasting of the craters are distributed in the surroundings of them. Accordingly, they contain some

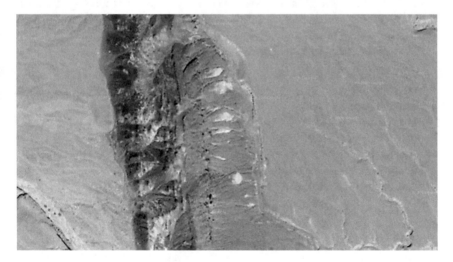

Fig. 6.14 Satellite image of Qaleh Hasanali volcano

materials of the older craters' walls. The basic rock fragments form the minority of the blasted materials compared to the acidic and intermediate ones. The basement of the area is mostly composed of pyroclastic rocks that have been cut by dikes.

Most of the materials thrown away from craters are of the Quaternary age like tuff and volcanic breccia composition.

The eastern, western and southern rims of the crater No. 9, the western rim of the crater No. 8 and eastern rim of the crater No. 10 are made of the abovementioned pyroclastic rocks. The tuffs and breccias have experienced a hydrothermal alteration resulted in a strong metasomatism. The tuffs include fragments and matrix that are rhyolitic, rhyodacitic, dacitic and trachyandesitic in composition.

Rhyolitic tuffs contain quartz and alkali feldspar as phenocrysts of which the later is altered to sericite. Trachytic tuffs and trachyandesites are of amphibole and oligoclases as phenocrysts. Their matrix is rich in iron oxides where ferromagnesian minerals have been altered into hematite and chlorite.

Dacitic tuff and rhyodacite have porphyritic texture and contain minor amounts of apatite and titanite minerals. The plagioclase minerals indicate a hydrothermal alteration.

6.1.7.2 Petrology of the Qaleh Hasanali Rocks

It seems that the ultrapotassic lava has erupted only from the crater No. 9, in the Qaleh Hasanali, Rayen area. Due to being several times deeper than the other craters and the throqing away the rock fragments too far from the crater No. 9. As a result, probably the intensity of the explosion occurred at this crater was more powerful compared to the other Quaternary craters of the region. The ultrabasic fragments of crater No. 9 can be found in the lowest to the highest layer of the sequence. The texture, matrix and sometimes chemical composition of the volcanic ultrapotassic bombs show wide variations.

The volcanic bombs can be divided into three categories:

1. **Less porous Granular Bombs**: As the name implies, these bombs are usually of less porous and grayish green or jade green in color. They contain olivine and phlogopite crystals.
2. **Hyalo-porphyritic porous bombs**: these bombs are scattered in northern parts of craters Nos. 8 and 9 and are green to dark gray in color. They are more abundant and sometimes contain coarse grains of phlogopite crystals.
3. **Porphyritic porous contaminated bombs**: The color of these bombs is lighter than above mentioned bombs and they contain some xenoliths.

The ultrapotassic rocks of Qaleh Hasanali, Rayen area, are dark gray in color and round or oval in shape found at crater No. 9. Some of these bombs contain small and large xenoliths of syenite, granite and rarely diorite and gabbro. The ultrabasic rocks of Qaleh Hasanali are porous and have granular or porphyric-microlithic texture. In the inner side of the pores and sometimes between the large crystals, small amounts of glass can be found. Microlite crystals are composed of clinopyroxcene and minor amounts of phlogopite accompanied by microcrystals of iron oxides. The ultrabasic rock-forming minerals of Qaleh-HasanalI, Rayen area, are as follows:

– Clinopyroxene is found as phenocryst and microlite. This mineral sometimes has poikilitic texture and contains phlogopite crystals. There are two types of clinopyroxene in bombs containing syenite xenoliths: some of them are colorless and homogen while others have green core and colorless margin. It does not seem that the abovementioned aegirine augite is derived from syenite xenoliths and has formed in an ultrapotassic magma resulting in forming a halo of salite around the augite minerals (Shishehbor 1993).
– Phlogopite is mostly automorph in the form of large single crystals and sometimes shows poikilitic texture and contains olivine. This mineral mostly occures as phenocryst and sometimes forms the microcrystals of matrix.
– Olivine is abundant in some lamproitic bombs of Rayen and in some others is rare. Olivine crystals are sub automorph and mostly surrounded by phlogopite and clinopyroxene crystals.
– The olivine crystals mostly are not decomposed and some of them have the characteristics of structural changes identifying that they are originated from the mantle.
– Apatite as small crystals are found within the phlogopite.
– Iron oxide and titanium appear as separate, cluster octahedr crystals, dendritic, or feather-like in glass matrix of lamproitic bombs.
– Analcime and haoline in glass matrix and ultrapotassic bomb pores are hardly recognizable.

6.1.7.3   Chemical Composition of Qaleh-Hasanali Rocks

The comparison of the average composition of ultrapotassic rocks of the Qaleh-Hasanali area with an average composition of lamprophyres indicate the same amount (%) of SiO_2, Al_2O_3, TiO_2, Fe_2O_3 and CaO in both rock types. The ultrapotassic rocks are depleted in terms of K_2O/Na_2O compared to lamprophyres. On the other hand, they are rich in MgO in comparison to the lamprophyres.

6.1.7.4   Pliocene Volcanic Rocks at Shahr-e Babak

The first phase of the Pliocene volcanic eruptions began by forming dacitic domes and lavas and completed by andesitic lava flows. The dacitic and andesitic lavas of Pliocene have passed through the volcano-sedimentary rocks of Eocene-Oligocene, and Miocene in northern Shahr-e Babak. The Pliocene volcanic activity is not so significant in Rayen area. This evidence indicates that the Pliocene volcanic activities have shifted to the western and southwestern of their initial location of occurrence.

The Pliocene volcanic rocks of the Shahr-e Babak region are most abundant in the following locations:

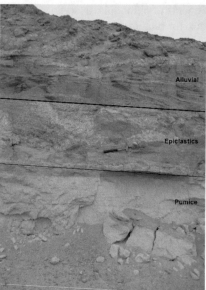

Fig. 6.15 Pumice of Anar plain, near Kuh-e Aj

(A)   **Dehaj Area**: This type includes dacite showing porphyritic texture in which
      hornblende and andesine are surrounded by a felsic matrix. In addition to
      dacite, quartz andesite with hyalo-porphyritic texture can be found in this
      area.

(B)   **Mozahem Area**: The main caldera, 5–7 km in dimensions, of the explosive
      volcanic activities is the Mozahem volcano, located in the northern Shar-e-
      babak. The volcanic activities began by pyroclastic explosions and continued
      by deposition of pyroclastic rocks accompanied by some sediments. Later,
      hornblende-bearing andesitic lava erupted. The andesite is overlain by younger
      pyroclastic rocks. The last phase of volcanic activity is the emplacements of
      dikes.

(C)   **Kuh-e Aj Area**: The volcanics of this area has been named based on two
      volcanic cones, Aj-e Bala and Aj-e Paein, meaning upper and lower Aj, respec-
      tively, which are found in Dehaj quadrangle. The cones have mainly andesitic
      lava flows and occasionally some dacitic eruptions have been recorded. The
      andesites of the Kuh-e Aj volcanos contain feldspar, hornblende, biotite, augite
      and minor amounts of Augite and quartz as phenocryst. Their matrixes are
      micro-poikilitic in texture with plagioclase as microlites (Fig. 6.15).

6.1.7.5 Kuh-e Mozahem Volcano, Northeastern Shahr-e Babak

Mozahem volcano is located 15 km of northeastern Shahr-e Babak in Kerman
province (Ahmadi Pourfarsangi 1993).

Its lava flows have covered an area of about 625 km^2. The volcano is located at the southwestern border of Urmia-Dokhtar volcanic belt and is hosted by the Hezar complex, sedimentary and igneous rocks, Upper Eocene in age.

The base of the pyroclastic cone is about 25 km^2 in diameter, and because of the morphological characteristics of the host rocks (Hezar complex), the erupted materials distributed mainly to the southern and western parts of the volcano. The height of the basement (Hezar Complex), at northeastern side, is 500 m higher than the southwestern part.

The volcano has a central collapsed caldera with very steep walls, up to 900 m in depth. The irregular circular shape of caldera is developed at the western edge of the volcanic cone and is 38 km^2 in diameter. Inside the caldera several small calderas, often more than 200 m in height, are found. The maximum thickness of pyroclastic sequence of Mozahem volcano is in the northern side of the caldera and they are well distributed at the southern and western sides.

The volcanic eruptions have resulted to the formation of following units:

(A) **Explosives eruptions**: the erupted materials include small pieces of host rocks, magmatic rocks such as dacite, rhyodacite, and pumice fragments which are embedded in a matrix composed of fine ash. Erupted materials are generally white in color, and covered by a red coat of the old soil. These materials are located at the base of the cone.

(B) **Breccia lava**: The breccia lava is only found in the central part of the caldera and contains angular fragments of rhyodacite and dacite that have the same matrix of former type. The breccias are formed by the contact of the magma and groundwater.

(C) **Pyroclastic falls**: They are found as an irregular ring around the caldera. They are very dense and have formed tall vertical walls. The magmatic rocks (andesitic) and host rocks are hosted by a matrix of ash. They are resulted from falling and welding of erupted loose clastic materials.

(D) **Pyroclastic flows**: The pyroclastic flows include very large to small andesitic bombs. The bombs were cemented by fine-grained ash matrix which is of a same composition as the bombs. The diameter of the bombs sometimes reaches to 6 m. The bombs are formed by collapsing and flowing of pyroclastic material. They should be categorized as pyroclastic ash flows and blocks.

(E) **Lavas**: The lavas are andesite and trachyandesite in type alternating in a sequence of falling and flowing deposits.

(F) **Intrusive bodies**: The Intrusive bodies are classified as three types.

- Acidic bodies intruded before the formation of caldera which are deeply altered;
- The dioritic dome-like bodies, dikes and sills intruded into the caldera or inside the pyroclastics; and
- Magnetite lamprophyric dikes that are so young.

Mechanism

The volcanic activity of Mozahem volcano could be described in three steps as follows:

1. **The first step**: During the upward movement of magma in late Cenozoic (Upper Pliocene), magmatic differentiation occurred and an acidic portion formed at the upper part of the magma. The acidic magma met the phreatic zone and as a result a phreatic eruption by which a column of eruptied materilas has formed. At the base of such column, eruptions occurred to form a thin well distributed layer.

    Then the lava breccia formed within the caldera after cooling of the raised magma; the granitic plutons have been simultaneously developed at the deeper parts. Afterwards, the volcano was inactive. This can be verified by the deposition of red paleo-soil layer at the top of the pyroclastics.

2. **The second step**: At the second stage, the andesitic magma ascents by a phreato-plinian eruption forming a huge column of pyroclastic materials. By deposition of these materials, the pyroclastics of the caldera and flows of the median facies are formed. These explosions accompanied by several relatively inactive periods in which the magma flowed away from the caldera. Due to the eruption of a huge volume of magma a collapse occurred at the central part of the volcano and afterwards the hydrothermal fluids altered the mentioned rocks.

3. **The third step**: After the formation of caldera at the last stage, domes, sills, and diorite dikes emplaced within the mentioned rock units.

Petrology and Geochemistry

Petrological and geochemical studies show that there is a very strong correlation between the various stages of the volcano evolution; so, most probably they are originated from a same magma. The mentioned magma, most probably, was a hydrous andesitic magma resulted in the formation of high -K calc-alkaline rocks.

The inclusions of basic granulite show that the magma is likely developed from partial melting of the lower continental crust.

Tectonic position

The volcano is located at the junction of two right-lateral faults, N–S Anar and NW–SE Shahr-e Babak faults. The activities of these faults resulted in the crustal decompression at the junction point and these phenomena facilitated the magma ascent (Nogol-e Sadat 1978). The volcanic activity has probably occurred in the Early Quaternary (Sabzehei 1994).

6.1.8   Quaternary Basalts

As mentioned above, the volcanic activities strongly started again in the Upper Miocene and Pliocene. Most of the volcanic activities are terminated in Plio. Quaternary by basaltic volcanic rocks. However, in some parts, the basaltic volcanoes of the Pliocene are not exposed and the Quaternary basalts are solely visible.

The Quaternary basalts of Iran are relatively well distributed and have been reported in many parts of the country. But the Quaternary basalts are more developed in two following areas:

– Azerbaijan and
– Eastern Iran

Many of these basalt occurrences have been described in Neogene/Quaternary part of the present book, so, only a few of them which are fairly large and have not been previously described are discussed here.

Azerbaijan young basalts
The areas of Azerbaijan where young volcanics can be found are as follows:
1—Ararat, 2—Maku, 3—Salmâs, 4—Ahar, 5—Neir, 6—Saveh, 7—Sarab and 8—Mahabad.

6.1.8.1   Ararat (Maku Area)

The Ararat mount is a stratovolcano which covers about a thousand square kilometers. The volcano is located at the junction of the major E-W and NW-SE fractures.

The Ararat, Nemrut, Suphan, and Tendurek volcanoes are parts of a NE-SW volcanic belt, 70 km in length. Also, there is a NW-SE volcanic belt in the area cut by the Aras River (Lambert et al. 1974).

In the Ararat region, the Quaternary sediments are covered by Lower Red Tuff, andesite, dacite and rhyodacite. Finally, the basaltic flows have covered the entire area (Fig. 6.16).

The young basalts at the surroundings of Ararat are easily distinguishable due to their dark color and showing flow texture (Figs. 6.17 and 6.18). These basalts have erupted as a fissure volcanism after the last glacial period and extend southwards of Ararat volcano. The Ararat volcanic rocks (except for the basalts) are divided in two types, yttrium-rich and yttrium-poor.

These two series consist of andesite, dacite and rhyodacite. The basalts of the area are generally poor in yttrium. The plagioclase phenocrysts of andesites and dacites are of more percentage of anorthite in yttrium-poor series compared to those of yttrium-rich. So, two kinds of plagioclase (labradorite and andesine) can be found. In addition, the dacite and andesite of both series are of ortho-pyroxene, clinopyroxene, magnetite and ilmenite. Hornblende has been altered to hematite in yttrium-poor dacites. There is no Quartz and alkali-feldspar as phenocryst (Lambert et al. 1974). Based on the

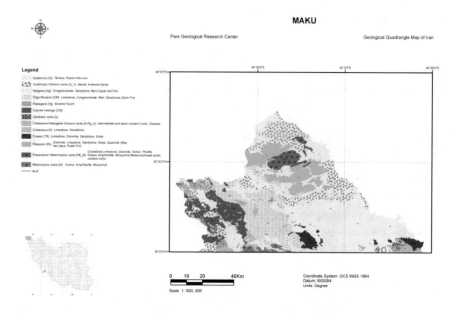

Fig. 6.16   Geological map of Maku

Fig. 6.17   A view of Poldasht lavas, Type aa

Fig. 6.18 The Ropy Quaternary basalts in the Poldasht region

Kuno (1959) diagram, the two series are divided as the calc-alkaline, and tholeiitic series (yttrium-poor). The tholeiitic series is originated from a differentiated andesitic calc-alkaline magma.

Radiogenic Sr ratio in dacites and andesites, 0.7042–0.7055, indicates their mantle source. Laboratory studies by Ringwood and Green (1972) have indicated that garnet and clinopyroxene can be crystallized at high pressures by this kind of magma. Such data encouraged Lambert et al. (1974) to think about a multi-stage magmatic evolution at the Ararat volcanoes. According to these authors, garnet and clinopyroxene have been crystallized of an andesitic magma originated from the mantle and contains 2% water. The magma has experienced a pressure of 15 K bars. The crystallization of minerals increases the water content of the magma. This mechanism makes a suitable chemical condition for amphibole crystallization, and at the same time, increases the ascent speed of the magma. High speeds of magma ascent led to upward movement in an adiabatic mechanism. In the first phase, at pressures between 8 and 12 Kb, the crystallization of amphibole has resulted in the formation of yttrium-poor series in response to the solid distribution coefficient of yttrium in hornblende which is equal to 5 or greater.

At the second stage, the water percentage raise again to about 2%. Some parts of water are released at the surface, and some other parts of the water have participated in the host rocks alteration. The reduction of residual water in the magma resulted in the upward and slow movement of magma and as a result the magma soon passed the amphibole stability field. In this case, instead of the amphibole crystallization, plagioclase and clinopyroxene are crystallized; so, an yttrium-rich series is developed. It should be noted that garnet, sphene and apatite have high yttrium solid distribution coefficient, but they are not observed in Ararat volcanic rocks. In case of creating an yttrium-poor series, it is necessary to count the garnet crystallization at high depths. Although the petrographic study of the rocks does not show any garnet, but there is a paragenesis of plagioclase, orthopyroxene and oxides. According to Ringwood and Green (1972), the oxides could be a pseudomorph of garnet.

The absence of spinel in volcanic rocks of Ararat is probably due to the existence of aluminum and titanium in the structure of pyroxenes.

Lambert et al. (1974) believe that huge crustal fractures along with the movement of micro plates have resulted in the volcanism at the Ararat area and the subduction theory cannot explain this event.

6.1.8.2   Maku Area

Maku is the most northern county of Iran. The outcrops of the volcanic rocks in Maku region are more widely-distributed compared to the other areas. The volcanic rocks are found in an area started from eastern Poldasht, 55 km of eastern Maku, and continued to southern Siahcheshmeh. They are of more exposures at three areas including Maku-Poldasht, Bazargan plain and the most northern embayment of the county. The volcanic outcrops are located at $44° 03' 13''$ to $45° 09' 06''$ longitude and $39°$ to $39° 46' 13''$ latitude.

In the Chaldoran region, these rocks found in the vicinity of Bazargan, Maku and Poldasht cities. In most cases, access roads to the outcrops are asphalted roads (Fig. 6.19).

The Maku green volcanic rocks are originated from the Ararat volcano. Pyroxene in porphyritic texture plays the major role as phenocryst and is mostly composed of clinopyroxene. About 10–20% of pyroxenes are identified as augite by measuring optic axial angle (2 V) (Kheirkhah and Emami 2010). They are altered to form biotite and rarely amphibole. Plagioclase is found as automorph to semi-automorph crystals in forms of column or fractured that constitutes the majority of phenocrysts. In the thin sections, the plagioclase shows polysynthetic twins and zoning structures. They display the extinction angle between 27 and 35°. So, the plagioclases are labradorite and andesine (48–62) in type. The grains are generally well preserved. In some cases, the central part of mineral is altered to sericite. The opaque minerals include magnetite and ilmenite.

About 40–60% of matrix is composed of plagioclase in the porphyritic samples. Pyroxenes are often found in the matrix, and in some cases constitute a majority of it. There is no olivine as matrix. Some spots of iron oxides were found in the matrix. The order of mineral crystallization in Maku's rocks could be noted as follows (Kheirkhah and Emami 2010):

In sub-volcanic Rocks (shallow in depth):

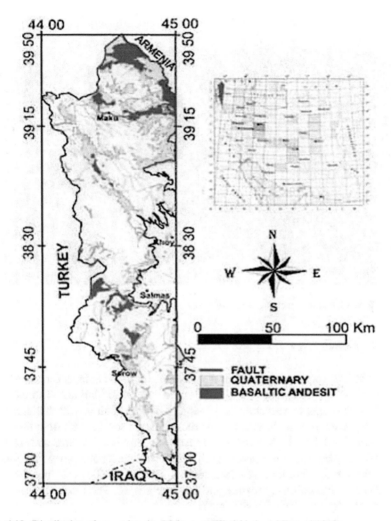

Fig. 6.19 Distribution of young basalts, Maku area (Kheirkhah and Emami 2010)

– Olivine + magnetite ± pyroxene + plagioclase.
– Olivine + magnetite + Pyroxene + plagioclase (phenocrysts).
– In porphyritic rocks:
– Pyroxene + plagioclase.

Denomination of rocks
Based on petrographic studies, the sub-volcanic rocks are named as diabase (micro-gabbro), porphyritic rocks (olivine basalt).

In the Maku region, on the Turkish border, on the way from Maku to Chaldoran, there are young basalts that have a special beauty (Fig. 6.20).

Fig. 6.20 A view of the prismatic basalts of Maku

6.1.8.3   Salmas Area

Young volcanic rocks are exposed in the western part of the Salmas City, at 44° 22′ 19″ to 44° 41′ 33″ longitude and 44° 59′ 44″ to 45° 17′ 21″ latitude. The outcrops consist of two major and several smaller target areas. The Salmas-Tazeh Shahr main road is the access road to the area and the rocks are exposed 3 km far from the Tazeh Shahr city. The Salmas basic lavas start from the western side of Salmas and continue to Iran-Turkey border. In the eastern part, they follow the topography of the area.

The thickness of basaltic lava flow increases to more than 600 m towards the west. The lava flow has columnar texture and irregular shapes. Their color is black and they are of large pores in hand specimens.

Texture: Their texture, in all samples is composed of fine grains and the matrix shows microlitic flow texture. Phenocrysts are about 20–30% of the whole rocks. They include:

1.   Olivine: This mineral which presents in all samples, generally, is automorphic to semi-automorphic and highly fractured. They are often larger than other phenocrysts, and based on optic axial angle (2 V), they are chrysolite which altered to iddingsite.
2.   Pyroxene: Pyroxene as well as olivine has automorphic to semi-automorphic form, green color and finer than olivines. The extinction angle is low. Because of the effect of alteration, the rim of grains is altered to form iron oxides. Due to their optic axial angle (2 V), they are Augite in type.
3.   Plagioclase: This mineral in the all samples is automorph and have polysynthetic twins. Some of them show crystal zoning and their twinning extinction angle is

about 27–35°. This implies that they are andesine to labradorite. Some of them are altered to sericite in the central part.

4. Opaque minerals: magnetite and ilmenite (less than 3% of the rock) are the main opaque minerals.

Matrix: the matrix is mainly composed of plagioclase and pyroxene. The plagioclase is mostly present as oriented microlites and they are often unaltered. The pyroxene minerals, augite in type, are automorph. They are comparable with plagioclases in terms of volume. Amphibole is also identified in some thin sections. The sequence of crystallization could be as follows:

Olivine + pyroxene + plagioclase + pyroxene ± amphibole + plagioclase

Denomination: According to petrographic studies of all thin sections, the abundance of olivine and presence of pyroxene as phenocrysts as well as parts of matrix, they are named as olivine-basalt.

6.1.8.4 Ahar Region

The Quaternary volcanic rocks in Ahar Region are exposed in the north and northwest of Ahar at 45° 41′ 20″ to 45° 56′ 35″ longitude and 38° 27′ 38″ latitude. The access road to the exposed rocks is unpaved Ahar-Alireza Chay road. In hand specimens, volcanic rocks of this area appear as dark to black rocks. Some of the rock samples are breccia and others are agglomerates.

The petrographic characteristics:
All the samples are porphyritic in texture that can be classified as below:

1. Porphyritic texture accompanied by microlitic matrix,
2. Porphyritic texture accompanied by microlitic-vitric matrix, and
3. Porphyritic texture accompanied by fine-grained matrix.

Phenocrysts constitute about 20–40% of the whole rock and include:

1. Olivine: It can rarely be found.
2. Pyroxene: The automorph to semi-automorph pyroxene crystals are usually found as coarse-grained, low-birefringence, fractured phenocrysts showing oblique extinction. Pyroxenes are augite in type and in some cases they have been altered to biotite; therefore, iron oxides are often found at the surroundings of phenocrysts and within the matrix. Also, the pores between the phenocrysts are filled by the secondary calcite.
3. Plagioclase: The minority of the phenocrysts is composed of plagioclase compared to pyroxene. The plagioclase crystals are usually automorph and unaltered. The plagioclase crystals are finer than pyroxenes and show polysynthetic twins as well as zoning. The extinction angle is measured and varies between 31 and 34%; so, they are labradorite type.
4. Amphibole: they are hornblende in type and show long and automorph crystals that are of parallel extinction.

5.   Opaque Minerals: Magnetite and ilmenite occur significantly in the rock and comprise about 5% of the whole matrix.

Matrix: More than 50% of the rock's matrix is composed of microlitic plagioclase crystals showing flow texture. Pyroxene is also a major phenocryst found at this rock and sometimes it is more abundant than plagioclase. Regarding to the paragenesis of minerals, the mineral crystallization in Ahar area may was as follows:

– Olivine + pyroxene + plagioclase
– Pyroxene + amphibole + plagioclase.

Denomination

Those samples contain abundant pyroxene (as phenocrysts and matrix) are called regular basalt, and due to the existence of hornblende in some samples and with respect to the composition of the matrix, some of the rocks are named hornblende-basalt to andesite-basalt.

6.1.8.5   Nir Region

These rocks are located at the western and southwestern parts of the Nir city, 37° 54′ 30″ to 37° 58′ 24″ longitude and 37° 54′ 30″ to 38° 08′ 16″ latitude. The best way to access the area is the Ardabil-Sarab main road. In this region, due to the smooth topography, the outcrops are widespread. The volcanic rocks are often of the vesicular texture in which small columns of plagioclase crystals are clearly visible in hand specimens. Outcrops are generally black in color and a thin cover of altered zone can be observed. Lava thickness is reported to be less than 20 m. Most of the samples contain pores and coarse phenocrysts of plagioclase. The rocks are fine-grained in texture, so the matrix can be classified as follows (Sayyar Miandehi 1993):

1.   Porphyritic texture accompanied by the flow-microlitic matrix (trachytic),
2.   Porphyritic texture accompanied by the pore bearing microlitic matrix and
3.   Porphyritic texture accompanied by vitric matrix.

Phenocrysts form approximately 25–40% of the rock's volume and include:

1.   Olivine presents in most cases. The crystals are often automorph, hexagonal and fractured. In some cases, it is completely altered to iddingsite (Sayyar Miandehi 1993).
     Olivine crystals sometimes comprise ten percent of the rock. Based on measuring optic axial angle (2 V) and other optical characteristics, they are chrysolite to hyalosiderite in type.
2.   Pyroxene: There are varied percentages of the rock's volume that is composed of pyroxene but in some cases pyroxene is the major phase of the phenocrysts. Based on the extinction angle is augite in type that has automorph to semi-automorph forms and is fine to medium in size showing inclined extinction.

In most cases, pyroxenes are unaltered and have one cleavage system. It is occasionally altered to amphibole, hornblende in type.

3.   Plagioclase: This mineral is always found as phenocryst in all samples, and in most cases it has no zoning. Generally, they are fine to medium in size which are slightly altered to sericite. The extinction angle is 28–32°. Therefore, the plagioclases are andesine and labradorite in type.
4.   Opaque Minerals: These minerals are mainly automorph magnetite and ilmenite crystals comprising 3–5% of the rock.

Matrix: The matrix constitutes only 5% of the rock's volume and is composed of lots of oriented columnar and unaltered plagioclase crystals. Pyroxene is also found, usually surrounded by plagioclase microlites. Olivine is also a part of matrix, but it is just observed in the rocks in which olivine phenocrysts can be found. The olivines of matrix are altered to iddingsite.

6.1.8.6   Nobaran Region, Western Saveh

The rifting and active volcanism at Quaternary, resulted in basaltic eruptions through the large faults and fractures of Nobaran area, Saveh. The main volcanic center is located in the western of Nobaran that continues to Dokhan village. The volcano is Stromboli and Hawaii in type and has a diameter of about one kilometer. Sometimes the phlogopite bearing basaltic xenoliths are found in the volcanic bombs. Around the cone of volcano, the basanitic lava has formed a basanitic platform of an area about 7 km^2 overlying the Quaternary terraces. The basanitic lava contains 15% nepheline and 15% phlogopite.

Regarding the presence of phlogopite and nepheline in the basanite two theories is mentioned (Mohammadi Siani 2005):

1.   The assimilation of carbonate rocks by basaltic magma can produce a carbonatic magma rich in phlogopite and apatite;
2.   Partial melting of a source rock rich in phlogopite is the second theory. As it is reported, there is a phlogopite basanite belonging to the Quaternary in the Eifel, Germany. The phlogopite is altered to apatite in Eifel, similar to Saveh's basanite. The Eifel basanite is of phlogopite bearing pyroxenite enclaves (Vosoughi Abedini 1977).

References

Abdi G (1996) Petrological study of northern Qorveh (Kurdistan) volcanic rocks. M.Sc thesis, Shahid Beheshti University

Ahmadi Pourfarsangi H (1993) Investigating the dynamism of volcanic activity in the disturbing mountain (Shahr Babak). Ministry of Science, Research and Technology—Shahid Bahonar University of Kerman

Ahmadi AR (2004) Petrography, petrogenesis and geochemistry o the metamorphic rocks of Tutak complex. M.Sc thesis, Shahid Beheshti University

Allenbach P (1966) Geologie und Petrographie des Damavand und seiner Umgebung (Zentral-Elburz), Iran. Abhandlung zur Erlangung der Wurde eines Doktors der Naturwissenschaften der Eidgenossischen Technischen Hochschule Zurich, 145 p

Arshadi S, Forster H (1983) Geological structure and ophiolites of Iranian Makran. Geodyn Project Iran R.G.S.I Rep 51:479–488

Babakhani A, Lescuyer JL, Riou R (1980) Explanation of Ahar quadrangle. Geological Survey of Iran, 123 p

Behrouzi A, Amini Fazl A, Amini Azar B (1997) Geological Map of Bostanabad, 1:100,000 Series, Sheet 5265. Geological Survey of Iran

Bout P, Derruau M (1961) Le Demavend. C N R S, Paris, Me´met Doc 8, 9-102

Brousse R, Lefevre C, Maury RC, Moein-Vaziri H, Amine Sobhani E (1977) Le Damovand: Un Volcan shoshonitique de la plaque iranienne. C.R. Acad Se. Paris, T 285:131–134

Christa E (1940) Ueber kristallisattion in magmatischen Gesteinen persiens. Min pet Mitt, 51

Darvishzadeh A (1987) The principles of volcanology. University of Tehran Publication, 299 p

Didon J, German YA (1976) Le Sabalan volcan pilo-Quaternaire de a'Azrabrijrn oaimnLra (Iarn) Edum gmoaogiqum mL pmLaogarphiqum du a'mdificm mL de son environment regional. These 3 eme cycle, Univ Scientifique et Medicale de Grenoble, France

Dimitrijevic MD (1973) Geology of Kerman Region. Gel Surv Iran, Report Number Yu/52, 334 p

Gansser A (1966) Ausseralpine ophiolite problem. Eclogae geologicae Helvetiae 52(2):659–680

Ghasemi Asl R (2003) Petrography, petrogenesis and geochemistry of the magmatic rocks in the East Harris (northeastern Tabriz). M.Sc thesis, Shahid Beheshti University

Ghorbani M (2003) Volcanological foundations with an attitude toward volcanoes in Iran, Pars Geological research Center

Girod M, Conrad G (1975) Les formations volcaniques recentes du sud de l'Iran (Kouh-e-Shahsasvaran), Donneés pétrologiques preliminaries, implications structurales. Bull Volcano; Ital., Da. 1975–1976, 39(4):495–511, Abstr. Angl.; BIBL. 1P. 1/2; 5 Illus.; 33 Anal Cent Geol Geophys, C.N.R.S., Montpellier, France

Irannezhad MR (1992) Petrology and volcanology of the Damavand Volcano. Shahid Beheshti University

Jang H, Kuresten M, Tarkian M (1975) Afor Nonograph post Mmsozoicvoacrnise in Iarn rnd iL's amarLion Lo Lhm subducLion of Lhm AfaoArabian underthe Eurasian Plate. Gel Surv Iran

Kheirkhah M, Emami MH (2010) The origin and evolution of Quaternary basaltic magmas in NW Azerbaijan (Burlan to Gonbad) using Sr-Nd studies. Geosciences 19(76):113–118

Kuno H (1959) Origin of Cenozoic petrographic provinces of Japan and surrounding areas. Bull Volcanol Ser 2(20):37–76

Lambert RSJ, Holland JG, Owen PF (1974) Chemical petrology of a suite of calc-alkaline lavas from Mount Ararat, Turkey. J Geol 82(4):419–438

Milton DJ (1977) Qal'eh hasan ali maars central Iran. Bull Volcanol 40(3)

Moein-Vaziri H (1985) Volcanisme Tertiaire et Quaternaire en Iran. M.Sc thesis, University of Orsay

Moein Vaziri H (1999) Petrography and geochemistry of ultramafic young zenolite basalt of Iran, examples of the peninsula Saray Bijar-Qorveh area, Damavand and Hasanali Rhine castles. The Second Conference Geological Survey of Iran, Tehran, Report 78, 23 p (in Persian)

Moein-Vaziri H, Amin Sobhani E (1977) Volcanology and volcano-sedimentology of the Sahand region. Tarbiat Moallem University Press, 59 p

Mohammadi Siani M (2005) Study of Quaternary volcanic rocks petrology in Taleghan region and their comparison with equivalent rocks in Qom-Saveh axis. Ministry of Science, Research and Technology, Shahid Beheshti University, Faculty of Science, M.Sc. dissertation

Montakhabi A (2012) Petrology of Sabalan volcanic rocks. M.Sc. thesis, Shahid Beheshti University

Mousavi SZ, Darvishzadeh A, Ghalamghash J, Vosoughi Abedini M (2014) Volcanology and geochronology of Sabalan volcano, the highest stratovolcano in Azerbaijan region. NW Iran Nautilus 128(3):85–98

Nogol-e Sadat MA (1978) Les zones de decrochement et les virgation structurales en Iran. Consequences des resultants de l'analyse structural de la région du Qom. PhD Thesis, 201

Nogol-e Sadat MA (1985) Les Zones de Decrochement et les Virgation Structurales en Iran. Geol Surv Iran Rep 55:259

Ovcinnikow A (1930) Outline of the geology of the Damavand region. Bull Naturalists Moseou, sect geol, 8/4

Rieben EH (1935) Contribution a la geologie de l'Azarbaidjan Persan. Neuchateh, these Sci Nat, 1–44

Ringwood AE, Green DH (1972) Crystallization of plagioclase in lunar basalts and its significance. Earth Planet Sci Lett 14(1):14–18

Sabzehei M (1994) Geological Quadrangle Map of Iran, No. 12, Hajiabad, Geological Survey of Iran, 1,250,000

Sayyar Miandehi (1993) Petrogenetic study of quaternary basalts of Azerbaijan. M.Sc. thesis, Shahid Beheshti University, 199 p

Shishehbor F (1993) Petrography, geochemistry and petrology of the Qaleh-Hosseinali lamproids. M.Sc thesis, Kharazmi University

Stocklin J (1974) Evolution of the continental margins bounding a former Southern Tethys. In: Geology of continental margins, Springer, pp 873–887, BIBL. 2P, 5Illus. U N Geol Surv Inst

Stocklin J (1975) On the origin of ophiolite complexes in the southern Tethys region. Tectonophysics 25:303–322

Vosoughi Abedini M (1977) Introduction to Petrographic and Petrological Research of basic Volcanics of Fourth Period in Azerbaijan Region. Report No. 69, Geological Survey of Iran

Wellmann H (1966) Active wrench faults of Iran, Afghanistan and Pakistan. Geol Runds 55:716–735

Chapter 7
Ophiolite Complexes in Iran

Abstract Ophiolite rocks of Iran are very significant due to the tectonic setting of Iran. So far, several research projects have been conducted at the academic and international levels in this regard. The author has visited all the ophiolite complexes and compared them with each other, reviewed the results of all the studies conducted and adapted them to the geology of Iran, and thus a logical classification of the ophiolites of Iran is provided. This chapter is like a mirror that reflects all the ophiolite features of Iran according to the studies that have been done on them so far. All ophiolite complexes of Iran are described in this chapter in terms of petrography, geochemistry, tectonic setting and geodynamics along with a geological map. They are also classified in terms of tectonic setting (according to structural units) and age (according to tectonics of Iran).

Keywords Ophiolite complexes · Geodynamic model · Geographic distribution · Iran

7.1 Ophiolite Zones and Ultramafic Rocks in Iran

There are widespread outcrops of ophiolite sequences and ultramafic rocks in Iran, nevertheless they can be generally divided into four main categories as follow (Fig. 7.1):

1. **Early Cretaceous-Paleogene ophiolites and ultramafic rocks:**
   Most of the rocks have the characteristics of ophiolite sequences related to the closure of the Tethys III (Neo-Tethys). Some of these rocks are as following:

   (a) The Kermanshah-Neyriz-Oman Axis,
   (b) Makran (southern Jazmurian basin),
   (c) Ultramafic, mafic rocks of the flysch zone in the Khash-Nosrat Abad-Birjand Axis,
   (d) Ultramafic and mafic rocks on north of the Darouneh Fault, Torbat-e Jam-Torbat-e Heydarieh-Sabzevar-Fariman,
   (e) Central Iran-Naien-Baft-Shahr e and

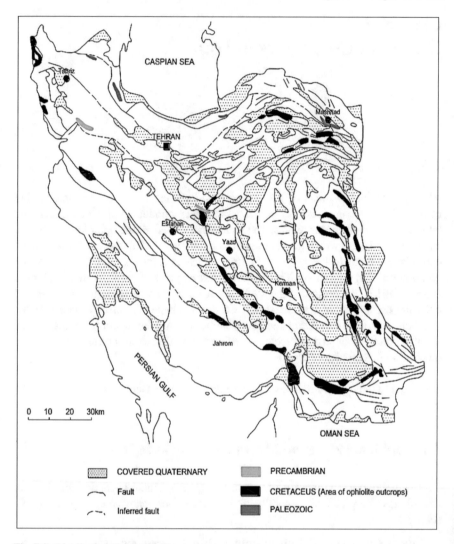

Fig. 7.1 Distribution of the ophiolites in Iran

 (f) Khoy-Maku.

2. **Upper Paleozoic mafic and ultramafic rocks**
  These rocks are represented by the metamorphic and non-metamorphic rocks
  in Fariman, Shanderman and Asalem. As it was previously discussed about the
  development of the Tethys II, these rocks originated in a continental rift zone
  which finally ended in forming an oceanic crust.
3. **Upper precambrian mafic and ultramafic rocks**
  It is described in Chap. 4 (i.e. Magmatic phases and their distribution in Iran).

4. **Ultramafic and mafic rocks associated with the great gabbroic intrusions**
These rocks were formed throughout a differentiation process and occurred in a large gabbroic batholith such as the Sarve ultramafic rocks in Urmia and the Masuleh gabbroic intrusions. They were previously attributed to the Precambrian. However, it is now believed that they have been formed during the Late Cretaceous-early Oligocene.

7.2 Structural Position of the Ophiolites in Iran

The Iranian ophiolites are parts of the eastern Mediterranean belt which extends from the Middle East and Dinarid-Helleni (Turkey, Troodos, Greece and Eastern Europe) to the eastern Asia (Pakistan and Tibet) (Bagheri and Stampfli 2008; Fig. 7.2).

The Iranian ophiolites can be classified into three categories as follows:

A. Structural geology and geodynamic mechanism,
B. Geographical distribution and
C. Time of emplacement.

There are similar characteristics among the three groups.

A. **Structural geology and geodynamic mechanism**

Many Paleozoic-Mesozoic ophiolite belts have been known in the Mediterranean region. In Iran, three ophiolite belts are recognized in terms of age and frequency as below (Lensch and Davoudzadeh 1982; Alavi 1991; Arvin and Robinson 1994). The Paleozoic ophiolite belt, limited in the northern Iran and the other two area, known as Triassic to Turonian Zagros-Oman-Baluchistan Belt (ophiolite belt of the outer Zagros axis), NW-SE trend (Stöcklin 1977; Alleman and Peters 1972); another one is formed in the eastern marginal part of the Central Iran micro-continent, mainly Late Cretaceous age, associated with the colored-Mélanges (Takin 1972). Based on the paleomagnetic data, the northern Iran was located in the margins of Great Gondwana Continent throughout Devonian-Carboniferous (Wensink and Varekamp 1980), and later the continent collided with the Turan Plate (Davoudzadeh et al. 1981; Sabzehei 1974). The Paleozoic ophiolites possibly confirm this suture zone (Stampfli 1978).

In general, it is assumed that the rifting zone has been occurred between late Triassic (Sabzehei 1974; Wensink and Varekamp 1980) and Jurassic (Haynes and Reynolds 1980), particularly in eastern and Central Iran. Gradually, with opening of the Neo-Tethys Ocean along the rifting zone, Iranian microplate was separated from the Gondwana Continent (Berberian and King 1981).

During the Mesozoic, the Lut Block probably began to counterclockwise rotate (Davoudzadeh et al. 1981). During the Cretaceous, the Arabian plate converged toward the Asian Plate and have caused the closure of Neo-Tethys and obduction of the ophiolites on the northern edge of the African continent (Ricou 1974) and shearing of the ophiolites of northeastern Iran around Lut Block. In general, the Mesozoic ophiolites have formed at a different period of time. The oceanic crust was

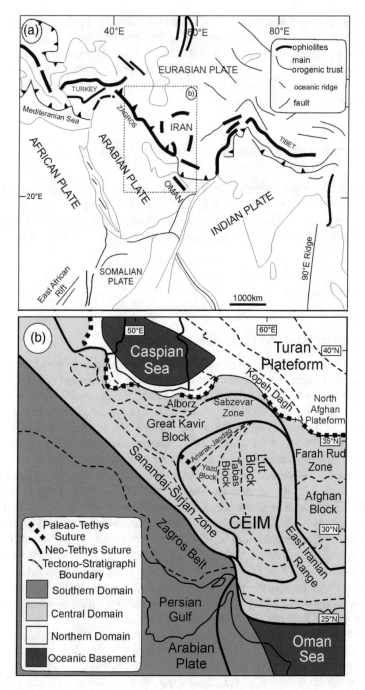

Fig. 7.2 **a** Distribution of the remnants of the Tethys Ocean in the Alpine-Himalayan Orogeny Belt.
b The Simplified geological map of the main tectonic units of the ophiolite belts in Iran modified
after (Bagheri and Stampfli 2008)

formed since Albian to early Paleocene and ophiolites have been emplaced in Late Cretaceous-Paleocene and the colored melanges were formed in Late Cretaceous-Oligocene.

The ophiolites of Sabzevar Zone were formed during Mesozoic and have been emplaced around the eastern Central Iran microcontinent (Alavi-Tehrani 1977; Lensch et al. 1977; Lensch 1980; Davoudzadeh et al. 1981) as small rifts (Stampfli 1978). There are six major complexes between the Lut blocks and Turan plate in the region. Except for the Sabzevar Complex, the other five complexes are incomplete (Lensch 1980). In the area, the Mesozoic ophiolites were followed by formation of the Paleocene Flysch and also the Island arc lavas flowed over the ophiolite complexes. It is noticeable that all activities were occurred in an extensional zone (Alavi-Tehrani 1978; Lensch 1980).

B. **Geographical Distribution**

The Iran ophiolites are divided into four groups (Alavi 1980) as follow:

1. The ophiolites of northern Iran along Alborz Mountain Range.
2. The Zagros trust (suture zone) ophiolites including the Neyriz and Kermanshah ophiolites which coincide with the Oman ophiolites and have been formed in the margins of the Arabian Plate.
3. The ordered ophiolites that have been formed in the basaltic columns in Makran, including Band-e Ziart, Darreh Anar and Remshk-Mokhtar Abad.
4. The Ophiolites and colored melanges represent the borders of the Central Iran micro continents (CIM), including the Shahr-e Babak, Naien, Baft, Sabzevar and Chehel Koureh. It is notable that the Yazd, Posht-e badam, Tabas and Lut Blocks constitute the Central Iran Micro continent (CIM) (Fig. 7.3).

C. **Age and time of emplacement**

The ophiolites of Iran are geochronologically divided into three groups as follows:

1. The late Proterozoic-early Cambrian ophiolites known to represent the isolated outcrops on the western edge of Central Iranian microcontinent. The ultra-mafic rocks have not considered in this classification,
2. The Pre-Triassic ophiolites, exposed in northern Iran with the same trend as the Alborz mountain range and
3. The Jurassic-Paleocene Ophiolites are more dominant and extended in Iran.

7.3 Distribution of Ophiolite Complexes and Ophiolitic Mélange in Iran

The ophiolite and ophiolitic mélange complexes are distributed widely in Iran. The major ophiolites zones are as below:

1. The Mesozoic-Paleocene ophiolites of the Central Iran and eastern Iran (the arc strip of Central Iran),

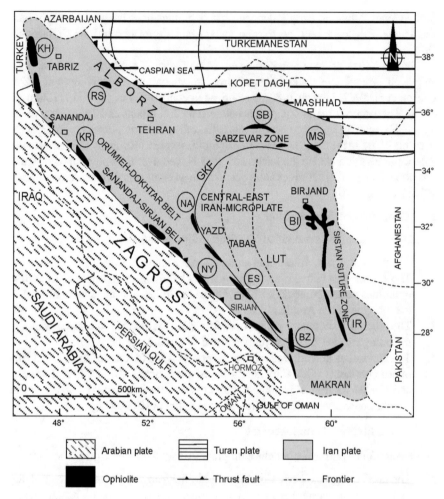

Fig. 7.3 Deformed and simplified map of Iran (Babazadeh and Wever 2004), which represents the major tectonic units (Lensch et al. 1984), the inner core (Yazd, Tabas and Lut; Şengör et al. 1988), and the position of the main ophiolites (Stöcklin 1977; Dilek and Delaloye 1992). Abbreviations (BI—Birjand; Bz—Band Ziarat; ES—Esfandaghehh; GKF—Great Kavir Fault; IR—Iranshahr; KH—Khoy; KR—Kermanshah; MS—Mashhad; NA—Naein; NY—Neyriz; RS—Rasht; SB—Sabzevar)

2.   The Mesozoic ophiolites of the southwestern Iran,
3.   The late Cretaceous-Paleocene ophiolites of Makran,
4.   The Cretaceous-Paleocene ophiolites of Zagros (Zagros-Oman Band) and the parallel ophiolites zone with them,
5.   The late Paleozoic-early Triassic ophiolites of north–northeastern Iran and
6.   The late Precambrian-early Cambrian Ophiolites and ultramafic rocks.

7.3.1 Central and Eastern Iran Ophiolites (the Circular Band of Central Iran)

A series of ophiolites rocks as a circular belt outcrop around the Lut Block, eastern Iran, and Central Iran. The complexes are as follows.

In the northern Central Iran, the ophiolite complexes of Torbat-e Heydarieh, Sabzevar and Oryan crop out in the north of the great Kavir-Darouneh fault. In the western Lut Block, the Naien ophiolites and the south and southwestern Kerman band of colored mélanges are exposed. In the eastern Lut Block, the N-S ophiolite band of the Birjand and eastern Iran outcrops and extends to the Khash region. Stöcklin (1974) considers the Maku-Khoy ophio-colored mélanges as a part of the Central Iran circular band ophiolites.

7.3.1.1   Northern Torbat-E Heydarieh Ophio-Colored Mélanges

A full sequence of ophiolite complex is exposed in the Asdabad to Robat-e Sefied (Razmara 1990). The complex's rocks are from the bottom to the top of the sequence as follows:

1. Basal amphiboles,
2. Metamorphic peridotites including: harzburgite, dunite, wehrlite,
3. Lerzolite, and pyroxenite. The peridotites have endured tectonic events, thus associated with plenty of fractures,
4. Cumulate rocks: Having the rock diversity from peridotite to gabbro and
5. Sheeted dikes and leucocraticrocks, tonalite to granite basalt and pyroclastic rocks associated with the metasomatic metamorphic rocks such as, rhodengite and amphibolite. The sedimentary portion of this sequence includes pelagic limestone, radiolarite and flysch sediments.

The chromite, talc, magnetite, asbestos, chrysotile, sepiolite and likely nickel deposits are considered in related to the ophiolites in the region.

7.3.1.2   Sabzevar Ophiolite Complex

The Sabzevar complex outcrops in an area about 2000 km^2 on the northern edge of the Central Iran Microcontinent (CIM) in the southern Miamay-Torbat-e Heydarieh fault and northern Sabzevar (Fig. 7.4). It is notable that the Sabzevar ophiolite complex is known as the largest multi-complex in northeastern Iran. All the ophiolite complexes are the boundary between Central Iran Micro-continent (CIM) and Turan continent (Kopeh-Dagh mountains). Consequently, the above ophiolite complexes have similar tectonic characteristics and features (Lensch 1980; Lensch et al. 1977; Soffel and Forster 1984; Şengör 1990; Şengör and Natal'in 1996).

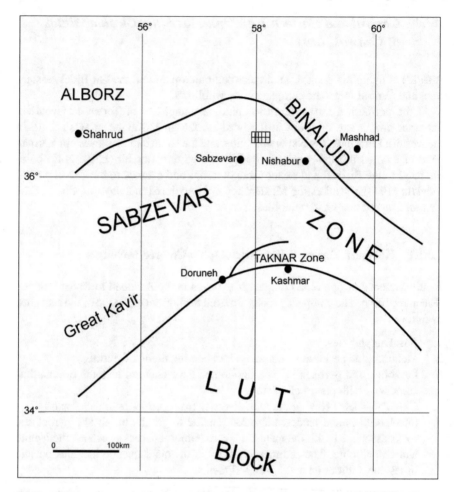

Fig. 7.4 Sabzevar Zone: confined between the Alborz and Binaloud mountains in the north, and Lut Block in the south (Pilger 1971)

According to Lensch (1980), six ophiolite complexes are exposed in the north of Doruneh fault and in the south of Torbat-e Jam-Shahroud Fault (Fig. 7.5). These complexes have their own unique characteristics of the geographical distribution and different ophiolite features (Fig. 7.5). For example, the Torbat-e Heydarieh ophiolite complex consists of the largest ultramafic and serpentinite rock units, whereas the Kashmar ophiolite complex includes more sheeted dikes. However, amongst the six complexes, the Sabzevar complex represents the most completed sequence of ophiolite rock units. The Sabzevar ophiolite complex considers being a portion of northern part of the Neo-Tethys Ocean, which was closed in Late Cretaceous (Lensch et al. 1977; Şengör 1990). Harzburgite rock forms the major portion of the ultramafic rocks of the Sabzevar ophiolite and is similar to the other classic ophiolite complexes

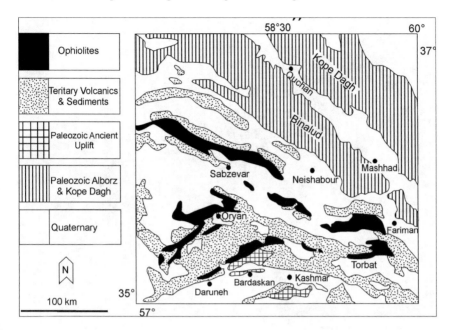

Fig. 7.5 Geological map of the northeastern ophiolites in Iran (Sahandi 1993)

of the world. All units of cumulate, micro gabbro and diabasic rocks of the Sabzevar ophiolite are formed as interlayers, between gabbro and basaltic-spilitic pillow lavas. The complex associates with sedimentary rocks, mostly pelagic limestone bearing minor radiolarite, are formed in Late Cretaceous. The metamorphic parts of the ophiolite have been originated from garnet and amphibolite rocks (Fig. 7.6). Based on Alavi-Tehrani (1977) and Lensch et al. (1977), the Sabzevar ophiolite complex is divided into various sections as below:

1. The Great Harzburgite and tectonite masses;
2. The layered series of ultramafic and the cumulate gabbro include diverse rocks' layers as lherzolite, plagioclase, bronzite, webstrite, norite and gabbro;
3. Acidic series such as quartz diorite and granophyre;
4. Dibasic dikes' series which represent as sills;
5. Pillow lavas which petrologically include a range of variation from lueco-basalt to spilite series and cratophyr. The enrichment and changing of the amount of sodium and potassium could be the result of hydrothermal alteration, and spilitification of origin composition (Sinton and Byerly 1980);
6. As previously noted, the major sedimentary rocks associated with the ophiolites are pelagic limestone and some radiolarite. Based on the paleontological evidence, these rocks with the volcano series have formed during the latest Cretaceous (Maastrichtian). The ophiolites associated with the red and green shale and occasionally flysch series; and

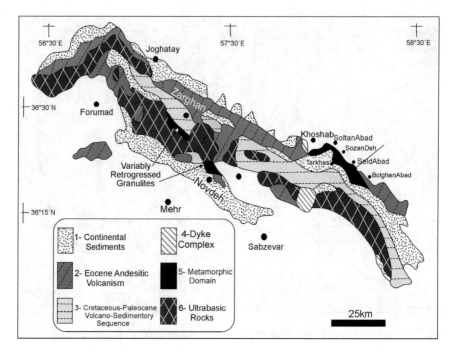

Fig. 7.6 Simplified geological map of the ophiolites in Iran (Lensch et al. 1977)

7.   Metamorphic rocks consist of two parts: (a) amphibolite and garnet amphibolite;
     (b) metamorphosed pillow lavas and glauchophan schist which represent the
     evidence of collision the Alpine orogeny and metamorphism.

In the Sabzevar region, besides the ophiolites, the colored mélanges units are
exposed; including almost all ophiolite complex units; accompany with some disso-
ciate and unknown rocks. The colored mélanges have formed along the major faults
and fractures. The general trend of the fractures and faults are east–west, coincide
with the trend of the geological structure of the region. It is believed they have
formed after Eocene and likely related to the Oligocene orogeny; because, the Early
and Middle Eocene sediments, are found as the main components of the colored
mélanges.

Kohansal et al. (2016) researches on the Forumad ophiolites, western part of
Sabzevar ophiolite, indicate a discontinuous, heterogeneous and thin crust, while in
the mantle section it is thick and relatively uniform.

The widespread areas of the harzburgite, clinopyroxene harzburgite, serpentinite and dunite, with low distribution of lherzolite, wehrlite, clinopyroxenite and
plagiogranite, are the main constituents of the mantle section. In the internal crustal
section of olivine gabbronorite–gabbronorite, clinopyroxene gabbro, amphibole
gabbro and microgabbro and in the external crustal section, in order frequency of
pillow basalt, tuffs, hyaloclastic breccia and sheet flew are important.

Field investigations show that main components of the oceanic crust have been formed from mantle peridotite and the molten/rock process was the main factor in the formation of lherzolite, wehrlite and clinopyroxenite.

Dunitic zone studies of the residual harzburgite transmission to dunite indicate a decrease in Mg# in olivine and an increase in Cr# in the spinel due to the incongruent melting process of orthopyroxene.

The measured equilibrium temperature between the olivine-spinel minerals for such a transition, showing the formation of automorph spinels from the types of vermicular to a minimum of 536–561 °C, indicates the greatest distance between the upraised mantle melts and the host peridotite in the dunitic channels. Maximum equilibrium temperature between olivine-spinel minerals is estimated for automorph spinels of 708–758 °C, which is related to the least distance of melting effect on peridotite in dunitic channels.

Mineral chemistry evidences, whole rock chemistry and petrogenesis of mantle and crustal rocks in the Forumad region indicate their origin from a heterogeneous mantle source with composition of spinel peridotite that caused by varying degrees of melting and the effect of the fluid flowing from the dipping slab, they have experienced a different enrichment, which occurred in the marginal arc basin tend to back arc basin of the suprasubduction zone.

Gabbros and basaltic pillow lavas are located in the domains of normal and enriched MORB that tend towards the arc. The study of geochemical trends along with generative characteristic in the crustal and mantle parts indicates their dependence on depleted MORB that have experienced some degrees of enrichment.

The absolute age of the U-Pb plagiogranites on zircon crystals is estimated to be 99.9 ± 0.92 million years old. Investigating the proportion of Th/U with the dependence of plagiogranites on the dunitic zones reveals their generation due to the heterogeneous melting of orthopyroxene. In this process, the mixing of silica-rich melts with melting from the dipping slab has played a major role in the production of plagiogranites in the Forumad region.

Geological and petrological studies in the Forumad area suggest the formation of oceanic crust in the basin with slow spreading speed as bellows:

1. The output sequence without mediation of sheet dikes and directly on peridotitic or gabbro rocks.
2. Superiority of pillow lavas versus sheet flows.
3. The presence of phyric texture in basalts with coarse plagioclase, clinopyroxene and olivine crystals.
4. Lack of sheeted dikes complex.
5. Gabbros do not form a continuous thick layer over the mantle rocks, but as small intrusive masses appear in the upper mantle.
6. Expansion of volumes of mantle rocks with composition of lherzolite amongst the host harzburgites rocks.

The study of paleogeography in the Forumad region suggests the formation of this ophiolite basin in the northern margin of the Central Iranian Microcontinent and in the suprasubduction environment, which was created by the subduction of the Neo-Tethys oceanic slab beneath the Central Iranian plate. The evolutionary process of this ocean basin began from a slow spreading marginal basin in the Early Cretaceous-Late Cretaceous (Albian-Cenomanian) boundary and gradually changed to the back arc conditions in the Late Cretaceous. In such a case, the composed main parts of the oceanic crust have formed through a mantle with spinel lherzolite in composition.

The course of the geodynamic evolution of the oceanic basin in the Forumad region began with the Pre-Albian rifting and continued with the opening and spreading of the oceanic crust in the Late Albian-Cenomanian. The subduction intraoceanic onset in the Late Turonian is accompanied by the formation of a tectonic mélange unit and ended up before Campanion. Subduction intraoceanic accompanied by the formation of the metamorphic sole with the composition of amphibolite and amphibole schist, which is related to the Late Turonian-Santonian event. Finally, the closure of the oceanic basin has occurred during the late Thanesian-Ypresian, causing discontinuity and unconformity between the Paleocene conglomerate and ophiolite rocks.

7.3.1.3 Naein Ophiolite and Colored Melanges

The Naien ophiolite is likely formed in Paleocene-Early Eocene (Davoudzadeh 1972). The ophiolitic area is in excess of 2500 km^3 (Fig. 7.7). Stahl (1911) conducted one of the primary researches in central Iran including the Naien and Anarak areas and later Huber (1978a, b, c) published the first systematic studies in central Iran on behalf of the National Iranian Oil Company. He carried out the systematic tectonic and stratigraphic studies of the Central Iran, including the sedimentary formations of the northern Naien.

De Böeckh et al. (1929) gave the name of "Persian Medium Mass" to Central Iran. Gansser (1955) has widely studied about Central Iran and for the first time used the term of "colored mélanges" for the ophiolitic rocks and represented a detailed description of the stratigraphy and geology in central Iran.

In fact, the Ph.D. thesis of Davoudzadeh (1969) is the most comprehensive study that has been done on the northern Naien rocks. Later a summary of his thesis published in report No. 14 provided by Geological Survey of Iran (GSI), Davoudzadeh, based on the geology of the northern Naien, divides the area into three major sections:

1. The colored mélanges, known as a mixture of ophiolites.
2. Limestone and radiolarites, expose throughout the middle part of the area with a trend of N-NW.
3. The Tertiary volcanic rocks in the western part of the region.
4. The Tertiary sedimentary formations in the eastern part of the region.

In the colored mélange's area, the composition of the ophiolites consists mostly of peridotite and serpentine along with a minor portion of pyroxinite and diabase. It is notable that peridotite and pyroxenite rocks have transformed to serpentine metamorphic rock through the metamorphism process, at different degrees. Despite, several

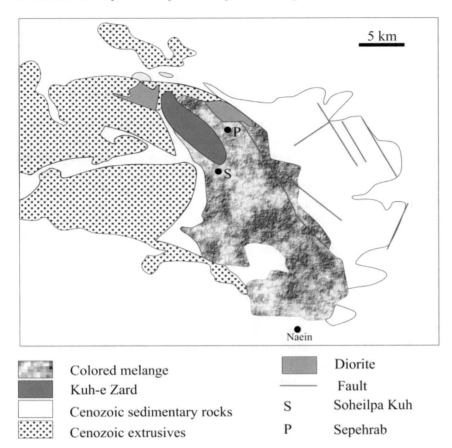

Colored melange		Diorite	
Kuh-e Zard		Fault	
Cenozoic sedimentary rocks	S	Soheilpa Kuh	
Cenozoic extrusives	P	Sepehrab	

Fig. 7.7 Simplified geological and tectonic map of the northern Naein

diabase, diorite and gabbro veins have exposed with small lenses of diopside and rhodengite formed in the ultramafic rock. The radiolarite limestone is as few meters to hundreds meters pieces. Some homogeneous mass of peridotite (each extends over several kilometers) exposed in the region. The Campanian-Maastrichtian limestone bearing planktonic foraminifers *Globotruncana* as thin layers of chert which is composed the oldest rock unit as well as the early Eocene Nummulite-Alveolina limestone and sandy limestone of the youngest sedimentary rocks units can be observed in the colored mélanges. The ophiolite units were attributed to the Paleocene-early Eocene. In the north of the colored melanges zone, there is a range known as Zard mountains which its major parts consist of the Paleocene-early Eocene *Globotruncana* and sandy limestone. A sequence of middle Paleocene basal breccia covers the Maastrichtian limestone without a clear unconformity. It proves that there was a gap of sedimentation before middle Paleocene. Also, the E-W trending faults that only occur in the *Globotruncana* limestone confirm the tectonic activity. The Paleocene tectonic phase were continued to early Eocene and the results are the formation of deep faults, landslides on the oceanic floor and the intrusion of basic and ultramafic rocks into the faults, and fractures zones. Eventually, all these processes end up forming the "colored mélanges". The early Cenozoic volcanic rocks in the western region are a small part of widespread volcanic rocks which occur all over the west Central Iran. The volcanic rocks are separated from colored mélanges zone by the major faults. However, the volcanic rocks are exposed in the southeastern and northern of the Colored Mélanges zone individually. The widespread bodies of andesite, dacite and pyroclastic rocks are extended in the volcanic zone. Nevertheless, in some area dengezite basalt, trachyandesite and porphyrite are exposed. Generally, the volcanic rocks are associated with less tuffs, crystalline tuffs, and rare calcareous tuffs. The Pyroclastic rocks thickness in excess of about 300–400 m. The acid volcanic rocks are slightly younger than the basic rocks, in the region.

In the eastern part of the region, an alternation of fine to coarse pyroclastic rocks with thickness of 3200 m "Akhureh Formation", covers the colored mélanges. The pyroclastic rocks confirm that the rocks were deposited with a relatively high rate of sedimentation in a gradually subsiding basin similar to a flysch-type depositional environment. The Akhureh Formation is assigned to the middle to late Eocene. However, its upper part is likely formed during the early Oligocene. The Lower Red Formation (450 m) which consists of the gypsum sandy marl and red sandstone cover the Akhureh Formation. The Qom marine Formation and the Upper Red formations (more than 1000 m), including the reddish gypsum marl, red sandstone and conglomerate locate on the Lower Red Formation.

The early Cenozoic intrusive rocks of diorite type, with 4–10 km length, in northern Naien are less widely-distributed. The NW-NE trending faults occur after Miocene and fragment the Naien area into several segments.

The major parts of the ultramafic rocks consist of harzburgite, lherzolite, enstatite, dunite and pyroxenite in the area. Harzburgite is the most common rock distributed in the region as large blocks with several km extensions (some blocks might be even, up to 5 km). The most important point is the rocks have been altered strongly, and mainly serpentinized. The diabasic and micro gabbroic dikes are rhodengitized and sometimes lawsonitized at different levels, and are dominant in the region. It should be noted that the metamorphism in the Naien region occurred as Amphibolite facies.

Jabari (1997) and Manouchehri (1997) conducted petrological studies on the northern Naien ophiolite for M.Sc. thesis. Jabari believed that the Naien Mélanges geochemically are poor of K2O. He also suggested that different terms of basic rocks represent the tholeiitic trend of the Island Arcs system which caused by partial melting of a deep source. According to Manochehri the ophiolite unit is similar to harzburgite type (HOT) unit. He believes that the rocks forming ophiolite unit are geochemically closed to the tholeiitic series. In his opinion, the comparison between basic units of the Naien region and Mid Oceanic Basalts indicate similarities and contradictions that make it skeptical and undocumented decision to present an absolute definition. But, on the basis of the available information, the Naien ophiolite unit is more or less the same as the Mid Oceanic Basalts (P-type).

7.3.1.4 Dehshir-Baft Ophiolite

The NW-SE Dehshir fault extends from the Naien and Dehshir areasin Yazd to Baft, Kerman area. In fact, the Dehshir-Baft fault is a part of the Naien-Baft fault. The Dehshir fault lies on the southern slopes of Shir Kuh Mountain, with almost the same trend of the major Zagros Fault. Also, the Dehshir-Baft fault is a part of Colored Melange structure. The north, northwest–south, southeastern Dehshir fault, in excess of 350 km, extends from south western Naien to Chahgoo, near Sirjan. The nearly vertical Dehshir fault is active and affects the Quaternary deposits. The lateral branches of Dehshir-Baft fault form the shear or fault zones of Touran Posht, Kheyrabad, Diz Run, Mohammadabad and Hushangabad.

The Dehshir fault separates the subsided Abarkuh from the ophiolite complex and younger non-ophiolite units. Along the Dehshir-Baft axis, many ophiolitic patches expose such as, Dehshir, Robad and Baft ophiolite (Makizadeh 1997; Noghreian 2004).

The Dehshir ophiolite includes the sedimentary rocks related to the ophiolitic sequence are associated with the metamorphic rocks. The ophiolite rocks outcrop 80 km to southwestern Yazd along the major fault of Dehshir-Baft is considered to be a part of the Central Iran circular ophiolite arc belt. Based on the geographical divisions, the Dehshir region is located in 70 km southwestern of the Yazd province and in 60 km of Taft Town, 31° 15′ to 31° 30′ latitude and 54° 30′ to 54° 45′ longitude.

Previous Studies

Geological Survey of Iran (GSI) initiated the researches in the region by providing the first geological map of the Dehshir-Baft Ophiolite Complex in 1983. Then, Amidi (1977) introduced the Surak Ophiolites in Surak area. Later, Noghreian (2004) classified the ophiolite as the plagiogranite units, ultramafic and sedimentary rocks with pelagic limestone, sandy clay limestone and chert. In addition, other researchers, such as, Makizadeh (1997), Shafaii Moghadam (2011) and Sultan Mohammadi (2009) conducted several researches.

Ophiolite Sequence of the Dehshir Region

The ophiolites of Dehshir dispersed in several parts of the region. It is believed that the activities of the Dehshir-Baft fault caused to form the numerous ophiolites disperses. It is classifying in three different areas.

1. The Surak region is located in the western Surak village between two nearly vertical faults with the north western-south eastern trend. Amidi (1977) introduced the area as the Surak colored melanges. In general, the ophiolites extension is limited as a narrow belt dispersed in Dehshir region. The fault activities have widely caused to crush and tectonize the area. The peridotites in the area consist of harzburgite, lherzolites, and pyroxenite which are serpentinized at different phases. The basalts and diabases are the most abundant crustal sequence, metamorphosed at greenschist facies, in fact, are spilitized. A sequence of cherts and pelagic limestone in the Surak ophiolites lay with

an unconformity contact on the mafic units. The sequence has mostly a fault contact with the Eocene volcanic units which formed by the Naien-Dehshir fault activities.

2.   Referring to the northern part of the Dehshir quadrangle, the complex has outcropped around the villages of Ardan, Dizan, Zolozarou, Aziz Abad, Ahmad Abad, and Rishkuiyeh. The Surak ophiolites are divided into two parts, in the northwestern and southeastern of the Ardan village: (a) a complex of meta basalts, meta cherts, serpentinite with diabase dikes, pillow lavas, and pelagic limestone is outcropped in the northwestern area; (b) An alternative sequence of dyke groups along with peridotites, pillow lavas, cherts, meta volcanic rocks and carbonates are exposed in the southeastern area of the Ardan village (Sharifabad).

3.   The southern part of the Dehshir geological map is connected to the Shahr-e Babak ophiolite sequence (Kamroud). A thick flysch sequence equivalent to the Akhureh Formation of the Naien ophiolites are exposed which consists of conglomerate, sandstone, greywacke and siltstone. The serpentinized peridotites of the complex include dunite, and harzburgite are depleted or poor of orthopyroxene.

Petrography of the Dehshir Ophiolite Complex

The Dehshir ophiolite complex's rock units include:

1.   The textures of peridotites are porphyroclast, and protogrannular with porphyroclast of olivine, orthopyroxene (partially altered to bastyt), and plagioclase with alteration of sericite-clay.

2.   Clinopyroxenite as dikes and sills have been found in the peridotites rocks with a coarse granular texture.

3.   Pegmatite gabbro, including clinopyroxene (altered to tremolite), and plagioclase (with sericite-clay alteration).

4.   Gabbro-anorthosites with a coarse grain intergranular texture, mainly consists of phenocrysts of plagioclase and clinopyroxene. In addition, they contain olivine phenocrysts, and fine grain of titano-magnetite and ilmenite. Plagioclase in the gabbro-anorthosite are much more abundant than the other type of gabbros and sometimes plagioclase is the only main mineral of the rock.

5.   Gabbro with fine to coarse granular texture mainly consist of sub-aphanitic, intergranular phynocryst pyroxene and plagioclase. Amphibole mostly presents in the rocks and occasionally is the main mineral component of amphibole-gabbro.

6.   The basalt with hyaloclastic texture consist of plagioclase and clinopyroxene appear as microlite and phenocryst. The secondary minerals such as chlorite, calcite and epidote are present, as well.

7.   Dolerite dikes which occur in pillow lavas porphyritic texture and vitric-microlitic or pillowtaxitic groundmass. They are fine grain silicate dikes and similar to the sheeted dikes in the complex. Phenocryst plagioclase (with

clay-sericite and epidote alteration) and some pseudomorph of mafic minerals (chloritizided completely) and titano magnetite are present.

8. Diabase dikes with intergranular texture; mainly consist of plagioclase phenocrysts (0.5–2 mm with clay-sericite alteration) and clinopyroxene (average size 0.5 mm, with chlorite alteration). The other minerals of the dyke's composition are amphibole (magnesium hornblende), titano magnetite and secondary minerals such as chlorite and prehnite.

9. Andesite-andesitic basalts: the rocks' sequence has a porphyritic texture with a microlitic and hyalopilitic groundmass. The main minerals which form the rocks' composition are plagioclase, amphibole (completely altered to chlorite or as small microlites in the groundmass) and with less content of pyroxene. Their secondary minerals are chlorite and calcite.

10. The accumulation of dikes' complex in a variety of andesite-basalts rocks: Mainly consist of plagioclase (with clay and epidote alteration), clinopyroxene (with clay alteration) and hornblende (0.8 mm) along with chlorite and quartz as the secondary minerals. The plagioclase phenocrysts, microlite and K-feldspar mainly constitute the dacite. The pseudomorph amphibole in the rocks has altered to chlorite and epidote and the groundmass is a composition of secondary minerals of amorphous quartz, chlorite and epidote.

11. Diabase dikes: They occur in the isotropic gabbroic rocks with an ophitic to sub-ophitic texture, mainly consist of plagioclase phenocrysts and clinopyroxene (altered dominantly to tremolite and less to actinolite).

12. Granite: The granite texture is granular mainly consist of K-feldspar orthose (strongly altered to clay, sericite and epidote), plagioclase (the main constituent minerals of granite veins in the Azizabad area) and anhedral quartz forms among the orthose mineral.

13. Metamorphic rocks: These rocks are associated with Dehshir ophiolites outcrop in the Zolozaro area and consists of amphibolite, actinolite schist, chlorite schist, quartzite, schist and epidote.

14. Sedimentary rocks: The major sedimentary units of the ophiolite complex are dominantly formed of the Upper Cretaceous pelagic and *Globotruncana*-bearing limestone and radiolarites, conglomerate, sandstone, greywacke and siltstone in the area.

7.3.1.5 Baft Ophiolites

The ophiolite belt in the south and southwestern Kerman (from southwestern Esfandaghehh to Hajiabad area) extends with an approximate 360 km length and 4–15 km width (Sabzehei and Berberian 1972).

The ophiolite complex consists of a series of the ultramafic rocks, gabbro, diabasic dikes, pillow lavas, massive basalts, basaltic sills, agglomerates, tuffs and the Late Cretaceous sedimentary rocks with gabbroic dikes which occur along the strike-slip of Dehshir-Baft fault. In fact, the ophiolites crop out along the Baft fault.

This NW-SE fault with a length of 96 km is located in the southern part of the Yashm Kuh crosses the Baft city. As a result of the activities of Baft fault, the ophiolite complex tumbled at different time in later than Late Cretaceous. The lateral branches of the fault have caused the displacement of the Neogene rocks. Accordingly, it seems that the last activity of the fault is related to the Neogene period. The distribution of the Baft ophiolite belt is fully shown on the Baft, Balvard and Chahar Gonbad (1:100,000) geological maps. It should be noted that the Baft ophiolite extends in the north–northwestern part and joins to the Shahr-e Babak ophiolites of which the south–southeastern part joins to the Esfandaghehh ophiolites.

For the first time, the Yugoslavia geologists initiated the geological studies on the Kerman region who prepared geology map of the area under supervision of the Geological Survey of Iran (GSI).

According to Dimitrijevic and Djokovic (1973), the Baft ophiolite complex mainly consists of a diversity of the ultrabasic basic to intermediate and acidic rocks.

The complex associated with sedimentary rocks of pelagic limestone cherts and sandstones types as well.

In the Chahar-Gonbad area where is the widest part of the ophiolite belt basically follows a particular order and mainly divided into three parts as follows:

1. Lower part: It includes the upper Turonian-early Coniasian sedimentary units mainly occur as sandstone, chert and limestone in various colors.
2. Middle part: It includes an alternate sequence of sedimentary and submarine volcanic lavas.
3. Upper part: This consists of a sequence of basic and ultrabasic rocks associated with pelagic limestone lenses. Based on the fossils found in the limestone, they were attributed to the late Cenomanian (Fig. 7.8).

Based on the geochemical data, the basalt and gabbroic rocks are tholeiitic type. Also, regarding to the mobility of trace elements they emplaced within the Mid Ocean Ridge Basalt (MORB). It should be noted that the Baft ophiolite melange resulted in formation of the tectonic activities and cold diapiric serpentinite (Dimitrijevic and Djokovic 1973).

According to Moeinzadeh Mirhosseini (2012), the Baft ophiolite complex which consists of ultrabasic (harzburgite, pyroxenite) and basic (gabbro, basalt and dolerite) rocks geologically has the characteristics of an ophiolitic mélange. The main volume of the ultramafic rocks has been formed by the tectonized and serpentinized of harzburgite. The gabbroic rocks generally occur as massive and dense bodies form the fine (micro gabbro) to coarse (pegmatite) grains rocks. The gabbroic rocks sometimes altered to rhodengite. The basalts which are associated with cratophyrs and clearly become spilitic are widespread as massive or pillow lavas in the region. Dolerites often are altered and occur as sheeted dikes all around the region, there are no sill in any part of the area. Generally, the contacts between different rock units are faulted. Results of the static metamorphism (oceanic crust metamorphism) metamorphose the ultramafic rocks to serpentinite, gabbro to rhodengite and basalts to spillitite. In the Baft area, the sedimentary rocks sequence consists of the Late Cretaceous limestone, silicified chert, *Globotruncana*-, radiolar-bearing limestone, respectively.

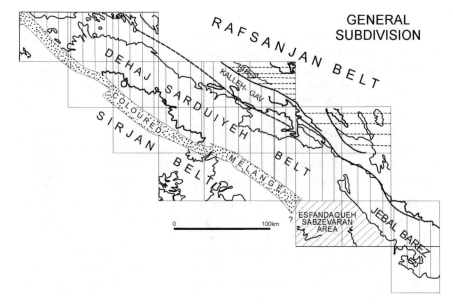

Fig. 7.8 General Geological Zoning Map of Kerman Region (Dimitrijevic and Djokovic 1973)

Based on geochemical data and analysis of the rocks, the ultrabasic rocks are as Komatiite (the "Komatiite series" is not adequate) and basalt, gabbro, dolerite are tholeiitic series.

In terms of replacement and formation of the ophiolite complex, it could be proposed that the same model as the "Red Sea Type".

Additional studies by Arvin and Robinson (1994) and Arvin and shokri (1997) indicate geochemically two types of basaltic lavas involved in the area as follows:

1. Tholeiitic basalts and
2. Transitional tholeiitic basalts.

They suggested that the basalts originated from mantle source and geochemically are intermediate of the island arc and mid oceanic tholeiiths.

Ophiolite Sequence in the Baft Region
Generally, this sequence is widespread in the area and as previously stated they are traceable on the three 1:100,000 geological maps of Balvard, Chahar-Gonbad and Baft. The Baft ophiolite sequences will be discussed in each map area as follows:

1. Balvard ophiolite zone: This zone consists of the Cretaceous serpentinite, peridotite, pillow lavas, massive basaltic lavas, agglomerate, breccia, tuff, trachybasalt, tuffaceous sandstone and limestone.

Serpentinite: They occur as allochthonous massive with a faulted contact which is associated with other units in the ophiolite complex. These rocks mostly have a fish scale texture.

Peridotite: The ophiolite complex consists of harzburgite and dunite.

In some cases, the rocks contain the chrome and spinel minerals. If the minerals are exposed with the orientation, the trend and plunge of lineation will be distinguishable. Basically, serpentinite and serpentinized harzburgite rocks occur on the north and northeastern of the Balvard quadrangle and southeastern of the Gughar village.

Isolated diabase dikes: They are rarely exposed in the mantle originated ophiolite sequence. They severely crushed, boudinaged and varied from highly to low rhodengenite.

Pyroclasts: They form the major rocks units in Balvard area consisting of breccia, agglomerate, tuffits, and basic to intermediate tuffs associated with the Late Cretaceous limestone and chert (unit K_2 on the map). The unit with a thickness of 300–400 m was considered as a lower (base) sequence (Mijalkovic et al. 1972).

1. The sequence of breccia and agglomerates (with hyaloclastics) contain nearly rounded to angular fragments of basalt, porphyry pyroxene basalts and porphyry andesite. A thick sequence of spilit, diabase and albitophyrs ($K2^d$ unit) is located on the sequence (Mijalkovic et al. 1972). In the area, the diabase as dikes cut the pyroclastics and spilite rocks. The spilites are highly vesicular and were filled by the secondary minerals calcite, quartz and less epidote and are characterized by amygdaloidal texture. The major part of the unit crops out in the west and northwestern in the Cheshmeh-Baghal and Kohan Siyah areas of the Balvard map.

2. Trachybasalt (unit B): The weakly vesicular (pores filled by calcite) basalts occurs with tuffits and turbidite ophiolites. It is believed that this basalt unit is younger, but discovery of thin layers of the pelagic limestone reject the idea (Mijalkovic et al. 1972).

3. Tuffaceous sandstone: It is a thick sequence associated with the Late Cretaceous limestone and chert lenses known as $K2^t$ Unit which occurs in west and southwestern of the Gughar Village. Obviously, all units of pyroclastics (tuffs, tuffits, agglomerate and breccia) with basaltic sills, the unit $K2^t$ and tuffaceous sandstone which overlies the massive basalt and pillow lavas. However, pyroclastic rocks are covered by the Late Cretaceous limestone mostly with a stratigraphic contact.

4. Limestone: The red limestone with volcanic breccia occurs around the Gughar village and is cut by the highly altered (kaolinite-clay) dyke which named as porphyritic diorite dikes in the Balvard quadrangle map.

5. Turbidite Series: The turbidite sequences occur noticeably widespread around the Gughar village.

There is an important factor that the Balvard-Baft ophiolites are compared to the Naein ophiolites based on their sedimentary sequences. The young sedimentary sequence in the Naein ophiolite complex consists of the middle Paleocene-early Eocene neritic limestone (lack of the early Paleocene deposition) which associated with the basal conglomerate are deposited on the ophiolites. Besides the limestone, some diabase sills occur with a tendency to IAT. It is similar to MORB pattern for the Rare Earth Element (REE) normalized to chondrite and the depletion of Nb element. On the other hand, the sedimentary sequence of the Balvard-Baft ophiolites contains the early-middle Eocene limestone associated with calc-alkaline to alkaline trachyandesite volcanic rocks. The other differences between two ophiolite sequences are the large volume of pyroclastic textured rocks in the Balvard-Baft ophiolite and absence of these rocks in the Naein and Dehshir ophiolites. Certainly, these two complexes have formed in two separate sedimentary basins.

2. Chahar-Gonbad Ophiolite region: The Chahar-Gonbad Quadrangle Geological Map is the continuation of northern trend of the Balvard Geological Map. The region is mountainous with many highlands such as Bid-Khan Kuh, Chehel-Tan Kuh, Panj-Kuh and Khersi-Kuh which are mainly covered by the Eocene volcanic rocks (Urmia-Dokhtar zone).

This ophiolite complex occurs in the south and southwestern Chahar-Gonbad which juxtaposes over the Eocene volcanic by a thrust fault. The ophiolites of the region contain a mixture of serpentinized ophiolite, breccia and agglomerate, basaltic lavas (spilites) and andesite, tuffs and tuffaceous sandstone with some interlayers of the Late Cretaceous pink-milky limestone. The ophiolites are named to as the "colored melanges" (CM) in the geological map.

3. Baft Ophiolite region: This is basically divided into two parts: (1) South–southeast of the Baft quadrangle; (2) Northwest of the Baft quadrangle. These parts of Baft ophiolite are scattered along the Baft-Esfandagheh fault. In addition to the old metamorphic units, the ophiolite sequences of Late Cretaceous is the oldest unit in the region.

Serpentines and serpentinized-harzburgite are the largest units which form the ophiolite. They are characterized by presence of the large porphyroclast crystals of orthopyroxene in rock sample. These minerals are elongated and the harzburgites occasionally are not affected by orthopyroxenes, resulted from the contamination of secondary magma, the surface foliation can be identified.

Severely serpentinized dunites which are occasionally associated with the films of depleted harzburgite occurs within the harzborgite unites. Pegmatite-gabbro as the small pockets found in peridotite. Isotropic gabbros as small stocks or larger masses have formed plentifully within the peridotite and serpentinite.

Pillow lavas are the other volcanic rock units which are of a tectonized contact with the ophiolite sequences in the area.

Balvard-Baft Ophiolite Units
The ophiolite rock units consist of:

1. Porphyric harzburgite with olivine phenocrysts surrounded by mesh-textured serpentine. Orthopyroxene (with an average size of 1–2 mm, and undulatory extinction), spinel minerals associated with chlorite and hydrous iron oxides.
2. Pegmatite gabbro: It contains phenocrysts of plagioclase, clinopyroxene (coarse variable sizes 0.5–5 mm) and a minor amount of olivine.
3. Isotropic gabbro: It includes phenocrysts of plagioclase (with clay and sericite alteration), clinopyroxene and magnetite. The chloritized amorphous amphibole (actinolite), iron oxide and sphene (titanite) occur in the area.

4. Pillow lava: This unit consists of 5.0–1 mm plagioclase phenocrysts (alter to sericite, clay and calcite) and clinopyroxene with hydrated iron oxide (palagonitic clays).
5. B Unit massive basalts in Baft quadrangle: They are with an intersertal and sometimes hyalopilitic textures which consist of plagioclase phenocrysts (1–4 mm, alter to clay and serisite). They have inclusions of clinopyroxene with fine-grained (albite) groundmass and associated with the secondary minerals such as, calcite and quartz.
6. Basaltic sills: This unit phenocrysts of plagioclase (with clay-serisite alteration) and clinopyroxene (1–5.1 mm), which is altered to chlorite and probably oralite. Matrix of the basalt consists of plagioclase microlites plagioclase associated with chlorite and iron oxide as accessory minerals.
7. Diabase dikes in peridotites: It is composed of phenocrysts of plagioclase, pyroxene (alters to chlorite and oralite), the opaque accessory minerals and secondary minerals are chlorite and epidote.
8. Diabase dikes with amphibole in isotropic gabbro: including phenocrysts of plagioclase and amphibole (actinolite) associated with quartz, chlorite, iron oxide and sphene (titanite).
9. Amphibolites in the Balvard-Baft ophiolite complex: These rocks occur follows:

    – Massive rocks usually form as bulk and compacted amphiboles along with the narrow quartz veins.
    – These rocks appear like gneiss with the alternating dark and light bands of amphibole, plagioclase and quartz, respectively.
      These two types of amphiboles form most the rocks (80–90%) which appear as anhedral and in some cases occur with inclusions of apatite. Also, Iron oxides form the accessory mineral of the rock. The second type of amphibolite is determined by the presence of poor undulatory extinction. The plagioclases are anhedral and alter to serisite completely. The epidote, have formed among the amphiboles crystals as fine grains or large crystals bands in the rock.

7.3.1.6 Eastern Iran

To describe the ophiolite complexes in this part of Iran, two Ratuk and Neh ophiolite mélange complexes are selected (Fig. 7.9). These complexes in terms of the nature and degree of mixture of the components are divided into three units:

A. Ophiolite unit: This unit includes different types of ultramafic, mafic and intermediate rocks and is characterized by disordering and being severely messy and the pelagic limestone contain the Cenomanian-Maastrichtian microfossils. Since, the geological structural events play a strong role to form the unit it could be called a "tectonic mélange".
B. Slightly metamorphosed flycsh unit: According to Tirrul et al. (1983), the unit is called "phyllite or phyllonite units" (the slightly metamorphosed Upper

Fig. 7.9 Main geological units of Sistan Suture Zone (Tirrul et al. 1983)

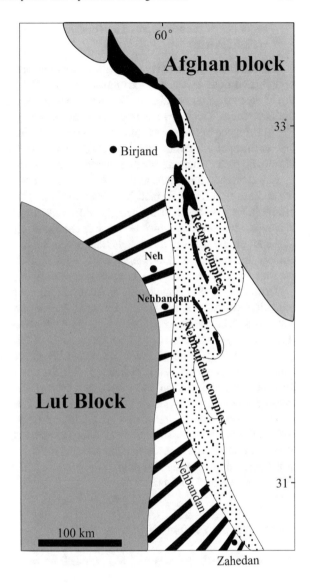

Cretaceous-Eocene turbidite sequences). The chlorite and muscovite growth related to the metamorphism phase have resulted in foliation which can be identified on the surface of rock units. Serpentinite in different sizes occur as thin interlayers in phyllite and locally forms a colorful mixture with pelitic cement in some locations.

C. Clastic marine sedimentary unit: The none-metamorphosed clastic marine sedimentary unit can be possibly attributed to the Paleogene age. Tirrul et al. (1983) believe that the unit is a cap for the mélange which consists of a turbidities sequence of sandstone, marl and clastic limestone with chert and even igneous rock fragments.

It should be noted that the ophiolite components form about 30% of Retuk and Neh ophiolite complexes. Most of the units are of an E-NW dip. The picks of pillow lavas, dip of layered gabbro, diabasic dikes and pelagic sedimentary rocks are generally at the same trend. The lower and upper contacts of the ophiolite complexes are faulted.

No evidence of the amphibolite metamorphic facies has been yet observed at the base of the ophiolite complexes. Harzburgite with tectonic texture occurs as dominant ultramafic rocks where dunite and lherzolite is less common in the area.

Based on chemical analysis, the ophiolite complexes are alkaline sodic type which the sodium was likely formed by metasomatism. The researchers believe that the pillow lavas that overlain by pelagic limestone, chert and radiolarite, are a good evidence of an eruption which occurs in a carbonate equilibrium level. The varieties of metamorphic rocks and minerals like, albite, sodic amphibole, quartz, actinolite, epidote and biotite or tremolite, chlorite, epidote and biotite indicate formation of the rocks in different pressure, temperature rates with green to blue schist facies.

Tirrul et al. (1983) consider the tectonic evolution of the Sistan Suture Zone as an individual structural unit in Iranian Plateau. They believe that the tectonic events have been occurred since Cenomanian to Campanian when two blocks of Afghan and Lut start to diverge with a relatively minor expansion of the oceanic floor between them. After completing the minor ocean in Campanian, the oceanic crust starts subducting under the Afghan Block and continues to late Eocene-early Oligocene. Meanwhile, the Sefidabeh flysch basin was closed and consequently, the Lut block collided to the Afghan block to form Sistan Suture Zone (Fig. 7.10).

The deformation of the zone was continued through the late Miocene by the collision of the Arabian-African plate to the triangle of Central Iran along the Zagros Major Thrust Zone. Then, the tension is transferred into the Sistan Zone. As a result of the main right-lateral strike-slip fault activities, the Sistan zone is cut and forms the current sigmoidal shape.

7.3.1.7   Birjand Ophiolite

The Birjand area in Eastern Iran is a part of the Sistan Crushed-Sutured Zone. The area with the NW-SE trend along with an area of approximately 800 km^3 is located in eastern Birjand City.

The first geological map quadrangle of Gezik scale 1:250,000 conducted by the National Iranian Oil Company in 1966. The geologists of the National Iranian Oil Company applied the stereoscopic technique and stratigraphy traverse sections to process the map.

Rahaghi (1976) published a detailed report with important results related to the microfossils in the area. Based on his efforts, the main units and structures of the Gezik quadrangle were identified and particularly gradual transition from the upper Maastrichtian to Paleocene-Eocene has been approved for the first time in the region.

Fig. 7.10 Tectonic
Evolution of Sistan Suture
Zone, N, 32^0 latitude (Tirrul
et al. 1983)

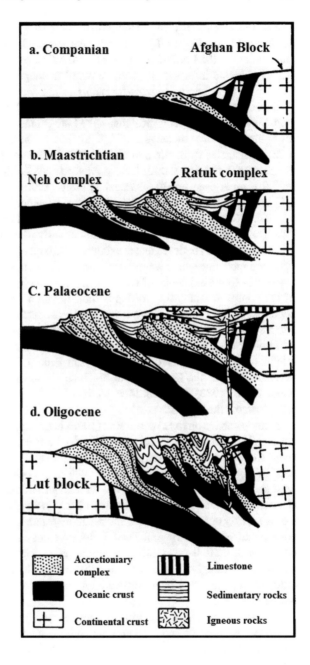

SOFIRAN geologist team (Dubois et al. 1976) has selected a fully exposed part of the Baran Mountain for oil prospecting. Although, Geological Survey of Iran (GSI) has conducted the Preliminary studies on the Central Lut Block and represented a lots of valuable information about "colored mélange" and its relation to the Late Cretaceous flysch and Cenozoic formation.

Based on the Sahlabad geological map (1:100,000) provided by collecting data from the surrounding geological maps, the Late Cretaceous age has been attributed to the ophiolite colored mélange and high pressure metamorphism which specifies with blue schist facies (Stöcklin and Nabavi 1973). The Gezik geological maps (scales of 1:250,000 and 1:100,000) have published by Geological Survey of Iran (GSI) in 1990, supervised by Eftekharnezhad et al. (1990). According to Stöcklin et al. (1972), the geological events caused to form irregular, disorder and messy mixture of the ophiolite colored mélange, which has emplaced in this part of Iran before the Paleogene and likely in Late Cenomanian or early Maastrichtian.

Ohanian (1983) believes that the southern Birjand ophiolite-melange is a complete ophiolite sequence which was possibly accumulated in an arch shape oceanic trench. He referes to the existence of flysch deposits which crop out in the north and west of the complex. According to this author, the arch shape ophiolite complex after deposition of has moved along a series of left-lateral strike-slip faults and rotated about 90 degrees. Stöcklin et al. (1972) in their initial statements in report no. 22 of GSI point out the process of subduction zone in the area is trending toward the west (Helmand block subducted beneath the Lut Block), but Maurizot (1990) believes that it is towards the East (the Lut block subducted beneath the Helmand block). Also, Tirrul (1983) based on their tectonic studies confirms the subduction zone trend towards the east.

A study carried out by Fotoohi Rad (1996) in the southwestern Drah, southeastern Birjand, led to discover and report of the eclogite facies metamorphic rocks and blue schist (for the first time in Iran). According to his studies the full sequence of different parts of ophiolite complex has been formed in the ophiolite mélange of the southwestern Darh, except for the basalts with sheet structure, dunite and dunite with chromite which is exposed in some ophiolite. A very clear prismatic alkalic-olivine basalt rock is outcropped in the most southwestern part of the region and nearly widespread across the region. The basalts have not endured any metamorphism. In other words, there is no evidence of alteration, except for some iddingsite euhedral olivine. Accordingly, it could be stated that the basalts are not associated with ophiolite mélange and have been formed later because they are much younger than the ophiolite.

The age of the Birjand ophiolite complex is pre- to late Cretaceous and is of the northwest–southeast trend. The complex has been covered unconformity by Cenozoic volcanic rocks. The colored mélange complex consists of a variety of rocks as follows.

1. Ultramafic and mafic rocks,
2. Sedimentary rocks and
3. Metamorphic rocks.

The above three groups of rocks are briefly described in below:

Peridotite, harzburgite, lherzolite and serpentinite: There are so many chrysotile and asbestos veinlets in serpentinite. Accordingly, the rock appears with a lattice texture. The Listvenite is abundant at the margins of shear zones of serpentinite bodies, along the faults and the adjacent intersecting faults. In some parts, the rodingitized dikes occur within the serpentinite-harzburgite bodies.

Flysch: Two types of flysch have been identified in this area which are barely distinguishable from each other. The Upper Cretaceous flysch occur in west and north of the region which consists of green shale, sandstone, siltstone and phyllite-shale and sometime, carbonized shale, like listvenite in the vicinity of the ultramafic rocks.

The upper part of the region more or less has endured metamorphism and contains the phyllite and shale beds. Generally, all rocks in the region, more or less, have endured a poor regional and tectonically metamorphism.

The metamorphic rocks mainly consist of phyllites and schist (epidote, chlorite, sericite and talc-schist). The metamorphosed spilitic rocks, gabbros (metagabbro) and calk-schist are amongst the Late Cretaceous metamorphic rock in the region.

Spillit is characterized by a pillow structure and maximum 2–5 m length.

7.3.2 Urmia-Maku Ophiolite

An ophiolite complex crops out in the southwestern Urmia to Maku, NW Iran, which extends towards the west to Anatolia, Turkey. Based on some studies, the Urmia-Maku ophiolite complex is not a part of the ophiolite ring band of Central Iran, but it is very similar to colored mélange zone in Central and Eastern Iran. In this area, ultramafic and radiolarite with diabase, tuff, shale and pelagic limestone are mixed and covered by Eocene thick flysch sediments.

7.3.2.1 Khoy Ophiolite Complex

This complex is not a part of the colored mélange ring of Central Iran. Actually it is exposed along the Khoy to Maku road.

On the border of Turkey and Iran, in the west of Urmia Lake, some ophiolite-radiolarite complexes with irregular forms are exposed and extend along the north–south towards the west to Anatolia, Turkey.

According to Alavi-Naeini and Bolourchi (1973) the Lower Cretaceous sediments mainly consist of the Aptian-Albian *orbitolina* limestone which unconformably over-lies the Jurassic sediments. The Upper Cretaceous ophiolite mélange juxtaposes against the Early Cretaceous limestone. Stöcklin (1974) believes that the ophiolite in northwestern Iran is very similar to the circular ophiolite mélange of Central Iran and consists of ultrabasic, radiolarite, diabase and abundant tufaceous volcanic rocks. This complex also contains shales which looks like flysch, pelagic limestone and often conglomerate. It is covered by the thick Eocene flysch sediments.

The Early Alpine Orogeny activities have caused to form the Khoy colored mélange and metamorphose to greenschist facies (Eftekharnezhad 1975). Also, the glaucophane schist occurs in the southwestern Khoy, around the Gheshlagh village.

According to Eftekharnezhad (1980), the Oshnavieh-Khoy ophiolite is a good evidence of the Late Cretaceous-Paleocene rifting zone, forming the deep oceanic basin and developing the Tethys Ocean. This ocean has been completely separating the western Zagros from the other parts of Iranian Late Cretaceous-Paleocene.

Hassanipak and Ghazi (2000) believe that the Khoy ophiolite represents an oceanic lithosphere that formed in the Neo-Tethys Ocean. They recognized two distinct types of geochemical basaltic flows in Khoy ophiolite:

1. The massive basalt with the same patterns of E-MORB and
2. Pillow basalts with the basic chemical composition which their trace elements patterns identify between E-MORB and N-MORB.

They also believe that results of comparison of the Khoy ophiolite with the other ophiolite complexes in Iran represent a clear evidence for a dominated tectonic relationship. In other words, these results indicate that the Khoy ophiolite and other ophiolite complexes in Central Iran such as Naein, Shahr-e Babak, Sabzevar, Chehelkureh and Band-e Ziarat have been formed at the same time, caused by closing the northwestern branch of the Mesozoic narrow sea surrounded the Central Iran microcontinent.

Based on studies of Ghazi et al. (2003) on hornblendes of gabbro and amphibolites indicate the age of 108–110 million years by applying the 40Ar-39Ar method.

The results of studies on the planktonic foraminifer microfossils indicate that they belong to late Cretaceous to middle-late Eocene. Based on these data, the authors suggest that the initial rifting and development of the oceanic crust of Khoy ophiolite had begun in the late Albian. However, the volcanic activities have continued to the Eocene.

According to the new field and laboratory studies (Khalatbari-Jafari et al. 2003), two ophiolite complexes of different ages are identified in the Khoy ophiolite as below:

1. Older multi-metamorphosed ophiolite: Age determination using 40 K-40Ar method on the amphiboles of the metamorphic rocks, indicates the Early Jurassic age for Khoy ophiolite complex,
2. The younger non-metamorphosed ophiolite: This ophiolite complex has been obviously formed during the Late Cretaceous.

It is believed that the ophiolite complexes have been formed in a tectonically active environment which typically is morb and super subduction types.

The results of detailed geological studies of Khalatbari-Jafari et al. (2004) on the Khoy ophiolite is presented in the geological map as below.

Accordingly, the final stage of ophiolite it forms in oceanic basin with slow spreading rate. To approve the above idea, they represent the sufficient evidence like,

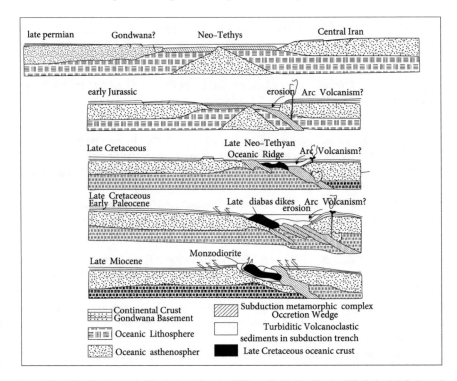

Fig. 7.11 The Geodynamic Evolution Model of Khoy Ophiolite Basin (Khalatbari-Jafari et al. 2004)

lherzolite mantle sequence, small volumes of gabbroic rocks, absence of diabasic sheeted dikes and abundant porphyric basalts in the volcanic sequence.

Khalatbari-Jafari et al. (2004) represent the geodynamic evolution of the Khoy oceanic basin (Fig. 7.11) as follows:

1. The Khoy oceanic basin starts to expand after the opening of the Neo-Tethys Ocean in Late Permian.
2. Since the Late Triassic to Late Cretaceous, while the Khoy oceanic basin was expanding, the subduction zone starts to form simultaneously along the Central Iran Block, in the eastern margin of Khoy ocean.
3. The final produced oceanic lithosphere which has never subducted and metamorphosed, forms the Khoy Cretaceous ophiolite complex.

The accumulation of turbidities pyroclastic rocks formed in the subduction zone, and masses of non-metamorphosed igneous rocks (gabbro and granite) intruded within the convergent metamorphosed complex;

1. The western margin of the basin in early Paleocene began to converge beneath the Late Cretaceous oceanic lithosphere accompanies with a series of isolated calc-alkaline diabasic dikes that cross cut all the Khoy ophiolite.

2.   Following the collision and folding, the calc-alkaline monzonite, sub-volcanic intrusive (Khalatbari-Jafari et al. 2004) intruded in the Khoy ophiolite during the Late Miocene, and cause to cover the Eocene-Paleocene sediments.

Petrology and geochemical studies which have been carried out on the Late Cretaceous rocks of Khoy ophiolite (Khalatbari-Jafari et al. 2006), represent the following conclusions:

1.   Petrographic studies have been confirmed by field data indicating the existence of two ophiolite complexes in the Khoy region.
2.   The Upper Cretaceous Khoy Ophiolite is formed in an oceanic ridge with slow spreading rate.
3.   Changing the various compositions of the extrusive rocks and layered gabbros indicate that the magma being injected alternatively and has caused to form a magma mixing within the origin magma chamber.

These authors also point out that the nature of Upper Cretaceous ophiolite lavas of the Khoy region is very homogeneous and is of T-MORB type. This indicates that these rocks have been formed by partial melting of the depleted mantle in the spreading zone of the ocean floor and that they possibly contaminated with one or more regional mantle plumes. Thus, the enrichment of ophiolite by LREE elements can be explained.

They also noted that the supera-ophiolite turbiditie series is accumulated on the Late Cretaceous ophiolite in the subduction trench along the southwestern margin of the Central Iran block. The subduction trench is feed by pieces of T-MORB volcanic rocks type from an oceanic source, in one hand, and from continental arc basalts on the other hand. However, the Khoy metaophiolite represents an oceanic basin slow spreading condition. The tectonites and porphyroclasts which are preserved in metamorphic slices indicate the terminate condition of shearing, particularly oceanic fracture zones.

7.3.3   Makran Ophiolites

According to the present author, the Makran ophiolites are definitely amongst the most complete ophiolite complexes in Iran. Researches carried out on this zone (e.g. Falcon 1974) indicate the following points:

1.   The Oman Sea crust is of the oceanic crust type (White and Klitgord 1976) which is subducted beneath the Lut Block (Falcon 1974).
2.   The speed of subducting is measured to be 5 cm/year.
3.   The Oman Sea crust is bended beneath the Makran baseal rocks at the beginning of the Jazmurian Basin and from the marginal southern side it slopes deeply under the Jazmurian Basin.
4.   The northern Jazmurian young volcanoes have been formed by melting of oceanic crust at a depth greater than 100 km.

5. Possibly during the Late Cretaceous to early Miocene, there was an oceanic trench at the present-day Makran and the Jazmurian basin is considered as its continental margin.
6. The Makran area is gradually subsiding.

Generally, there is not any kind of magmatism activities in the Makran subduction zone, but considering the general geology of the region, the volcanoes in the northern area are evidence of the special structural features in the Makran zone.

These volcanoes which are located between two subsided zone, i.e. the Lut Basin in the north and Jazmurian basin in the south (with EW trend), have been active during the Pliocene-Quaternary. Thus, the Makran ophiolites are the result of the subduction zone activities which still going on.

According to Arshadi and Forster (1983), the Makran region can be divided into two zones, northern and southern Makran. These two zones are separated by a narrow continental crust zone, 250 km length.

The northern zone can be considered as a real ophiolite mélange complex with a complete sequence of an ophiolite complex. The above-mentioned authors report the high-pressure and medium-temperature metamorphic rocks that include glaucophane, lawsonite, pumpellyite, albite, chlorite and titanite in the northern margin of the complex (Fig. 7.12).

The fossils found in the sedimentary rocks confirms the Jurassic-Paleocene for the tectonic ophiolite mélange complex. Arshadi and Forster (1983) believe that this complex originated in an oceanic rift and simultaneously mixed with other deposits. However, the Upper Cretaceous-Paleocene pelagic sediments in this complex indicate that they formed during Paleocene. Although, the final age of the ophiolite

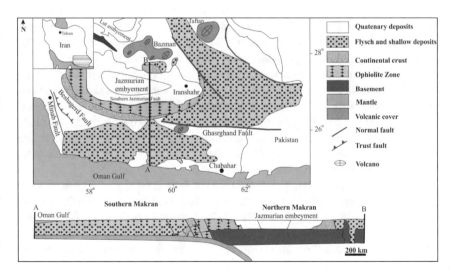

Fig. 7.12 Geotechnical and schematic cross section of the Makran based on seismotectonic map of Iran (Berberian 1976); Report of the geology of Eastern Iran (Geological Survey of Iran 1980); (Coleman 1981); Fenouj and Iranshahr geological map (Arshadi and Forster 1983)

emplacement is known as the Cretaceous and Paleocene. Nevertheless, based on Haynes and Reynolds (1980) the age of hornblende mineral in an amphibolites associated with the ultramafic of western Makran ophiolite identify approximately 170 ± 5 Ma by 40Ar-39Ar Method. According to Haynes and Reynolds (1980) as amphibolites associated with ophiolite, in Iran, Greece and Yugoslavia have the same age, so possibly the first collision of Tethys (Neo-Tethys) oceanic crust have occurred during the Middle Jurassic.

7.3.3.1 Kahnouj Ophiolite Complex

The Kahnouj ophiolite complex is located in the most westerly part of the Makran zone and western Jazmurian basin (as a part of the map 1:250,000 of the Minab quadrangle and the map 1:100,000 of the Kahnouj). Tectonically, the complex is located at the conjunction of the three main tectonic zones; i.e. Sanandaj-Sirjan, Makran and Lut Block (Stöcklin 1968). The Kahnouj ophiolite complex is one of the main rotational structure zones in Iran (Nogol-e Sadat 1978) which is located between two main N-S trending Sabzevaran and Jiroft faults.

In fact, it is exactly the units that McCall (1985) called them different names as: Band-e Ziarat for plutonic units, diabase dyke complex for sheeted dikes unit and Darrehanar Complex for the volcanic unit with its sedimentary cover rocks (Fig. 7.13).

There are different ideas about how the Kahnouj ophiolite belt has been formed. Stöcklin (1974) and Knipper et al. (1986) suggest that during the Triassic, the Neo-Tethys Ocean has been formed between Gondwana, Iran and Afghanistan continental blocks where have already separated from Gondwana and collided to the Eurasia continent. Stöcklin (1974) believes that in fact the Central and Eastern Iranian ophiolite belts indicate several small oceans which are the branches of big Neo-Tethys Ocean. Contrary, Berberian and King (1981) presented a different idea implying that the northwest, central, eastern and southwestern Iran ophiolite belts are the remnants of a small, narrow ocean which are resulted from a rift zone similar to the Red Sea rift type. They noted that during the lower Paleozoic-Middle Triassic, multi-branches of rifts developed by the divergent movements. In this model during the lower Mesozoic a small ocean open between the Lut block and Sanandaj-Sirjan zone, a later closed during the Late Cretaceous-Paleocene because of northward movement of the African-Arabian plates (Glennie et al. 1973).

McCall (1985) and Hassanipak et al. (1996) believe that different sequences of the Kahnouj ophiolites complex are syngenetic and have been formed in the Cretaceous-Paleocene. According to Mccall and Kidd (1981), the Bajgan-Dourkan Zone located on the south of the internal Makran Ophiolite belt indicates the presence of an arc basin extended towards the Sanandaj-Sirjan Zone. The Bajgan complex and its sedimentary cover (complex Dourkan) was a part of the narrow continental crust where shallow platform sediments are deposited during the Cretaceous to Paleocene. At the same time, on both sides of this uplifted micro-continent, the deep water sediments were deposited. He believes that the colored mélanges on the southwest of the

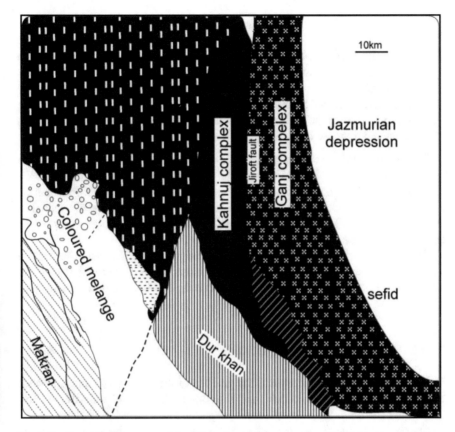

Fig. 7.13 Geological structures of the Kahnouj ophiolites complex with neighboring complexes, modified after (McCall 1985)

micro-continent developed at the margin of a deep large oceanic trench connected to an active subduction zone, up to Early Paleocene (McCall 1985). The Kahnouj ophiolites formed along a diverged oceanic rift in a marginal basin, since the Neocomian to early Paleocene.

However, Sabzehei et al. (1993) believed that the Kahnouj ophiolite is a polygenitic complex which has been formed by three individual magmatic sequences and a series of mylonitic and saussuritic rocks as follows:

1. The Precambrian-Paleozoic gabbro cumulates which consist of trectolite, olivalite, feldspar predotite and melagabrro at the base and gradually towards the top transfer to olivine, diallage, gabbro, noriteandferro-gabbro. This author believes that the attributed age of the complex is incorrect.
2. There is a Late Cretaceous subvolcanic and volcanic series which include sheeted diabasic dikes, basaltic pillow lavas and the Upper Cretaceous pelagic limestone. According to Sabzehei, the unit intrudes into the layered gabbro

and causes partial melting and recrystallization of rocks to form the coarse amphibole-gabbro.

3. An Upper Paleocene-Eocene sodic plagiogranite complex intruded into the above semi deep volcanic series. Sabzehei et al. (1993) believe that the complex is the result of the anatexis of the old hydrated gabbro.

4. At Early Cenozoic, a young set of susuritic and mylonitic rocks forms during the process which the hydrothermal fluids intrude into the old magmatic rocks and metamorphosed them, along the faulted zone.

Sabzehei et al. (1993) emphasis the tholeiitic nature and low-K content of all above magmatic series, represent the Inversion Tectonic Theory, to explain the formation of these rocks. They believe that the volcanic and subvolcanic rocks form on the widespread old gabbroic rocks of oceanic crust, while the rifting event occured in the Late Cretaceous. Then, during the Late Paleocene Laramide Orogeny, these rocks endure the folding activities by the inversion tectonic system.

However, the Makran ophiolite belts and their northeast extension indicate a marginal basin development in Cretaceous and close in Late Cretaceous-Paleocene. Desmons and Beccaluva (1983) believe that an arc-basin system existed in this part of the region.

7.3.3.2 Southern Iranshahr Ophiolites

The Iranshahr ophiolitic mélanges is the junction of the north–south and east–west borders of the Lut Block in Baluchistan. The complex includes all units of the ophiolite sequence associated with sediments, porphyric pillow lava (spilitic basalt, sanidine trachyte) and upper lava. In addition to the above rocks, some blocks of granitic rocks, amphibolite facies metamorphic rocks (hornblende, biotite, garnet and andesine) or green schist facies (chlorite, albite and epidote) extended few meters to one kilometer. These rocks are called as "exotic rocks" of the ophiolites. Various interpretations are expressed regarding to their origin likely the origin of rocks are older than the Permian base rocks of the Central Iran or Lut Blocks.

The limestone was deposited on basic lavas and is interpreted as "simont sediments". The massive amphibolites consist of the plagioclase, hornblende and brown garnet (Reinhardt 1969). However, the existence of granite associated with biotite-gneiss and the amphibolite among the exotic rocks, is considered to be a part of the Mesozoic platform and simont sediments.

The Iranshahr ophiolitic melange is an olistostrom complex which includes sedimentary and igneous rocks on a sedimentary matrix. Also, Olistostolite is metamorphosed in zeolite greenschist facies or glaucophane and lawsonite blue schist facies. Three type colored complexes of colored mélanges as, ophiolitic melange, olistostrom, and sedimentary-volcanic sequence are exposed together in the region.

7.3.3.3 Fanuj-Mascutan Ophiolite Complex

The complex around the Lut Block with a WE trend is covered by the Cretaceous-Paleogene andesitic volcanic rocks and interlayered turbidite sediments, in southern Baluchista (Desmon 1977). The ophiolite complex includes the rock units as follows (Fig. 7.14):

1. Harzburgite serpentinized in some locations to form antigurite and lyzardite. The gabbroic dikes cut the rock unit.
2. The cumulative layered sequence of alternating layers of wehrlite, troctolite, olivine norite, leucogabbro and anorthosite.
3. Clinopyroxene-hornblende gabbro contains troctolite lenses, which are crosscut by the diabase dikes.
4. Massive and pillow lavas with hyaloclastic and associated sediments.
5. Basic sheeted dikes with the ophitic and interstitial texture (Houshmand-Zadeh 1977; Delaloye and Desmons 1980; Arshadi and Forster 1983).
6. Massive and pillow lava along with hyaloclastic and other associated sediments.
7. Hornblende amphibolite partially changes to lawsonite and glaucophane schist.
8. Granitoid, plagiogranite, trondhjemite to tonalite.

In the southern region, the ophiolite sequence converted to tectonic mélange which consists of ophiolites and turbidites. In most southerly region, the Makran turbidity rocks indicate the Cenozoic continental margin basin (White and Klitgord 1976; Farhovdi and Karig 1977; White 1976).

The contact between different lithological units, intrusive and metamorphic rocks is actually disturbed by the fault activity. Based on some studies (e.g. McCall and Kidd 1981), a new concept was released as follows:

1. The ophiolite is transitional type and is likely formed in a marginal basin and
2. The colored mélanges have only the tectonic characteristics.

There are two volcanic series:

1. Ophiolitic lava (lower lava) (Desmon 1977),
2. Andesitic lava (upper lava) associated with the flysch-type sediments.

7.3.4 Zagros Ophiolite Complex

During the early Triassic, the Neo-Tethys Ocean was developed due to the separation of Iran from the Arabian plate. During the Late Cretaceous, the Neo-Tethys has been closed and Iranian plate began to move towards the Turan plate. Consequently, the ocean between Iran and Arabia start to be closed and the plates collided. These were finally resulted in formation of the Zagros Mountain Range. Later, when the Red Sea start to open in the Miocene, pushed back two plates (Iran and Arabia), and the result is to intensify the collision of two plates. The final complete closure of the Neo-Tethys Ocean has been occurred during the Pliocene.

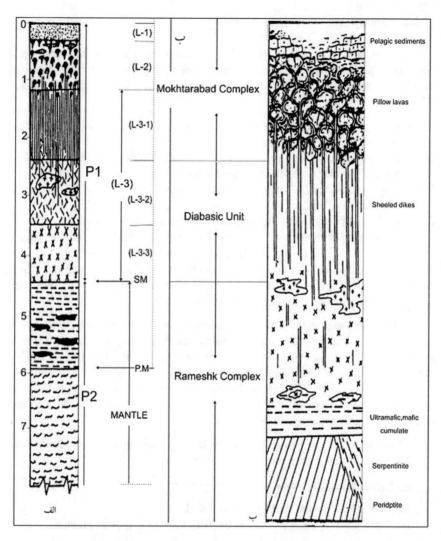

Fig. 7.14 Schematic Cross Section of a Complete Ophiolite Sequence (A) and the Ophiolitic Sequences in the Makran Region (B) (Arshadi and Forster 1983). L-1 = the Sedimentary Part, L-2 = the Volcano Part, L-3-1 = Subvolcanic part, L-3-2 = Massive Gabbro, L-3-3 = the Layered Gabbro, P.M = Petrological Muhu, SM = Seismic Moho, = P1 = Crustal Segment and the Mantle ophiolitic sequence

In the northern part of the Zagros Thrust Zone, a series of ultramafic, mafic and radioralite rocks are exposed. The formation of this series was ascribed to the oceanic crust and trench basins.

Some geologists suggest that the presence of the Zagros ophiolite and colored mélange around the Lut Block is related to a rifting event. The ophiolites in western

Iran are track down along the High Zagros or shear zones. The High Zagros ophiolite-radiolarite belt mainly formed tectonic unites in the Neyriz and Kermanshah regions and gradually thrusted on the north Zagros carbonate sediments with a continental platform.

In the Fars region, the ophiolite unit is unconformably covered by a coral reef limestone (the Late Cretaceous Tarbur Formations). The unconformity contact indicates the ophiolite has obducted along the Zagros belt in Late Cretaceous.

The Oman and Zagros ophiolite-radiolarite sequences in terms of the stratigraphy and structure geology are more uniform. The basal section of the complex is clearly exposed in Neyriz, Fars. According to Stöcklin (1974), the complex covers the Cenomanian-Turonian clayey limestone (Sarvak Formation), without any disconformity or faulting. It includes three sections as below:

Lower part or sedimentary part: The base of ophiolite-radiolarite sequence has been described as a specific lithological unit called the Zagros Radiolarite Unit which is a mappable unit. Ricou (1971) described this unit as "Pichakun Formation" in Zagros and is known as "Hava Sina group" in Oman. This unit includes a series of consecutive thrusts, generally with flake structure. The lower part consists of the red radiolarite, associated with the red, green to gray layers of cherts that encompasses very thick crystalline limestone rich in fossils. Turbiditic rock types are also formed within the complex.

Middle part or mélange: This part of the ophiolites is located on the sedimentary unite of the lower part. The thickness is less than the other two parts includes non-host sedimentary and igneous (ultrabasic, diabase and spilite) rocks, that scatter in a sedimentary matrix.

Upper part or ophiolite: At the base of this section an ophiolite masse, with several meters thickness is exposed, and overlain by a set of crystalline limestone blocks. The crystalline limestone slightly metamorphosed, and apparently looks to be allochthonous.

7.3.4.1 Kermanshah Ophiolites

The Kermanshah ophiolites are exposed in the Sahneh-harsin areas (Braud 1970; Moghadam and Stern 2011) and consists of three separated zones.

First outcrop: The first outcrop is mostly exposed in the Sahneh, northeast of Kermanshah, and composed of cumulative, peridotite, pyroxenite and harzburgite. It is covered first by gabbroic rocks and then by lava. There is no mélange structure.

Second outcrop: The second outcrop located in the southeastern of the first outcrop, the Organa area. Unlike the previous outcrop, this is highly tectonized. The tectonic pressure caused to mix the ultramafic rocks and limestone associated with calcareous thin layers and radiolarite flakes.

Third outcrop: There is a serpentinite massive along with the sheets of biological pyroclastic re-crystallized limestone of Triassic in Harsin area. Ricou (1974) introduced an ultramafic and mafic complex as a part of trusted rock unit near Kermanshah City, and south of Sahneh as follows:

1. Harzburgite, serpentinite (chrysotile) are cut by gabbro and diabasic dikes.
2. Homogeneous gabbro, diorite and tonalite.
3. Sheet dikes compose of basalt to andesite.
4. Cenozoic intrusive rocks which include diorite, gabbro and pre-Miocene troctolite.
5. Volcanic rocks, as interlayers within the flysch sediments (Motamed 1987).

The ophiolite complex's composition is radiolarites, lava, limestone, and serpentinite. The low content of non-compatible elements, Nb, Zr, P, Ti shows clearly geological environments of tholeiitic island arc (Ewart 1976).

7.3.4.2   Neyriz Ophiolite

Studies of radiolarite–ophiolite of Neyriz (Ricou 1974) and Oman (Glennie et al. 1973) indicate that subsiding, and rifting events have formed within the continental crust during Triassic.

These geologists have studied the Upper Triassic radiolarites and indicated that these fossils are the indications of the earliest life deposited during the Triassic. Therefore, it is accepted that the ophiolitic-radiolarian complex of Oman-Zagros, is a result of an oceanic trough formed by rifting with a too steep slope basin. But there are different opinions on the age, and the size of the trough and its formation mechanism.

The Neyriz radiolarites (Pichogn Formation) which include radiolarian cherts, and fossiliferous turbiditic limestone cover the shallow limestone of Late Cretaceous with unclear contact. The radiolarites content plenty of Paleozoic and Mesozoic fossils, apparently without the younger Turonian fossils (e.g. Moghadam et al. 2014).

Located in the Late Cretaceous limestone at the above and the radiolarites with older fossils at the below of the complex offers two possibilities:

1. The fossils in radiolarite sediments are in situ.
2. Most of the fossils in the turbidity limestone and radiolarites are detrital and transported.

If the fossils are in situ, so, it should be accepted the radiolarites are displaced and transported horizontally far about several hundreds of kilometers, and there must have been at least one trough since the Early Triassic, and also the trough has been far away from the sedimentary basin as today.

If the radiolarite fossils are detritus and transported, so there should be a shallow subsidence basin in the margin of Arabia Platform in Cretaceous. Consequently, the radiolarites have been deposited in a shallow basin.

The theory of in situ radiolarite is assumed mostly based on the characteristics of the Zagros and Oman Eugeosyncline System. Therefore, Falcon (1967) and Stöcklin (1968) believe that the radiolarites have been formed in an intercontinental trough with severe depression, on the edge of the Arabian Plate. During the Late Cretaceous and later the ophiolite emplaced within the radiolarites sediments.

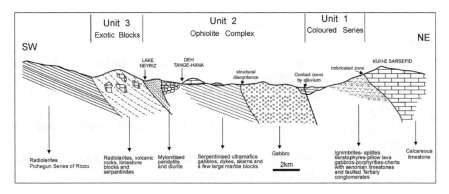

Fig. 7.15 Cross section of the northwest Neyriz ophiolites with identified units 1, 2 and 3

Regarding the forming modes of emplacements of radiolarian and ophiolites, in situ or transported, all the researchers agree with the idea that the formation of ophiolite complex has ended in Campanian or early Maastrichtian. It is because, the Maastrichtian shallow limestone (Tarbur Formation) covers the radiolarites and ophiolite complex unconformably. Therefore, it is now accepted that the Arabian and Iranian plates jointed again with a narrow strip of an oceanic crust in the pre-Maastrichtian time.

Based on the field observations and the characteristic of structural geology, Arvin (1982) divided the Neyriz area into three sections.

1. The colored series include spilite, keratophyre, ignimbrites, less alkaline gabbro, porphyrite, chert, sandstone, and Late Cretaceous limestone.
2. Ophiolite complex consists of harzburgite, tectonites, and layered gabbro, small quantities of dunite, lherzolites and pyroxenite.
3. Exotic fragments include chert, massive dolomitic limestone mass, altered peridotite, spilite, alkali-basalts, amphibolite, green schist, quartzite, rodengtinite, tuffite, red limestone and thin dolomitic limestone layers. (Fig. 7.15).

The Neyriz ophiolite complex is divided by Sarkarinezhad (1989) into the following units:

1. Ultrabasic units including dunite, chromite, webstrite, and clinopyroxenite,
2. Gabbroic units,
3. The sheeted dikes and basaltic which include diabasic and basaltic dikes. Also, a small amount of plagiogranite varies from diabase to upper gabbros and
4. The units of tectonized harzburgite and lherzolites with intrusive foliation.

7.3.4.3 Esfandagheh Ophiolites

The Esfandagheh ophiolites is located on the northern Hormuz Arc, approximately 150 km north of Hormuz Island, along the axial plain of the Hormuz Arc. The ophiolites of the area are along the following three branches (Houshmand-Zadeh 1977).

1. Northeast branch or the Ashin branch cut across the Mahabad-Manoj zone which includes large elliptical peridotites (generally dunite) formed as fault breccia features. This tectonite dunite lenses contains chromite layers which are extractable economically.
2. A branch in excess of about 100 km with the northwest trending parallel to Zagros.
3. The central branch trending southwest its direction changes to the north–south trend, and continues towards the south up to Makran ophiolite.

Sabzehei (1974) described three major rock units in the Esfandagheh as follows:

1. **Ultramafic and mafic units:** The ultramafic and mafic rock units are located along the major faults zones and usually their formation has occurred during more than one phase. These unites are associated with Mahabad-Manoj zone with faulted contacts zone. The units form as a great mass of the serpentinite dunite, serpentine, and rarely layered gabbro.
2. **Volcanic and associated sediments:** Volcanic and its associated sedimentary rocks are diabase, tuffs, tuffaceous breccia and pillow lavas. Some diabase dikes form sheet complexes and the pillow lava rocks are covered by *Globotruncana* limestone. Generally, the sedimentary rocks associated with volcanic rocks are radiolarite, pelagic limestone and flyschs.
3. **Mélange**: This mélange is a mixture of the Jurassic limestone, basic rocks, ultrabasic, volcanic, and related sedimentary rocks. Generally, they are located alongside the ophiolite and cratonic rocks.

The metamorphic rocks associated with the Esfandagheh ophiolite and mélanges are mostly volcanic and its related rocks which are metamorphosed. Sabzehei (1974) identifies three phases of developing the metamorphic rocks (Fig. 7.16). The first phase consists of green schist including, quartz albite, epidote and almandine which are followed by the second phase of the development of high-grade phases of glaucophane, valestonite and jadeite. The greenschist of dynamic metamorphism is third phase.

Tectonic: Since the real flysch sequence in Esfandagheh has a considerable thickness. So, they could be formed in a triple junction of divergent system. Sabzehei (1974) proposed a rifting zone similar to the Red Sea rift system for the formation of Esfandagheh ophiolites, ophiolite mélange and related sediments in a large trough basin.

Geochemistry: The cumulate magma of the Esfandagheh ophiolites are tholeiitic type (Sabzehei 1974) and rock's geochemical data indicate that the chemical composition of ultramafic rocks of Esfandagheh is similar to the other metamorphosed harzburgite of the world's ultramafic rocks (Alavi-Tehrani 1979).

Age of the Zagros ophiolites: The mechanism of emplacement of the Zagros ophiolite, particularly its age is not yet well known. The Zagros ophiolites as large banded masses, cover the radiolarites. The contact between the sedimentary series

Fig. 7.16 Esfandagheh ophiolitic melange (Sabzehei 1974) [A = Eocene, B = conglomerate, C = Almost non-metamorphic rocks that gradually locate on the metamorphic units. The series include, pillow lavas, flysch, radiolarite, red argillite, and occasionally limestone, D = Pelagic limestone, E = Pillow lavas, F = Multiphase metamorphic series of the basaltic volcanic rocks (tholeiitic basalt), diabase, red argillite, chert and layered sandstone, greywacke and small amount of limestone, G = Tholeiitic basalt facies include, eclogite, amphibolites, garnet, prasinite and uvarovite), H = Layered gabbros and I = Dunite, harzburgite, wehrlite, layered pyroxenite with unknown basal part]

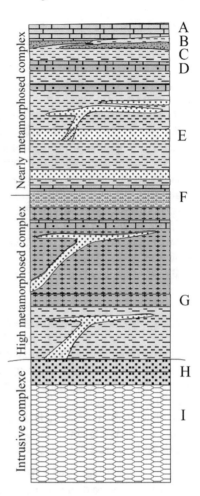

and ophiolite masses is as a narrow mélange strip with typically Permo-Triassic exotic fragments. The exotic limestone fragments have caused to complicate the original connections between the ophiolite rock units. There is little evidence to determine the age of the ophiolite. It seems that the age of the ophiolite is Late Cretaceous. It is only because of the last formation of basaltic activities such as pillow lava, and tuff breccia which rarely are interlayered with siliceous shale, radiolarite, and occasionally pelagic limestone. Alleman and Peters (1972) determined the age of tuff formations of the Oman Hawasina ophiolite about 90 ± 6 Ma. by using P/Ar technique. Probably, the Zagros ophiolites age could be the signs of the Oman Hvasyna Ophiolite. Although the age of the Zagros ophiolites is known as Late Cretaceous. It should be noted; further studies are needed to prove the age precisely. However, the emplacement age of the Zagros ophiolite is varied from Jurassic to Cretaceous.

7.3.5 The North–Northeastern Iran Ophiolites (Late Paleozoic-Early Triassic)

The north and northeastern Iran ophiolites like the other ophiolites such as Virani, Mashhad (Ab and Bargh areas), Talesh in Masuleh area and Gorgan schists are the indications of remnants of the ophiolite suture zone of Paleo-Tethys Ocean. The ophiolites are formed by the collision of Iran and Tourna plates (Alavi 1991). Here is described only the Virani ophiolites as follows.

The Alborz mountain thrust faults trend north–south direction. It is due to the subduction of Iran Plate (oceanic crust between Iran and Turan) beneath the Turan plate with a northern dip. There are several normal faults which formed by the gravity mechanism and are mostly thrusted.

Also, in the Binaloud zone, there are some thrust faults towards the south, that formed by subduction of the Paleo-Tethys oceanic crust beneath the Turan crust. However, both Alborz and Binaloud thrust systems have been developed by the collision of the Iran and Turan plates.

7.3.5.1 Torbat Heydarieh Ophiolites

In the northern Torbat Heydarieh and southwestern Fariman several E-W trending ophiolites and mélanges are exposed. The ophiolites melange of north Torbat Heydarieh with approximately 3000 km^2 according to the Fig. 7.17 has a nearly complete sequence.

Although, the northern ophiolites of Torbat Heydarieh is highly blended, but in some cases the following sequences are identified by compiling different parts of the sequence from base to the top (Darvishzadeh 1991):

1. Basal amphibolites with limited distribution,
2. Metamorphosed and large spread peridotite.

The peridotite sequence contains dunite, harzburgite, wehrlite, lherzolites pyroxenite and serpentenite. The highly-tectonized features such as lineation and fractures are their obvious characters.

Cumulate rocks: The composition of cumulate rocks is ultrabasic, intermediate to basic, which begin by peridotite (peridotite contains chromite) and ends by gabbro. The light and dark pyroxenite layered in the middle of complex indicate a transitional stage and the gabbro occur as layers and massive with pegmatite texture.

Sheeted dikes complex: They are mainly composed of diabase sheeted dikes and have the massive structure on the upper part. Their chemical composition is calc-alkaline type (Darvishzadeh 1991).

Leucocratic rocks: The rocks composition are typically tonalite to granite which have appeared as dikes and thin layers.

Extrusive rocks: These rocks are composed of a complex of spilite basalt, tuff, agglomerate and Lapilli tuff. They are calc-alkaline rock's type.

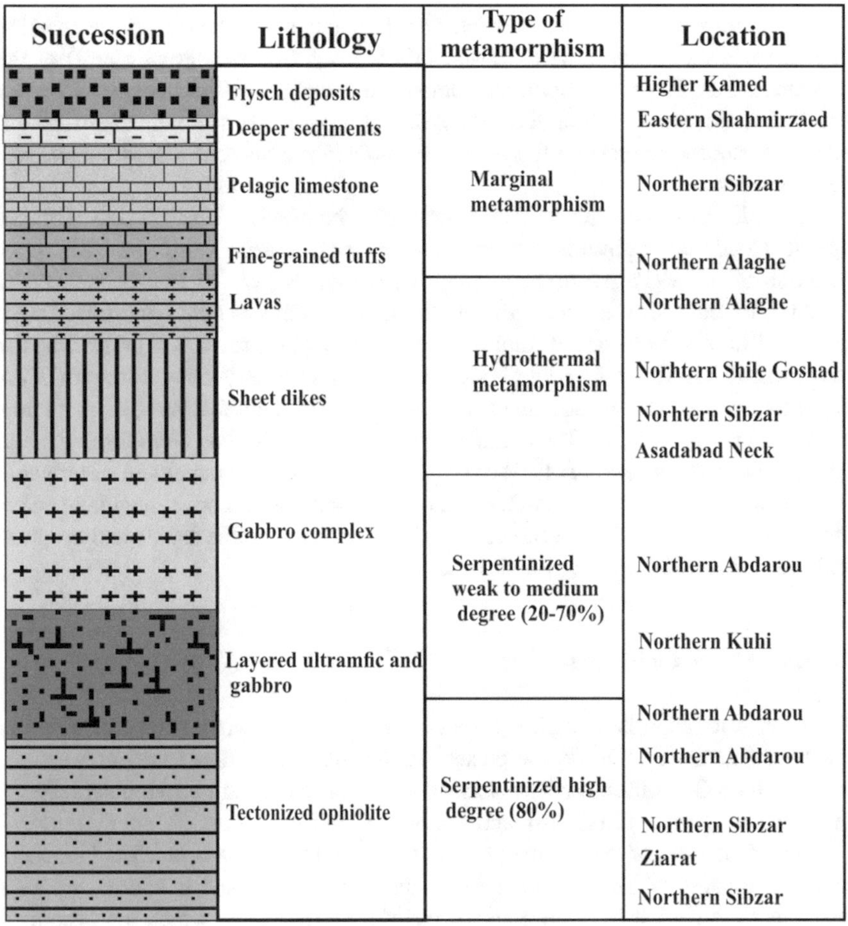

Succession	Lithology	Type of metamorphism	Location
	Flysch deposits		Higher Kamed
	Deeper sediments		Eastern Shahmirzaed
	Pelagic limestone	Marginal metamorphism	Northern Sibzar
	Fine-grained tuffs		Northern Alaghe
	Lavas		Northern Alaghe
	Sheet dikes	Hydrothermal metamorphism	Norhtern Shile Goshad
			Norhtern Sibzar
			Asadabad Neck
	Gabbro complex	Serpentinized weak to medium degree (20-70%)	Northern Abdarou
	Layered ultramfic and gabbro		Northern Kuhi
			Northern Abdarou
	Tectonized ophiolite	Serpentinized high degree (80%)	Northern Abdarou
			Northern Sibzar
			Ziarat
			Northern Sibzar

Fig. 7.17 Main rock units of the northern Torbat Heydarieh ophiolite (Razmara 1990)

Sedimentary rocks: The sedimentary rocks mostly contain pelagic limestone with radiolarite and *Globotruncana* foraminifer. Also, there is a very thick flysch sequences are located on the ophiolite complex.

Metamorphic rocks: The orogeny and tectonically movements caused to develop amphibolites, rodingite and crystalline limestone (marble) within the ophiolites series.

Geochemistry: Based on the rock's composition diagram (Razmara 1990), the cumulate series are separated from metamorphosed peridotites and the original magma of cumulate series is calc-alkaline. Also the basalts are tholeiitic type of Oceanic Islands Arc. The amount of trace elements in basalts indicates the basalts originated directly from partial melting of the upper mantle (Razmara 1990).

Age determination: Darvishzadeh (1991) believes that based on the location of radiolarites and limestone on the of pillow lavas (top stratigraphy unit) of the northern Torbat Heydarieh ophiolite complex and the similarities between Sabzevar and Torbat Heydarieh ophiolite complexes, their age are Campanian-Maastrichtian. Also, the Eocene-Oligocene flysch in the southern region confirms the above age determination.

Tectonic: Based on the structural relation which exists between the northern Torbat Heydarieh ophiolites, their tectonic environment could be rift zones (Darvishzadeh 1991). At the beginning, the rift zones were as transformed faults around the Central Lut microcontinent. Then, the Gondwana continent moved right laterally towards the Eurasian continent. Consequently, the result was converting the transformed faults to a wide, deep and expanding rift zone. Probably, some part of the upper mantle has risen up as a diaper and caused to form ultrabasic rocks, metamorphism and partial melting. The cumulate sets and basaltic pillow lavas formed in the oceanic environment and finally the rift gradually closed by convergent of orogenic movements in Late Cretaceous-Paleocene. The result of the above magmatic activities is the metamorphism at the green schist and amphibolite facies and erupting andesitic tuffs which have covered the area.

7.3.5.2 Virani Ophiolites

The Paleozoic ophiolite complex has a whole structural sequence in comparison to the other Paleozoic ophiolite complexes. So, it is described more in detail.

The Virani Ophiolite complex is located in the northern Binaloud Mountain, 20 km far from southwestern Mashhad, and southern Virani village.

The oldest rocks of the Virani ophiolite are related to Precambrian. The Paleozoic rock's characters of Virani are similar to the Central Iran, and the Jurassic rock's characters are similar to Alborz zones. Beside the rock units which are described above, there is a series of ultramafic rocks with cherty bands, quartzite, and slate that are discussed as follows.

Those Iranian geologists who have studied the magmatism of Iran believe that the above ophiolite sequence is the remnants of an ocean which exists between the Gondwana and the Eurasia continents, Paleo-Tethys Ocean. After the collision of the continents (plate to plate) the remnant trace of the oceanic floor remains as ophiolite suture zone.

There is several elongated and banded layers northwest–southeast direction which looks like the sedimentary sequences. However, these layers are not stratigraphic deposits, they are the fold axis of different eroded folding deposits of various original sources. The eroded folds are tight and elongated forms. So, because of the apparent topography of the folds to survey the geo-traverse perpendicular to the fold axis is a so confused task.

General petrology: The ophiolites complexes are two separate units in south and southwestern Mashhad and both of them are surrounded by a series of metamorphosed sedimentary rocks which is parallel to the general trend of banded layers. The

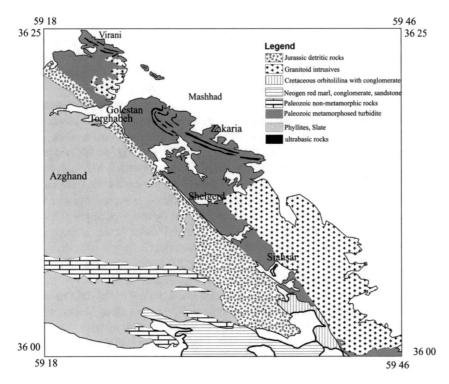

Fig. 7.18 Simplified geological map of southern Mashhad. The zones are separated by the Sangbast-Shandiz fault zone (modified after Majidi 1978)

general trends of the units are northwest–southeast direction, parallel to the Alborz zone (Fig. 7.18).

The Virani ophiolite complexes are mostly composed of mafic and ultramafic sequences as an ideal ophiolite sequences. But, the Mashhad ophiolite complexes (in Ab and Bargh area) contain relatively more volcanic and sedimentary rocks. The ophiolite complex consists of peridotite, pyroxenite, isotropic gabbro, and basalts. The meta cherts interlay with thin bedding marbles together with meta-basalts sequences. The layered gabbro and sheeted dikes are not found in the region.

1. The peridotite rocks mainly include:
2. Donites have been wholly serpentinitezed,
3. Wehrlite of which CPX has been replaced by amphibole,
4. Lherzolite with cumulate texture and Mg rich olivine form as rounded shape minerals,
5. Harzburgite with a lots of OPX,
6. Course-grained gabbro which includes OPX, CPX, Plag and hornblende. The primary minerals have been replaced by actinolite and chlorite minerals and
7. The basalts have relatively erosion and exposed as pillow lava flow.

The cherts and slates are exposed alternately. The metamorphosed pelagic lime-stone (marble) is originated at deep-sea basin. The marble are alternated layers with the metacherts. The metamorphic temperature is 300 °C. So, it is low to intermediate facies. The rock's metamorphic minerals approve the type of metamorphic facies.

Metasedimentary Rocks: The Mashhad and Virani ophiolites both are surrounded by meta-sedimentary rocks. The main meta-sedimentary rocks are; slates, phyllites, schists, marbles, calcareous conglomerates limestone and olistostrome. Alavi-Tehrani (1977) believes that all of these sediments are considered as a deep-water flysch type. The slate and phyllite in many areas are well stratified and particu-larly the slate layers demonstrate the stratification. Also, graded- and cross-bedding forms indicate a turbiditic environment for sandstone and mudstone formations. In other words, the turbidities facies are the sign of a turbidity complex, associated with the metasedimentary rocks in the area. It is reported by many geologists that at the present time, there are several tectonically active turbidities' environments similar to the above turbidity set in different parts of the world.

Several exotic fragments are among the turbiditic complex basins that have not been transferred from the outside, but they displaced from the different parts within the basin. The transferred fragments have the same composition as the surrounding sediments or probably of different compositions. The slate unit of Virani turbiditic deposits content pieces of marble, which exceed of 20 m thickness. The marble particles contain crinoid in thin section. So, what does the sediment variety means in the sedimentary rocks of the ophiolite complex?

It could be concluded that the crinoidal limestone is deposited in a shallow-water continental shelf. On the other hand, the contact between the limestone fragments and the other rocks are gradual. Therefore, we can suggest that the sedimentary deposits are olistostrome.

There are several carbonate layers among the meta-sedimentary sequences which are not similar to the regular stratums and seem to be breccia limestone particles. So, it is better to call them metamorphosed limestone conglomerate with occasionally well-bedded strata in some area. These layers have a thickness of 5–10 m and sometimes detectable up to 200–300 m, but are frequently hidden beneath the sediments. The limestone does not have the characteristic of the deep environments such as trench, and likely to be formed in a continental shelf basin, in which the turbidite deposits are formed.

Clastic rocks: The clastic rocks actually are quite volcanic complex contain series of tuffs and lapilli-tuff that formed alternately (Mashhad ophiolites formed in the axial zone of a synform anticline). The tuffs have two structural characteristics. First they are formed at inner zone of a closed synform anticline and second they are of thrust-faulted contact with the other sediments. So the tuffs should be deposited in another area by air fall mechanism and transferred latter to the basin, to thrust over the metasedimentary deposits. Regarding to the rate of lapilli-tuff deformation, the stress and strain index could be identified.

Age of Virani Complex (Paleo-Tethys remnants age): As it is well known, the remnants parts of Paleo-Tethys are intruded by granite and granitoid (Alavi 1996). Recently, the age of these rocks is known to 120–146 Ma. by Potassium-Argon method. Majidi (1978) classified the Mashhad Granite in three groups (see Chap. 1) and identify their age 211–246 Ma. So, the Early Cretaceous age (120–146 million years) for the granite cannot be acceptable because:

1. The metamorphosed ophiolites' rocks are covered uncomfortably by a series of sedimentary rocks with a basal conglomerate. Some residue plants of Late Jurassic to Early Triassic are found in sedimentary series which consists mainly of shale and sandstone. It is noted that the above sediments are not metamorphosed. In the conglomerate some fragments of ophiolites, ultramafic, and particles of intruded granites in the ophiolite complex are found. So obviously, the early Cretaceous age for the granite is doubted and should be followed for some other interpretation.

2. Intruding the granite into the metamorphic complex has caused to form contact metamorphism and porphyrobasts of garnet, sillimanite, andalusite in schists. Based on microscopic thin sections study, the andalusite crystals are cut by crenulation cleavage during the folding. Indeed, the sedimentary sequences with basal conglomerate equivalent to the Shemshak Formation have not been deformed. As a result, the intrusion of granite should take place earlier or same time as the Early Triassic deformation. This fact is applicable at least for some part of the granite. Therefore, all the Mashhad granites are not the same origin, and are not intruded simultaneously. The Potassium-Argon age determination identifies that some of these granites intruded before the end of Triassic and some others in Cretaceous. So, certainly the first group intrudes into the second one, and cut them. Except crinoid fossil that have been found in Mashhad metamorphic rocks, there is no other fossils. Nevertheless, Ruttner (1984) reported a crinoid fossil with a Permian age in northern Fariman rocks which are the same age as Mashhad metamorphic rocks. Also, Eftekharnezhad and Behroozi (1991) identified some fossils of Sakmarian age in the northeastern of Fariman, so the age of the complex could be generally Permo-Triassic.

The relation between Mashhad and Fariman rock formations is not well known, and there is no certain evidence to approve the same age of both Mashhad and Fariman rocks formations, and also no precise geological map for the area. Thus, the age of the above fossils is not sufficient to identify the Mashhad complex's age.

Replacement of the Virani complex: As mentioned previously, the Virani ophiolitic complex is the remnants of Paleo-Tethys Ocean which basically consists of a set of residues ophiolites, metasedimentary, volcanic and clastic rocks. The composition of ophiolite quite is the same as oceanic crust. Majidi (1978) believed that the Virani ophiolites have tholeiitic composition similar to the Mid-Oceanic Ridges.

The depositional mechanism of meta-sedimentary rocks of Virani, as a turbidity complex, is similar to the present active turbidity's environment. There are several olistolite rock fragments which form olistostrome, in fact we deal with a deep flysch type containing the allochthonous boulders. Based on the following reasons the Paleo-Tethys oceanic crust tectonically moves towards the north under the continental crust of Turan plate, consequently, the turbidity sediments gradually are deposited along the Turan subduction basin.

1. The thrust mode is dominated in ophiolite contact, with meta-sedimentary rocks, everywhere.
2. The direction of the thrust towards the edge of the Iran continent, and the stress direction are northeast–southwest.
3. The metamorphisms are usually low metamorphic facies indicating temperature about 300 °C.
4. The volcanic and pyroclastic rocks are essentially composed of tuff and lapilli tuff which situated at the core axis of a fold.
5. The Paleo-Tethys oceanic crust was located between two continents of Iran and Turan.

During the subduction, some parts and segments of the oceanic crust pull apart, and locate in the subduction basin, so the results are fault contacts with flysch or turbidites. The separated segments eventually thrust on the flysch and turbidities. The thrust segments have moved slowly towards the edge of the Turan plate. In addition, the continental shelf carbonates have been pulled away possibly by seismotectonic activities through the submarine valleys and located in the turbidity area. The oceanic crust segments thrust on the trench's sediments, and fold (which may be done with metamorphism), then joint to the continental plate that is getting close to the trench zone. Eventually the jointed complex, thrust on the Iran plate, and tolerates the tectonically deformation. Obviously, all the deformation is done within the continent crust at lower temperatures when the rocks' temperature was lower than before. As it is known, the result of the subduction phenomena forms the side edge volcanic arcs. The arc volcano's ash can be deposited in the basin, and then be driven tectonically on the previous sedimentary series. The entire volcano-sedimentary complex with a contact fault is traceable at the core axis of a syncline fold. At the beginning of the collision of the continents (Iran and Turan plates), the ophiolite complexes connect to the edge of the Iran continent. In fact, still the collision continues and the evidence of the collision are proved as thrust zone that is remained in the Alborz Mountains with sloping to the north and the moving direction towards the south.

According to the paleo-geographical evidence of the general tectonic activities in the region, the arcs' pyroclastic deposits and the turbiditic deposits prove an oceanic trench environment. Also, their geochemical compositions identify the tholeiitic origin for the rocks, and ocean origins for the ultramafic complexes (Majidi 1978).

Identifying the stress factors associated to the fold's axis, is the best theory and mechanism to explain the obduction phenomenon.

The arcs' activities are studied and confirmed mostly by Kazmin and Kulakov (1969) and other geologists in Caucasus, east of Caspian Sea, Aghdarband, and Afghanistan, while there is no report to prove the volcanic activity features related to the volcanic arc activities in Alborz region.

Emplacement of the Virani complex starts at the beginning of the collision and continues for a long period of time. It takes a long time, because the closure of the basin is not an immediate event and for closuring the basin that needs to be continued for a long time to be completed. Turbidite and flysch facies require an oceanic trench to form as accretionary complex at the growing edge of Turan Plate, then after reaching Iran Plate to Turan Plate the turbidities and flysch sediments situate at the margin of the Iran continent.

7.3.5.3   Asalem-Shanderman Ophiolites

The base rocks of the Asalem-Shanderman ophiolite complex are metamorphosed with long strips of serpentinite. Although, the contact of the complex with nearby rocks are tectonite. But may be developed simultaneously with the formation of Paleo-Tethys ocean floor (Stampfli 1978).

The Asalem-Shanderman ophiolites is located in the Guilan province, southwestern Caspian Sea. Most geologists believe that these ophiolites complex and its related metamorphic rocks are related to the Mashhad ophiolite that occurred during Hercynian Orogeny (late Carboniferous).

Asalem metamorphic complex: The Asalem which ophiolite is located just west the Syahbyl village, Asalem City, consists of green schist, gneiss, amphibolite and serpentinite and also is covered by the Jurassic sediments. The Jurassic sedimentary rocks consist of deformed detritic sedimentary rocks and tuffaceous sediments with basic lava layers.

Shanderman metamorphic complex: Shanderman village is located in 40 km south of Asalem and Shanderman Metamorphic Complex in the west of the village. The Shanderman complex's relationship with the Jurassic and Paleozoic rocks are faulted (Haghnazar 2009). Its petrological characteristics are more or less same as the Asalem Complex including actinolite, garnet, muscovite Zoisite or fine grain gneiss, and amphibolite. Moreover, there are highly crushed lenses of serpentinite ophiolite within the complex. The length of the lenses in exceeds 20 km, and width of 1.5 km or more in south of the Shalrah village.

The metamorphic intensity of complex varies green schist to amphibolite facies (Clark et al. 1975). Also retrogressive metamorphism has occurred and the garnet and actinolite minerals were metamorphosed into the chlorite schist rocks.

The ultramafic rocks usually have metamorphosed to antigorite and serpentinite. The light color layers within the serpentinite rocks are related to the formation of rodingite (the rodingite lenses or discontinuous layers is formed by advanced alteration of gabbroic rocks). According to its mineral composition (hydrogrossular, prehnite, titanite, and apatite), it is inconceivable that this rock be formed simultaneously with the serpentinization and the calcic metasomatism in crushed zones. Amongst the Triassic conglomerate near Gorgan river, and Agholar Village, several rounded rock fragments of serpentinite have been found that approve the metamorphic phenomena (Clark et al. 1975).

The Asalm-Shanderman ophiolite complex and its associated serpentines' age is assumed to be Paleozoic. Their current position of the complex were attributed to the thrusting and shearing solid stage. The contact of Asalem-Shanderman ophiolite complex with Shemshak deposits and Cretaceous sediments is thrust fault, in which partial of Shemshak Group thrusted on the Asalm-Shanderman complex.

However, the age of these two metamorphic and ultramafic complexes are not certainly known. Nevertheless, the ophiolite segments of the complexes probably are the remnant of oceanic crust of the Hercynian or Early Cimmerian (the compression stage of Middle Triassic) when the Paleo-Tethys oceanic floor is closed.

The results of detailed micropaleontology studies (Eftekharnezhad 1991), indicate that the previous age interpretation are somewhat inconsistent with the new data. According to the recent age determination data, the ophiolite complexes are younger than Hercynian event. The new controversy interpretation causes to be recommended an alternative explanation based on associated with the mobilistic tectonics of the Alps and Himalaya geodynamic problems. Davies et al. (1972) believe that the age of the southwestern Caspian ophiolites are Precambrian in age. However, the recent years re-examination of the Precambrian age determination data of Asalem and Shanderman compleces and also finding the organic substances in the complexes' rocks, the attributed Precambrian age is questionable.

Systematic studies of the geological map of Khalkhal, conducted by Eftekharnezhad et al. (1990). Most of the geodynamic model which has been presented is based on the Davies interpretation. So, the recent important results about the age of Asalem-Shanderman metamorphic ophiolites complex (Eftekharnejad et al. 1989) suggest to take into consideration the revision of previous age interpretation of the complexes. Davies et al. (1972) in the report of Bandar-Anzali quadrangle (1:250,000), refer the age of the partial complexes to the Precambrian metamorphism, and the other parts to the non-metamorphic Jurassic formations. However, Eftekharnezhad et al. (1992) referred to Bozorgnia that according to the age determination of some microfossils in the ophiolite complex, seven fossils belong to Permian and one to Carboniferous uncertainly.

To attribute some part of metamorphic complex to Shemshak Formation by Davies et al. (1972) are not so irrational. Because, around the Zendaneh village in the Khalkhal highlands in metamorphic rocks, shunkinite shale and even in some metamorphic rocks the trace of plants are detectable. But the crystallized limestone contains the Permian fossils. In general, the Permian sediments are different from Permian epicontinental carbonate of Iran Paleozoic Platform sedimentary facies. These sediments are exposed nearby the western area.

The general characteristics of basic and highly serpentinite ultramafic mass of Asalem-Shanderman ophiolite have similar characters to the age and facies of Mashhad ophiolitic rocks. The new information about the similarity relationship of geodynamic phenomena between Southwestern Caspian Sea ophiolite, and the folding Alps-Himalaya Mountains, causes to think about the same origins and mechanism of their formation. So, proofing the new age for Asalem-Shanderman ophiolite and metamorphic rocks, the Hercynian event as a factor of deformations of southwestern Caspian Sea as well as in northeast Khorasan is rejected. The southeastern Caspian Sea magmatism is related to an oceanic convergent system in upper Paleozoic Era (Permian-Carboniferous). Therefore, it is not associated with the characters of Palo-Tethys Ocean (Early and Middle Paleozoic in age) that folded and cratonized in Hercynian event.

Recently, the contradiction phenomenon about the Paleo-Tethys relation to the Pamir-Hindo Kush, northern Tibet, and southeastern Asia have been discussed by many geologists, Stöcklin (1974), Belov et al. (1986), Şengör et al. (1988) and Şengör (1990).

Eftekharnezhad and Behroozi (1991) provide the new data about Mashhad and southeastern Mashhad ophiolites to improve the lack of information, about the relation of southern semi-oceanic basement with Paleo-Tethys genesis.

In recent offshore drillings, in the northern coastal region of Azerbaijan, at a depth of about 5000 m after the sediments of Mesozoic, a series of metamorphic rocks have been identified as absolute age 296–342 Ma. which is compatible with the Permo-Triassic time. Generally, it is comparable with southwestern Caspian Sea Permian ophiolites and metamorphic rocks. This theory supports the idea implied the existence of remnants of basement in the second Paleo-Tethys in the Caspian Sea. Thus, that may be part of the upper Jurassic volcanic rocks formed the basement of the South Caspian Sea. Also, the same characteristics ophiolite belt could be followed in the northern mountain chain of the southern Black Sea (Şengör et al. 1988) which fits perfectly with the remnants of Alpine Meso-ophiolites.

The Svan-Karabakh ophiolites of Caucasus suture zone with a general trend can be observed in the southern Aras and northern Meshkinshahr and probably continue to ophiolites of Asalem-Shanderman. Nevertheless, the new data, recognized the Asalem-Shanderman ophiolites with a different feature from Swan-Gharabakh ophiolites.

According to the geological and geophysical studies conducted in Republic of Azerbaijan, it is thought that the Vndam highland, Bagher, Talesh areas and some part of south Caspian Basement rocks has been lifted up within two faults. The depth of the elevated segment is about 7000–8000 m.

7.3.6   Ophiolites and Ultramafic Rocks of Late Precambrian to Early Cambrian

As already mentioned, because of the metamorphism or erosion of the late Precambrian to early Cambrian rocks often do not have any significant exposures in Iran. Although, these rocks show the oceanic crustal rocks, but often do not reveal all the characteristics of an ophiolite sequence, so calling them as old ophiolites is not accurate. These rocks are widely exposed in the Anarak and Takab area (it was described in the Takab and Anarak magmatism chapter) related to continental rift zone to form the oceanic crust in some areas, such as ophiolite zone of Takab, Anarak, Jandaq, Bayazeh and Posht-e badam. The Naein ophiolite has been described in latter chapter.

7.4 An Overview on the Petrography and Petrogenesis of Ophiolites in Iran

The results of field studies, mineralogy, lithology and geochemistry of ophiolites have resulted to revise the ophiolite sequences of Iran as a polygenetic origin type as follows:

7.4.1 Main Ultramafic and Mafic Body with Metamorphic Texture Consists of Three Main Parts

1. Alternation of dunite and harzburgite and lesser chromite as the base of the mass body.
2. Alternation of wehrlite, vebstritet-lherzolite, clinopyroxenite and olivine, webstrite and less chromite in the middle section of the sequence, which in many texts is introduced as transitional zone,
3. The order of the gabbro part is metagabbro, feldespatic peridotites, troctolite, formed at the lower part; anorthite and norite in the middle part; and ferrogabbros, loco-gabbro and ferro diorite at the upper part.

There is no detailed information available on the formation of ultrabasic rocks, but most likely, they are the oldest unit of the ophiolite complexes. The emplacement of these different masses in the ophiolite series is certainly related to tectonic forces to separation the fragments of oceanic upper mantle.

In some areas such as Esfandagheh (Sabzehei 1974), it is believed that such huge masses of ultrabasic is the segregation of a basic magma and they all have endured different deformation stages to form layering, as specified in the series of Bushveld and Stillwater areas. But, Alavi-Tehrani (1977) mentioned that consideration of generalized concepts for all ultrabasic rocks of the Iranian ophiolite complexes in Iran can be questionable.

7.4.2 Basic Mass Body

The basic mass body consists of isotropic gabbro, pillow lava and sheeted dike which generally indicates the activities of basaltic magma. Timing of the magmatic activities and also the field observations have signified that the feeder channels of magma cut the old mafic and ultramafic rocks, and in many cases the hydrothermal factors are affected on the host rocks (previous rocks). The field evidence indicates that intruded basaltic magmas (the secondary gabbro) in the upper part of the previous mafic and ultramafic rocks (primary gabbro) crystallized to form an isotropic gabbro.

Consequently, the root of diabase dikes' protolith (the secondary gabbro) disappear within the isotropic gabbro mass.

These relations show that the intruded magmas on the basis of similarity of density between gabbros and basaltic magmas were crystallized as isotropic gabbro masses, or sheeted dikes and pillow lavas at the upper part isotropic gabbro. These intruded magmas lesser have cooled within the dunite, harzburgite and transitional zone.

The geochemical composition of Sabzevar ophiolites indicates a tholeiitic magma contributed to the formation of cumulate gabbro series (Alavi-Tehrani 1977). The tholeiitic gabbro magma is followed in some parts of Iran like Esfandagheh (Sabzehei 1974). Nevertheless, there is not enough evidence to generalize the above phenomenon in the other parts of the ophiolites in Iran.

Due to the tectonic and crushed nature of the ophiolites in most areas of Iran (except for the south Jazmurian), the geological structures and age relation between the above mentioned complex and the other main components of ophiolitic complexes is unknown. For their generation to the ultrabasic masses, different ideas presented. Sabzehei (1974) mentioned the extensive sheet gabbro in Esfandagheh is the final result of differentiation of a basic magma which the ultrabasic series are produced at the first stage of activity. Therefore, he believes that the basic and ultrabasic rocks are generated from a single magma.

The interpretation of the gabbro and peridotites which contain plagioclase in most ophiolites areas, particularly in the Eastern Mediterranean ophiolites could be related to the phenomenon of partial melting of the upper part of mantle and oceanic crust. The result of this phenomenon can produce a basic magma that through of differentiation will formed feldspathic peridotites and sheet gabbro in the upper part of the mass. Then, in the upper part of the mass, the acidic rocks such as tonalite and plagiogranite are segregated from the upper part of the basic magma (Eftekharnezhad et al. 1990; Coleman 1971). In general, the frequency and wide distribution of basic and ultrabasic sheet rocks in ophiolite complexes is less than basic and ultrabasic rocks with tectonically texture. The geochemical studies of acidic rocks such as plagiogranite and tonalite associated with ophiolite complex, for example in Sabzevar, and Jazmurian, indicate that these rocks could be resulted from the last crystallization process of a basic magma (Alavi-Tehrani 1977).

Another interpretation regardless to the above explanation for the formation of the layered rocks and perhaps in a simpler manner they could be formed directly by the injection of younger tectonically masses. The technical mass is also extracted from a previous basic magma.

7.4.3 Diabase Dikes and Microgabbroic Rocks

The diabase and microgabbroic rocks are the abounded layered rocks associated with ophiolite complex in the most ophiolite regions in Iran. The sheeted structures are more abundant and extended. Nevertheless, sometimes it is necessary to follow the tectonic or alteration phenomena by detailed field study and control to find out the chilled margins and sheeted structures in diabase zone as well.

The main outcrops of diabase and micro-gabbroic rocks are Sabzevar, Nehbandan, Birjand, Torbat-e Heydarieh and Jazmurian (southern Kahnouj and Fanuj) regions. The contiguous connection of sheeted diabases with mostly ophiolite pillow lavas particularly in terms of similar specification of petrographic and geochemical nature is clearly identified.

These rocks in some areas where are not deformed tectonically, cross cut the gabbro, and feldspathic cumulate peridotites series (southern Kahnouj and regions). Thus, their contact with the ophiolitic rocks are specified. The relative age of their formation is Late Cretaceous to Paleocene. The evidence of metamorphism in these rocks can be followed in some so area such as Khoy, Torbat-e Heydarieh, Sabzevar, and southern Jazmurian. The metamorphic temperature is varying from mild to green-schist facies and occasionally is more common to amphibolite facies.

Arvin (1982) refers the lack of sheet dikes in Neyriz to a factor that distinct these ophiolites from the other parts of the world ophiolites series such as Oman ophiolites.

Despite the clear situation and the exact contact relationship of the sheeted diabase with the main parts of ophiolite complex. Nevertheless, the position of lenticulars micro-gabbro dikes that are abundant in ultramafic masses is uncertain. The uncertain factors are the lack of age relation and the rock's magma composition. We must also determine whether the sheeted diabase beside the ophiolite complexes are related to ophiolite complex and an oceanic crust. As some believe it develop in a Mid-Ocean Ridge and the diabase originate in an oceanic floor spreading or according to other authors these diabases are formed in an Island Arc system (Miyashiro 1973).

Detailed chemical analysis of these rocks shows a tholeiitic magma origin; although, in some cases because of the rock alteration, tend towards the calc-alkaline composition (Alavi-Tehrani 1977).

7.4.4 Volcano-Sedimentary and Associated Sedimentary Rocks

The volcanic rocks which consist of different types of lavas and tuffs associated with sedimentary rocks are other major constituents of the most ophiolite complexes in Iran. The name of the complex is volcano-sedimentary ophiolite complexes. Because of the volcanic lavas and tuffs are adjacent with a variety of pelagic sedimentary rocks widely. The sedimentary rocks are mostly micritic and pelagic limestone and radiolarite which consist of a significant percentage of the rocks in some areas. Clastic and pyroclastic rocks scatter and occasionally without certain contact over-lain the complex. The volcanic rocks as pillow lava are the most common features and massive structures also can be observed within the volcanic rocks. Despite the metamorphism, the petrological composition of ophiolites is varied from loco to spilitic basalts. The main characteristic of the above mentioned rocks in the ophiolite complex is the degree of regional metamorphism affecting the rocks. The facies of metamorphism is varied from epithermal to mesothermal phases (green schist to prehenite-pumpellyite facies) and sometimes dramatic changes occur in mineralogy, texture and structure above the volcanic series. Hyaloclastite and ophiolitic tuffs including vitric, crystalline tuff and andesite are abundant volcano-sedimentary rocks in the ophiolite complexes.

The presence of intercalation of pink pelagic limestones and foraminifer-bearing sediments laying within the layers of lavas and ophiolite tuffs define the age of the volcano-sedimentary formations to the Late Cretaceous (Maastrichtian) in the most ophiolite complexes in Iran.

Geochemical characteristics of volcanic and diabase units of the ophiolite complexes in different region of Iran generally show a calc-alkaline nature to generate the diabase, and volcanic rocks in the ophiolite complexes.

7.4.5   Acidic and Intrusive Rocks

The intrusive masses that in most cases cut off the basic and ultrabasic rocks have been formed during the Late Cretaceous. The plagiogranite are the major intrusive and granular rocks of the ophiolite complexes. Despite their less development, they are significant in terms of abundance and origin. They tend to form so small b or in some cases as dikes and thin veins (maximum up to half a meter thickness) which are associated with the diabase rocks and spilite lavas. The above mentioned rocks of the ophiolite complexes are extended in the Sabzevar, Torbat Heydarieh and Jazmurian regions, and their genetic relationships are also known. Also, the acid intrusive rocks and other intrusive rocks such as scatter and vine forms of albitophyre (albitite) and ktatophyre. However, all of them are in relation to the last phase of the crystallization and differentiation of a mafic magma, which at the first step peridotites contain plagioclase, gabbro and diabase and finally developed the acidic rocks are developed.

Some authors believe that the ultramafic and ophiolite gabbro consists of two types, tectonites (including dunite, harzburgite and chromitite) and cumulate (including wehrlite, lherzolite, webstrite layered gabbros). These two parts are petrologically detachable by a petrological connection named petrological Moho. Based on this idea, the tectonites belong to mantle, and cumulate to basic lavas and dikes to the crust.

Sabzehei (1995) argued that this assumption about the ophiolites of Iran is incorrect and all the ophiolites of mafic and ultramafic bodies with metamorphic texture are generated together.

Based on the studies of Sabzehei (1995) on the Neyriz, Esfandagheh, Kahnouj, Fenouj and Sabzevar ophiolites, with no deformed sequences of mafic and ultramafic rocks, are characterized as follows:

1. The different segments of the mafic and ultramafic bodies have an original connection and there is no major tectonic connection. It should be noted that the subsequent tectonic activities occasionally distorted and disconnected the primary sequences. Nevertheless, the field study of tectonic slices and their relationship with respect to the marker beds such as layers which the clinopyroxene and plagioclase minerals form for the first time, it has been possible to reconstruct the primary row and

2. It has been proved that the plastic deformation phenomena are not only occurred in harzburgite dunite, but also up to the whole sequence of gabbro. Certainly, it can be said the whole complex is essentially a cumulate unique which the metamorphism and deformation cause to change their igneous texture into the metamorphic texture. Thayer (1960) believe that the layered igneous texture (in stable zones) is comparable to the metamorphic texture similar to compare the metamorphic and sedimentary rocks.

The thickness of the different parts of the sequence is not well known but it could be stated that:

1. All constitutes of the ultramafic sequence is much more than the mafic sequence in many ophiolites zones of Iran.
2. The thickness of the upper sequence of gabbro rocks such as, loco-gabroo, ferodiorite, quartz diorite and granophyre are insignificant.
3. The amount of serpentinite ultramafic rocks such as dunite and harzburgite is more than the other components in color mélanges. Thus, based on the all above evidence the primary ophiolites' magma must be an ultramafic magma.

There is certain evidence that shows the differentiation flow of Komatiite magma have caused to create a sequence of ultramafic rocks like harzburgite, dinaite, wehrlite, gabbro and occasionally granophyre in the Paleozoic aulacogen of Iran. Collecting and processing all the analytical data by Minpet Software program indicates the following facts:

1. A series of komatiite peridotites to basalts and eventually theoleites rich in Mg and slightly calc-alkaline rocks have been differentiated from the primary komatiitic magma.
2. Based on the Irvine and Baragar (Irvine and Baragar 1971) diagram, the general trend of total FeO increases in tholeiitic rocks. Sabzehei (1995) by processing the data of 484 samples of mafic and ultramaphic rocks with metamorphic texture and applying the Irvine and Baragar diagrams shows that the fractionation trends in the complex are very similar to the trends derived from komatiite data processing aulacogen of Paleozoic in Iran. It seems that the magma which creates the ultramafic and mafic complex are likely to be generated from the same magma of komatiitic ultramafic and mafic lavas of Iran Paleozoic aulacogen. Based on the filed relation which shows the komatiitic metamorphic rocks of the lower Paleozoic and upper Precambrian mafic and ultramafic rocks are directly located on the mafic and ultramafic with metamorphic textures of the main ophiolite complex. So, the original magma of the main ophiolite complex is the

same magma which creates komatiitic mafic and ultramafic rocks, and it named ophiolitic magma (Bruun 1962).

The author has visited the most features of the mafic and ultramafic complexes of Iran that have been explained in this book. However, the Sabzehei's idea about the komatiitic magma may not be accurate about the Zagros and the ophiolite arc zone of Central Iran.

7.5 Geodynamic Model

Relating to the formation and emplacement of the Iran ophiolite complexes, most researchers believe that they are actually the fossilized remnants of the primary lithosphere of oceanic crust, which are formed in the extensional oceanic (rift) zones. After development and moving towards the subduction zone, instead of depositing at the deep mantle, they obducted along the continental margins, such as back island arc basins. Hassanipak and Ghazi (2000), Stöcklin (1968) and other researchers classify the most of Iran ophiolite complexes into two groups based on their ages.

1. The ophiolites related to the Paleo-Tethys oceanic rift (Paleozoic),
2. The ophiolites related to the Neo Paleo-Tethys oceanic rift (Mesozoic).

Nevertheless, Sabzehei (1995) provided a new perspective about the different ages of the ophiolites of Iran. He believed that different parts of the complexes form at different stages and not necessarily to be the same age and forming simultaneously. Based on the all above discussions, the main conclusion is as follow.

The multi-generation of the Iranian ophiolites is the result of several eruptions of an individual magma. The composition of ophiolitic magma is ultramafic and its geochemical characteristics is able to generate all ophiolites sequences from ultramaphic (komatiiticperidotite) to melagabbro (komatiiticbasalt) according to Yensin Diagram (Jensen 1976). Eventually, the tholeitie rich in Mg (norite and gabbro norite), tholeitie rich in Fe (Ferro gabbro) and the acidic rocks are developed through differential.

On the other hand, the study of the Paleozoic aulacogen of Iran shows that their arrangement within the ophiolite zones should be proposed in a rift environment, then the diapers are formed by the activities and partial melting of the mantle (>70%). The formation of ophiolite magma, differentiation magma flow and replication of these phenomena occur in different geologic periods. In this model, the central part of the aulacogen changes completely into an oceanic crust, by effecting of the influx of ophiolite magma, consequently, its sialic parts are destroyed. By repeating the influx of magma in the central part of the rift or aulacogen, have caused to change the composition of this part mainly to ultramafic. The ophiolites tectonic emplacement and the formation of color mélange are interpreted by inversion of Paleozoic Iran's aulacogens. According to the proposed model and the characteristics of the following geological tectonics, a simpler interpretation of ophiolites could be expressed as follows:

1. It is not necessary to create a complicated and large oceanic pattern to explain the formation of Iran's ophiolites, then, for destructing the created oceanic floor to reinvent a subduction zone, which there is not strong evidence to prove that.
2. The issue of justification multi-generational Iranian ophiolites is more accurate than the other ideas particularly when the evidence proves that primary magma of basalts and pillow lavas is most likely the same as the ultramafic magma.
3. The ultramafic and mafic complexes are generated from a single ultramafic magma, and then metamorphosed. Thus, it is not needed to refer some part of magma to the mantle and the other part to the crus. But, the entire magma is the ophiolite type. The deformations and metamorphism of the ophiolite magmas occur during the inversion and closing the aulacogen zones.
4. The multi-generational ophiolites represent the portion of the aulacogens of the old continental crust which entirely is dead, disappeared and changed to a new oceanic floor. Consequently, the mafic and ultramafic lavas are mostly placed at the marginal zones of aulacogens as interlayers of turbidities. Just fewer deposits mainly include pelagic chert and pelagic limestone occurs at the new ocean's floor.
5. The serpentinization of the ultramafics at the central part of aulacogens increases the uprising ability of diapirs. During the inversion of these ultramafics, the aulacogens as diapirs are injected into the all ophiolite sequences and have formed the colored mélange complex.
6. For explaining how the ophiolitic magma erupted into the oceanic floor basin, it can be cited that only some part of ophiolitic magma erupts into the basin to form the basaltic pillow lavas and the other portion of the ultramafic magma for any reason—which is derived from the magmatic segregation, crystallized within the old crust. In this case the composition of the ultramafic intrusions such wehrlite diapirs within some ophiolite complexes such as Oman ophiolite (Nicolas 1989; Nicolas et al. 1990) is not unusual, and probably related to the ophiolitic magma, that basic lavas and sheeted dikes are generated (Sabzehei 1995).

Sabzehei (1974) stated that the layered gabbro ophiolite of the Kahnouj, Esfandagheh, Neyriz, Kermanshah and Saqqez ophiolite zones have the following specifications (Amani 2000):

1. The rhythmic and cryptic layering similar to the stable region of continental gabbro,
2. Having repeated folds,
3. Lack of microscopic foliation, because of the simultaneously functions of the convection currents of the asthenosphere below the magma chamber, and the crystallization of magma that cause to deform the minerals.
4. They have a significant metamorphic foliation, due to the lack of lineated of deformation of minerals, and simultaneous crystallization and deformation phenomena in a semi-solid state (Nicolas 1989; Boudier et al. 1989). In fact, the deformation phenomena affected to the magma, due to the final stages of crystallization plenty of plagioclase, pyroxene and olivine as a thin layer, surrounded

the basaltic magma. All above researchers believe that the small amount of basaltic magma attracts and releases all the stresses resulting from deformation and have caused to protect the internal basaltic crystals against the deformation phenomena.

The lithological order of ophiolite complexes from the old to new periods are as follows:

1. Dunite, lherzolite, harzburgite with signified layered structure of magmatic and metamorphic foliation.
2. Alternation of pyroxenite (clinopyroxenite, webstrite), dunite and wehrlite at the end of the sequence, mela-gabbro and trictolite form alternatively with pyroxenite.

Therefore, the complex of dunite, harzburgite, lherzolite, pyroxenite, wehrlite, mela-gabbro and troctolite normally converted to layered gabbros which their original composition is neurite. According to Sabzehei (1974, 1995) the complex are interdependent of petrological, mineralogical and magmatic evolution and originated from a spinel-plagioclase lherzolite, spinel magma which have been derived from an ultramafic magma. There is so many evidence in the Sanandaj-Sirjan Zone that indicate the ultramafic magma erupted directly from a komatiitic or picritic composition in the lower Paleozoic. The lower Paleozoic ultramafic lavas in Sanandaj-Sirjan zone are the best evidence to prove this phenomenon.

The geological studies in the Sanandaj-Sirjan zone, Sabzevar, Hajiabad, Neyriz and Kermanshah, have shown that the mentioned ultramafic-gabbro complexes are located below the oldest metamorphic rocks and their ages have been approved by many Iranian geologists to the lower Paleozoic (Ordovician-Precambrian) and possibly Precambrian. The ultramafic and gabbroic complexes names as the lower complex.

The diabase dikes and plagiogranite cut and intrudes the lower complex in the Neyriz, Esfandagheh, Hajiabad, and Kahnouj ophiolite complexes.

The intruding of basaltic and granitic magmas mainly occurred in Cretaceous, particularly in the late Cretaceous, causing re-crystallization and metamorphism of the lower complex and removing the rock foliation, renewing the texture of the lower Complex. Sometimes, particularly in the gabbros, the texture changes into granoblastic to xenoblastic. The primary traces of rootless isoclinal and superposed folds, or disconnected folds occur in the outcrops of gabbro mass. However, due to the hydrothermal alteration the microscopic granoblastic texture occurs, and the minerals recrystallized fully by a factor of heating in a statistical situation. So, based on the above evidence the following facts are identified:

1. The ophiolites of Iran consist of Paleozoic and Mesozoic ophiolites complexes. The Mesozoic ophiolite composed of pillow lavas, sheeted dikes, plagiogranite and some specified gabbro, while the ultramafic-shetted gabbro complex belongs to lower Paleozoic or probably Precambrian.

2. These two ophiolite sets occur in certain geological zones of Iran, in which
    the most magmatic activities took place through the upper mantel in different
    periods.
3. The best justifying pattern of forming the ophiolite complexes is raising hot
    mantle diapirs repeatedly in different periods of forming the intercontinental
    rift type, similar to the Red Sea Rift (Sabzehei 1974).
4. Finally, according to Sabzehei (1974) applying the term, refers to low level
    gabbro to the layered gabbro of Jazmurian-Kahnouj and the other layered
    gabbros of Iran ophiolites as McCall and Kidd (1981) mentioned in the report
    of east Iran project, absolutely is not true.

7.6   Metamorphic Rocks of the Iranian Ophiolite Complexes

Three different metamorphic rock types could be distinguished in the Iran ophiolites:

1. A group of metamorphic rocks in Iran that is not related to the ophiolites and
    the oceanic crust, having the other original sources. They are mostly older than
    ophiolite and emplaced mostly tectonically. These group of metamorphic rocks
    are mica schist, and sometimes, marble, amphibolite and gneiss.
2. The metamorphic rocks that form after the formation of ophiolites. The meta-
    morphism is associated with metasomatic (alteration) without identified defor-
    mation, but the main factor for deformation is hydrothermal fluids that cause
    chemical deformations at a large scale. As a result, serpentinite, talc, rodengite
    (formed by the alteration of gabbro) and magnesite have formed. These alter-
    ations exacerbated during mélange and mixing that caused to segmented and
    disturbed of ophiolite complexes.
3. This metamorphism created after formation and solidification of ophiolites.
    That is ocean floor metamorphism type and almost equal to hydrothermal
    metamorphism that occurred as static.

The ultrabasic rocks endured serpentinization. The results are to provide lizardite,
and chrysotile. The basic rocks such as, gabbro, diabase along with the serpentinization endured rodingitization. Consequently, vesuvianite, grossolar and chlorite are
composed of the main mineral compositions of the rocks. On the other hand, there are
some ideas implying that the spilitization phenomenon is the results of the spiliteting
of lava that associated with this type of metamorphism. Some of amphibolite and
green schist without foliation can be associated with such metamorphism.

A group of dynamic metamorphic rocks are associated with ophiolites which are
related to the regional metamorphism. Possible evidence of geology, petrology and
geochemistry approve their dependency to the ophiolite complex. Some of these
dependent rocks are green schist, blue schist and occasionally amphibolite and
amphibolite-garnet. These rocks are the result of the metamorphism of dibasic rocks,
pillow lavas and tuffs of the ophiolites. Sometimes the carbonate materials and rocks,

and volcanic rocks of ophiolite complexes, respectively, have been metamorphosed into marble and amphibolite. This type of metamorphism should be occurred after the Cretaceous tectonic phases.

The above-mentioned metamorphism is related to the Alpine orogenic phases (post-Cretaceous) which particularly have severe dynamic effects in a wide regions. The results of this metamorphic phase are green to blue schist facies and occasionally up to amphibolite facies in a series of pyroclastic sediments, in south of Birjand, Sabzevar and western Iran (Alavi-Tehrani 1979). The effects of this type of metamorphism (oceanic floor metamorphism) impacts mostly on the hydrothermal metamorphic rocks. The results of the metamorphism in the ultrabasic rocks are abundant veins of antigorite in the high pressure and low temperature conditions. These veins cross cut the most chrysotile and lizardite minerals which formed during the previous metamorphism. The stepwise metamorphic of the alpine in basic rocks, especially in diabase and pillow lavas can be followed dramatically. Typically, the type of the metamorphism is advanced and active progressively. Initially, during this metamorphism, the pressure increases versus the temperature and have affected to the pillow lavas to from perhinite-pumpellyite to blue schist facies and resulting in the recent case, caused to form a dramatically deformation and alteration in the primary rocks. The outcrops of pillow lavas are remarkable in southern Sabzevar and Jazmurian regions. In late orogenic phase, which the temperature increases over the pressure, the pillow lavas and diabase have endured the metamorphism from green-schist to even in some places up to amphibolite facies. The volcano-sedimentary rocks with the volcanic series such as pelagic limestone, more or less, metamorphosed to the crystalline limestone and marble in Sabzevar and Torbat Heydarieh region.

Amongst the metamorphosed rocks associated with ophiolite we can mention those of the Torbat Heydarieh and Khoy regions in which the dibasic rocks are changed into amphibolite and limestone into marble. There is also evidence of such metamorphism around Mashhad that is compatible with the above ultramafic-mafic-sedimentary metamorphosed complex. However, it has been formed at the low temperature facies.

References

Alavi M (1980) Tectonostratigraphic evolution of the Zagrosides of Iran. Geology 8(3):144–149

Alavi M (1991) Sedimentary and structural characteristics of the Paleo-Tethys remnants in northeastern Iran. Geol Soc Am Bull 103:983–992

Alavi M (1996) Tectonostratigraphic synthesis and structural style of the Alborz mountain system in northern Iran. J Geodyn 21:1–33

Alavi-Naeini M, Bolourchi H (1973) Geology of the Maku area Geol. Surv Iran Rep No. Al

Alavi-Tehrani N (1977) Geology and petrography in the Ophiolite range NW of Sabzevar (Khorasan/Iran) with special regard to metamorphism and genetic relation in an ophiolite suite. H T 9(SI):1–47

Alavi-Tehrani N (1978) Ophiolite complexes in Iran (results and conclusions). Geol Surv Iran, Internal Report

Alavi-Tehrani N (1979) Ophiolitic rocks of Iran (in Persian). Geol Surv Iran

Alleman F, Peters T (1972) The ophiolite-radiolarite belt of the north Oman mountains. Ecolgae Geol Helv 65:657–698

Amani K (2000) Petrology of Saqqez granitoid. M.Sc. thesis, Tehran University

Amidi SM (1977) Etude géologique de la région du Natanz-Surk (Iran Central) stratigraphie et pétrographie. PhD thesis, Geol Surv Iran, Rep. 42, 316 p

Arshadi S, Forster H (1983) Geological structure and ophiolites of Iranian Makran, Geodynamics Project in Iran. R.G.S.I. Rep. 51:479–488

Arvin M (1982) Petrology and geochemistry of ophiolites and associated rocks from the Zagros suture. Neyriz, Iran

Arvin M, Robinson PT (1994) The petrogenesis and tectonic setting of lavas from the Baft ophiolitic mélange, southwest of Kerman, Iran. Can J Earth Sci 31(5):824–834

Arvin M, Shokri E (1997) Genesis and eruptive environment of basalts from the Gogher ophiolitic mélange, southwest of Kerman, Iran. Ofioliti 22:175–182

Babazadeh SA, De Wever P (2004) Radiolarian Cretaceous age of Soulabest radiolarites in ophiolite suite of eastern Iran. Bulletin de la Société géologique de France 175(2):121–129

Bagheri S, Stampfli GM (2008) The Anarak, Jandaq and Posht-e Badam metamorphic complexes in central Iran: new geological data, relationships and tectonic implications. Tectonophysics 451(1):123–155

Belov AA, Gatinsky YG, Mossakovsky A (1986) A precis on pre-Alpine tectonic history of Tethyan paleooceans. Tectonophysics 127(3–4):197–211

Berberian M (1976) An explanatory note on the first seismotectonic map of Iran, a seismotectonic review of the country. Contribution to the seismotectonic of Iran (Part III)

Berberian M, King GCP (1981) Towards a paleogeography and tectonic evolution of Iran: Reply. Can J Earth Sci 18(11):1764–1766

Boudier F, Nicolas A, Ceuleneer G (1989) De l'accretion océanique á la convergence, le cas de l'ophiolite d'Oman. Bull Soc Géol France 8:221–230

Braud J (1970) Les formation au Zagros dans la region de Kermanshah (Iran) et leurs rapports structure. Compt Rend 271:1241–1244

Bruun P (1962) Sea-level rise as a cause of shore erosion. J Waterw Harb Div, Am Soc Civ Eng 88

Clapp FG (1940) Geology of eastern Iran. Bull Geol Soc Am 51(1):1–102

Clark LD, Cannon WF, Klasner JS (1975) Bedrock Geologic Map of the Negaunee SW Quadrangle, Marquette Co., Michigan. U.S. Geol Survey Geol Quad Map, G-Q-1206

Coleman RG (1971) Plate tectonic emplacement of upper mantle peridotite along continental edges. J Geoph Res 72:1212–1222

Coleman RG (1981) Tectonic setting of ophiolite obduction in Oman. J Geophys Res 86:2497–2508

Darvishzadeh A (1991) Geology of Iran. Neda Publication, Tehran, p 901

Davies RG, Jones CR, Hamzepour B, Clark GC (1972) Geology of the Masuleh, Sheet 1:100,000; Northwest Iran. Geol Surv Iran, Rep 24:110

Davoudzadeh M (1969) Geologie und petrography des Gebietes Nordlich von Nain, Zentral- Iran, Mitt. Geol Inst Eidg Tech Hochsch Univ Zuerich, Che 98:40

Davoudzadeh M (1972) Geology and petrography of the area North Naein, Central Iran. Geol Surv Iran, Rep 14:92

Davoudzadeh M, Soffel H, Schmidt K (1981) On the rotation of the Central-East Iran microplate. N Jb Geol Palâont Abh 3:180–192

De Böeckh H, Lees GM, Richardson FDS (1929) Contribution to the stratigraphy and tectonics of the Iranian ranges. The structure of Asia. Methuen, London, pp 58–176

Delaloye M, Desmon J (1980) Ophiolites and melange terranes in Iran: a geochronological study and its paleotectonic implications. Tectonophysics 68:83–111

Desmon J (1977) Ophiolites and mélange in the Tchehel Kureh sheet (Western Zahedan, SE Iran). Geol Surv Iran; Inter, Rep 16

Desmons J, Beccaluva L (1983) Mid-ocean ridge and island-arc affinities in ophiolites from Iran: Paleographic implications: complementary reference. Chem Geol 39(1–2):39–63

Dilek Y, Delaloye M (1992) Structure of the Kizildag ophiolite, a slowspread Cretaceous ridge segment north of the Arabian promontory. Geology 20:19–22

Dimitrijevic M, Djokovic I (1973) Geological map of Kerman region, scale 1:500,000. Geol Surv Iran, 1 sheet

Dubois P, Mazelet P, Ricateau R, Bozorgnia H, Moshtaghian A (1976) Rapport de Mission de Terrain, Dasht-e Lut, Block No. 12, Sofiran (unpublished report)

Eftekharnejad J, Ghorashi M, Mehrparto M, Arshadi S, Zohrehbakhsh A, Bolourchi A, Saidi A (1989) Geological map of Tabriz-Poldasht, No. B1 & B2, 1/250,000 scale. Geol Surv Iran, Tehran

Eftekharnezhad J (1975) Brief description of tectonic history and structural development of Azerbaijan, Field excursion guide, No. 2. Note A Sym, Geodynamic of Southeast Asia, Tehran, pp 469–478

Eftekharnezhad J (1980) Explanatory report for the Mahabad quadrangle map 1:250,000. Geol Surv Iran Geological Quadrangle, No B, 4

Eftekharnezhad J (1991) Geodynamic significance of recent discoveries of ophiolites and the Paleozoic rocks in NE, Iran (including Kopet-Dagh). Abh Geol B.A. Wien, pp 89–110

Eftekharnezhad J, Alavi Naini M, Behroozi A (1990) Explanatory text of the Gazik quadrangle map 1:250,000. Geol Surv Iran

Eftekharnezhad J, Asadian O, Mirzaee AR (1992) Geological map of Khalkhal-Rezvanshahr, sheet 5765, 1:100,000 scale. Geol Surv Iran, Tehran

Eftekharnezhad J, Behroozi A (1991) Geodynamic significance of recent discoveries of ophiolites and Late Paleozoic rocks in NE Iran (including Kopeh-Dagh). Abh der Geol Bund 38:89–100

Ewart A (1976) Mineralogy and chemistry of modern orogenic lavas, some statistics and implications. Earth Planet Sci Lett 31:417–432

Falcon NL (1967) The geology of northeast margin of the Arabian basement shield. Advan Sci 24(119):31–42

Falcon NL (1974) An outline of the geology of the Iranian Makran. Geogr J 140(2):284–291

Farhovdi G, Karig DE (1977) Makran of Iran and Pakistan as an active are system. Geology 5(11):664–668

Fotoohi Rad GR (1996) A study of petrology, petrography and geochemistry of ophiolite mélange of north west of Drah region (southeast of Birjand) with a view on the region's economic potential. Master thesis, Department of Geology, Tarbiat Moallem University, Tehran, 230 p

Gansser A (1955) New aspects of the geology in Central Iran. Proc, 4th World Petrol Cong

Ghazi AM, Pessagno EA, Hassanipak AA, Kariminia SM, Duncan RA, Babaie HA (2003) Biostratigraphic zonation and 40Ar/39Ar ages for the Neo-Tethyan Khoy ophiolite of NW Iran. Paliegeogr Palaecl Palaeoecol 193:311–323

Glennie KW, Boeuf MGA, Clarke MH, Moody-Stuart M, Pilaar WFH, Reinhardt BM (1973) Late Cretaceous nappes in Oman Mountains and their geologic evolution. AAPG Bulletin 57(1):5–27

Haghnazar, S. (2009). Petrology and geochemistry of mafic rocks in Javaher Dasht area, East Guilan Province, North of Iran, Ph.D. thesis, Shahid Beheshti University

Hassanipak AA, Ghazi AM (2000) Petrology, geochemistry and tectonic setting of the Khoy ophiolite, northwest Iran, Implications for Tethyan tectonics. J Asian Earth Sci 18:109–121

Hassanipak AA, Ghazi AM, Wampler JM (1996) Rare earth element characteristics and K-Ar ages of the Band Ziarat ophiolite complex, Southeastern Iran. Can J Earth Sci 33(11):1534–1542

Haynes SJ, Reynolds PH (1980) Early development of Tethys and Jurassic ophiolite displacement. Nature 283:561–563

Houshmand-Zadeh A (1977) Ophiolites of south Iran and their genesis problems (unpublished). Geol Surv Iran. http://www.usbm

Huber H (1978a) Geological Map of Iran, 1 sheet. Geol Surv Iran

Huber H (1978b) Tectonic Map of Iran, 1 sheet. Geol Surv Iran

Huber H (1978c) Geological Map of Iran, 1:1,000,000 with explanatory note. North-East Iran, Expl and Prod Aff, Tehran, 1 sheet

Irvine TN, Baragar WRA (1971) A guide to the classification of the common volcanic rocks. Can Jour Earth Sci 8(5):523–548

Jabari A (1997) Geology and petrography North of Nain. Unpublished M.Sc. thesis Isfahan University, Iran (in Persian)

Jensen LS (1976) A new cation plot for classifying subalkalic volcanic rocks. Ont Div Mines, Geol Rep 66:1–20

Kazmin CG, Kulakov VV (1969) Quelques traits de la structure tectonique de I'Iran et de I'Afghanistan. Bull Moskov, Ispytatelej prirody, Otd Geol, Sun 44(2):61–67

Khalatbari-Jafari M, Juteau T, Bellon H, Emami H (2003) Discovery of two ophiolite Complexes of different ages in the Khoy area (NW Iran). Comptes Rendus Geosci 335(12):917–929

Khalatbari-Jafari M, Juteau T, Bellon H, Whitechurch H, Cotten J, Emami H (2004) New geological, geochronological and geochemical investigations on the Khoy ophiolites and related formations, NW Iran. J Asian Earth Sci 23(4):507–535

Khalatbari-Jafari M, Juteau T, Cotton J (2006) Petrological and geochemical study of the Late Cretaceous ophiolite of Khoy (NW Iran) and related geological formations. J Asian Earth Sci 27:465–502

Knipper A, Ricou LE, Dercourt J (1986) Ophiolite as indicator of the geodynamic evolution of the Tethyan ocean. Tectonophysics 123:213–240

Kohansal R, Ghorbani M, Pourmafi M, Khalatbari Jafari M, Omrani J (2016) Petrology and petrogenesis of pillow lavas of Forumad area, Northeast Iran. Sci Q J, Geosci 26(101):147–158

Lensch G (1980) Major element geochemistry of the ophiolites in northeastern Iran. In: Panayiotou A (ed) Ophiolites. Proceedings to International Ophiolite Symposium. Geological Survey Department, Ministry of Agriculture and Natural Resources, Republic of Cyprus, pp 389–401

Lensch G, Davoudzadeh M (1982) Ophiolites in Iran. Neues Jahrbuch für Geologie und Paläontologie-Monatshefte 5:306–320

Lensch G, Mihm A, Alavi-Tehrani N (1977) Petrography and geology of the ophiolite belt north of Sabzevar/Khorasan (Iran). N J Min Abh 131:156–178

Lensch G, Schmidt K, Davoudzadeh M (1984) Introduction to the geology of Iran. N Jb Geol Palaont Abh, 155–164

Majidi B (1978) Etude pétrostructural de la région du Mashhad (Iran), Les problemes des métamorphites, serpentinites et granitoides Hercyniens. Thése Universite Scientifique et Medical de Granobel France, 277

Makizadeh MA (1997) Petrology and geochemistry of Dehshir Ophiolites with special insight on the dependent hydrothermal alteration process (running and listing). M.Sc. thesis, Isfahan University

Manouchehri S (1997) Petrography and petrology of the North Naein Ophiolites. M.Sc. thesis, Shahid Beheshti University

Maurizot P (1990) Explanatory text of the Shahrakht QuadrangleMap, Scale 1:250000. Geol Surv Iran, Quadrangle L7

McCall GJH (1985) Explanatory text of the Minab Quadrangle, Map; 1:250,000; No. J13. Geol Surv Iran, Tehran, 530 p

McCall GJH, Kidd RGW (1981) The Makran, southeastern Iran: the anatomy of a convergent plate margin active from Cretaceous to present. In: Legget J (ed) Trench-fore Arc Geology, Geological Society, vol 10. Special Publication, pp 387–397

Mijalkovic N, Cvetic S, Dimitrivic MD (1972) Geological map of Balvard, 1/100000 Series, Sheet

Miyashiro A (1973) The Troodos ophiolite complex was probably formed in an island arc. Earth Planet Sci Lett 19:218–224

Moeinzadeh Mirhosseini SHA (2012) Geochemistry and petrogenesis of lamprophyric Dikes in the Hur-Norkhahr Village of Kerman, 6th National Geological Conference of Payam Noor University

Moghadam HS, Stern RJ (2011) Geodynamic evolution of upper Cretaceous Zagros ophiolites: formation of oceanic lithosphere above a nascent subduction zone. Geol Mag 148(5–6):762–801

Moghadam HS, Zaki Khedr M, Chiaradia M, Stern RJ, Bakhshizad F, Arai S, Ottley CJ, Tamura A (2014) Supra-subduction zone magmatism of the Neyriz ophiolite, Iran: constraints from geochemistry and Sr-Nd-Pb isotopes. Int Geol Rev 56(11):1395–1412. https://doi.org/10.1080/00206814.2014.942391

Motamed A (1987) Sedimentology, Tehran University Press

Nicolas A (1989) Structure of ophiolites and dynamics of oceanic lithosphere. Kluwe, Dordrecht, p 367

Nicolas A, Peters TJ, Coleman RG (1990) Ophiolites genesis and evolution of the Oceanic lithosphere. Proceeding of the Ophiolite Conference, Oman, 7–18 January 1990. Kluwe Academic Publishers, Dordrecht

Noghreian M (2004) Comparison of ophiolites of the North Naein with Dehshir Ophiolites. 8th Conference of the Geological Society of Iran

Nogol-e Sadat MA (1978) Les zones de decrochement et les virgation structurales en Iran. Consequences des resultants de l'analyse structural de la région du Qom. PhD thesis, 201

Ohanian T (1983) The Birjand ophiolite: an intercontinental transform structure, Eastern Iran. Geodynamic Project (Geo-traverse) in Iran, Rep 51:239–245

Pilger A (1971) Die zeitlich-tektonische Entwicklung der iranischen Gebirge, Pilger

Rahaghi A (1976) Contribution a l'etude de quelques grands foraminiferes de 1' Iran. N.I.O.C. Publication N. 6

Razmara M (1990) Geology, petrology and mineral potential of ophiolite melange of North Torbat Heydarieh (Asadabad Area). M.Sc. thesis, University of Tehran

Reinhardt BM (1969) On the genesis and emplacement of ophiolites in the Oman mountains geosyncline. Schweizerige Mineralogishe und Petrog- raphishe Mitteilungen 49:1–30

Ricou LE (1971) Le Croissant ophiolitique peri-arabe, un ceinture denappes mises en place au Cretace superieur. Rev Geogr Phys Geol Dyn 13:327–350

Ricou LE (1974) L'etude géologique de la région de Neyriz (Zagros iranien) et l'évolution structurale des Zagrides [The geological evolution of the region of Neyriz (Iranian Zagros) and the structural evolution of Zagros]. PhD thesis, Universite Paris-Sud, Orsay

Ruttner AW (1984) The Pre-Liassic basement of the Eastern Kopeh-DaghRange. N Jb Geol Paläont Abh 168(2–3):256–268

Sabzehei M (1974) Les Mélanges ophiolitiques de la région d'Esfandagheh (Iran méridional). Étude pétrologique et structurale, interprétation dans le cadre iranien, Université Scientifique et Médicale de Grenoble

Sabzehei M (1995) Layered ultramafic- mafic komatiitic lava flow and their bearing on the genesis of Iranian ophiolites. 30th Inter Geol Vongr, Bijing, China Abstract

Sabzehei M, Berberian M (1972) Preliminary note on the structural and metamorphic history of the area between Dowlatabad and Esfandagheh, South-East Central Iran. Geol Surv Iran, Int Rep 30 p and 1st Iranian Geol Symp, NIOC, 1973 (abst)

Sabzehei M, Roshanravan J, Amini B, Eshraghi SA, Alai Mahabadi S, Seraj M (1993) Geological map of the Neyriz quadrangle, H-11, 1:250,000. Geol Surv Iran

Sahandi M (1993) Geological map of Ophiolites in the North East of Iran

Sarkarinezhad K (1989) Petrology and geology of ophiolite Neyriz-Hashtjin. 8th Earth Science symposium, Geol Surv Iran, pp 35–39

Şengör AMC (1990) Plate tectonics and orogenic research after 25 years: a Tethyan perspective. Earth Sci Rev 27:1–201

Şengör AMC, Altincr D, Cin A, Ustömer T, Hsu KJ (1988) Origin and assembly of the Tethyside orogenic collage at the expense of Gondwana Land. In: Audley-Charles MG, Hallam A (eds) Gondwana and Tethys. Geological Society, vol 37. Special Publications, London, pp 119–181

Şengör AMC, Natal'in BA (1996) Palaeotectonics of Asia: fragments of a synthesis. In: Yin A, Harrison, M (eds) The tectonic evolution of Asia. Rubey Colloquium, Cambridge University Press, Cambridge, pp 486–640

Shafaii Moghadam H (2011) Petrogenesis of the volcanic sequence of Kamyaran ophiolites (Kermanshah). M.Sc., Earth Faculty, Shahid Beheshti University

Sinton JM, Byerly GR (1980) Geochemistry of abyssal oceanic magmas. Supplement to: Sinton JM, Byerly GR, Silicic differentiates of abyssal oceanic magmas: evidence for late-magmatic vapor transport of potassium. Earth Planet Sci Lett 47(3):423–430

Soffel HC, Forster HG (1984) Polar wander path of the Central-East Iran microplate including new results. N Jb Geol und Paläontol Abh 168:165–172

Stahl AF (1911) Persien, in Hanlbuch der regionalen Geologie. Heidelberg (winter). Hft 8, 5, 6, 46, 5 figs, 2 Maps

Stampfli GM (1978) Etude géologique générale de l'Elburz oriental au S de Gonbad-e-Qabus, Iran NE: thèse présentée à la Faculté des sciences de l'Université de Genève, Cooperative d'imprimérie du Pré-Jérôme

Stöcklin J (1968) Structural history and tectonics of Iran; a review. Amer Associ of Petr Geol Bulletin 52(7):1229–1258

Stöcklin J (1974) Evolution of the continental margins bounding a former Southern Tethys. In: the geology of continental margins. Springer, pp 873–887, BIBL. 2P, 5Illus. U N Geol Surv Inst

Stöcklin J (1977) Structural correlation of the Alpine ranges between Iran and central Asia. Anonymous. Livre a la memoire de Albert F. de Lapparent (1905–1975) consacre aux Recherches geologiqes dans les chaines alpines de l'Asie du Sud-Ouest. Soc Geol Fr MemHors-Ser 8:333–353

Stöcklin J, Eftekharnezhad J, Hushmandzadeh A (1972) Central Lut reconnaissance, East Iran. Geol Surv Iran, Rep 22:62

Stöcklin J, Nabavi MH (1973) Tectonic map of Iran, 1:2,500,000. Geol Surv Iran, Tehran Offset, Press, Rep 31:100

Sultan Mohammadi A (2009) Geochemistry and petrogenesis of the ophiolite melange in the Marvast, northwest of Babak city. M.Sc. thesis, Shahid Beheshti University

Takin M (1972) Iranian geology and continental drift in the Middle East. Nature 235(5334):147–150

Thayer TP (1960) Some critical differences between Alpine-type and stratiform peridotite- gabbro complexes. Inter Geol Congr, 21st Sess. Copenhagen 13:247–259

Tirrul R (1983) Structure cross-sections across Asiakforeland thrust and fold belt, Wopmay orogen, District of Mackenzie. Geol Survvey Can, Pap 83(1b):253–260

Tirrul R, Bell IR, Griffis R, Camp V (1983) The Sistan suture zone of eastern Iran. Geol Soc Amer Bull 94(1):134–150

Wensink H, Varekamp JC (1980) Paleomagnetism of basalts from Alborz: Iran part of Asia in the Cretaceous. Tectonophysics 68:113–129

White RS, Klitgord K (1976) Sediment deformation and plate tectonic in the Gulf of Oman. Earth Planet Sci Lett, Netherl 32(2):199–209

White S (1976) The effects of strain on the microstructures, fabrics, and deformation mechanisms in quartzites: Royal Society of London Philosophical transections. Ser Math Phys Sci 283:69–86

Chapter 8
Metamorphic Rocks of Iran

Abstract In this chapter, all metamorphic rocks of Iran have been studied and evaluated from the perspective of metamorphic facies, protoliths and their relationship with magmatic and orogenic phases. The distribution of metamorphic rocks is described along with the geological map and the grade of metamorphism. The causes of metamorphism and geodynamics of all metamorphic rocks are classified according to their structural zones, in stratigraphic order. Metamorphic rocks are also classified into orogenic phases according to their structural zone:

– Pan-African orogenic phase in Late Precambrian
– Caledonian orogenic phase
– Metamorphic rocks of Hercynian phase
– Metamorphic rocks of Cimmerian phase
– Laramide metamorphic rocks
– Young Alpine metamorphic rocks.

In each phase, the distribution of metamorphic rocks, petrography, grade of metamorphism and geodynamics are described in detail. This chapter also evaluates contact-metamorphism in relation to intrusive masses.

Keywords Metamorphism in Iran · Metamorphic zones of Iran · Metamorphic rocks of Iran · Metamorphic facies of Iran · Metamorphic phases of Iran

8.1 Metamorphic Phases in Iran

Generally, the metamorphic phases coincide with the orogeny and magmatism phases in Iran. Regarding to these fact, the metamorphic phases can be divided as follows:

© The Author(s), under exclusive license to Springer Nature Switzerland AG 2021   387
M. Ghorbani, *The Geology of Iran: Tectonic, Magmatism and Metamorphism*,
Earth and Environmental Sciences Library,
https://doi.org/10.1007/978-3-030-71109-2_8

1. Late Precambrianis the oldest metamorphic phase in Iran. It has resulted in a low to intermediate regional metamorphism. Even scarcely, in some areas, the amphibolite facies has been reported to be occurred at the Late Precambrian.[1]

2. During the Caledonian tectonic phase, epeirogenic movement took place in Paleozoic of Iran in, caused changes in the climate and sedimentary environments. Anyway, no well-known metamorphic rock units are reported to be related to the Paleozoic in Iran.

3. During the collision due to the closure of the Paleo-Tethys Ocean (the Tethys II) that has coincided with the final stages of Hercynian and probably the initial stages of the Early Cimmerian orogenies in Late Paleozoic-Early Triassic, a metamorphism phase has occurred in northeastern, north and northwestern parts of Iran.

4. The Early Cimmerian metamorphic phase (Middle Triassic-Late Triassic): The Sanandaj-Sirjan zone and some parts of the Central Iran have been affected by this metamorphic phase where the Paleozoic sediments deposited in a platform environment were metamorphosed up to greenschist and amphibolite facies.

5. The Middle Cimmerian metamorphic phase (latest Early Jurassic-Middle Jurassic): There is an evidence of metamorphism caused by this phase in Sanandaj-Sirjan and Central Iran zones that has resulted in a wide metamorphic area at the Sanandaj-Sirjan zone.

6. The Late Cimmerian metamorphic phase: It has caused metamorphism of the Sanandaj-Sirjan and Central Iran zones.

7. The Laramide metamorphic phase: This phase is traceable in ophiolitic belts and the Sanandaj-Sirjan zone. The Late Cretaceous and early Cenozoic orogenic movements have resulted in the weak regional metamorphism forming schists and glaucophane schist in ophiolitic complexes throughout Iran.

8. The last orogenic phase in Iran which occurred during the Eocene to Oligocene, has resulted in the metamorphism of the Eocene rocks of Alborzin zeolite facies.

9. In addition to the metamorphic phases consistent with orogenic phases, there are also some metamorphic events associated with large batholiths (it does not mean contact metamorphism). The metamorphic rocks that previously were attributed to the Precambrian could be mentioned as the metamorphic core complex. The attributed age of this type of metamorphism has been changed later. They are related to high temperatures and intermediate pressures.

10. The high-temperature and low-pressure contact metamorphism caused by intrusions is relatively abundant in Iran. It can be added to the above mentioned phases.

---

[1] There are several rock units in the Takab and Central Iran zones older than late and early Precambrian which are metamorphosed up to amphibolite facies. The present author believes that they are equivalent to the rocks older than the Late Neo-Proterozoic.

8.2 Metamorphic Zones in Iran

The metamorphic zones in terms of their position in structural units of Iran (Fig. 8.1) are described as follows.

8.2.1 Sanandaj-Sirjan

Generally, the largest metamorphic zones are in the Sanandaj-Sirjan Zone. This zone is mostly known as the magmatic-metamorphic Sanandaj-Sirjan Zone. Though, in many previous studies, the metamorphic rocks of this zone were attributed to Precambrian but such a metamorphism just can be found in the lower parts of the Soltanieh Formation and/or older rock units. Although having low grade in metamorphism, the main parts of the metamorphic rocks of this zone have metamorphosed during the late Paleozoic-Mesozoic.

In geology of Iran, the Sanandaj-Sirjan zone which is the most active structural zone of Iran is mentioned by other names such as Esfandagheh-Marivan (Nabavi 1976), Urmia-Esfandagheh (Takin 1971) and Manujan-Marivan (Houshmandzadeh 1977). This zone is as an elongated NW-SE magmatic-metamorphic belt that can be observed from Urmia to Sirjan and is almost 1500 and 150–250 km in length and width, respectively. It is confined to Zagros main fault in southwest and Urmia-Dokhtar volcanic belt in northeast (Tillman et al. 1981). The Sanandaj-Sirjan zone is characterized by complex deformations of metamorphic rocks, deformed and plenty of undeformed igneous rocks of Mesozoic (Mohajjel and Sahandi 1999). Mohajjel and Sahandi (1999) divided the Sanandaj-Sirjan zone in to five structural subzones with different types of deposit (Fig. 8.2).

1. Radiolarite subzone;
2. Bisotun subzone;
3. Ophiolite subzone;
4. Completely deformed subzone and
5. Marginal subzone.

In the mentioned subzones, following metamorphic areas are discussed that most of them are of Mesozoic.

A.

1. Southern Sirjan;
2. Northern Dehbid-Surian;
3. Northwestern Saman;
4. Aligoudarz-Azna axis;
5. Baba Ali-Galali axis (surroundings of Almogholagh mount) and
6. Divandarreh-Saqqez axis.

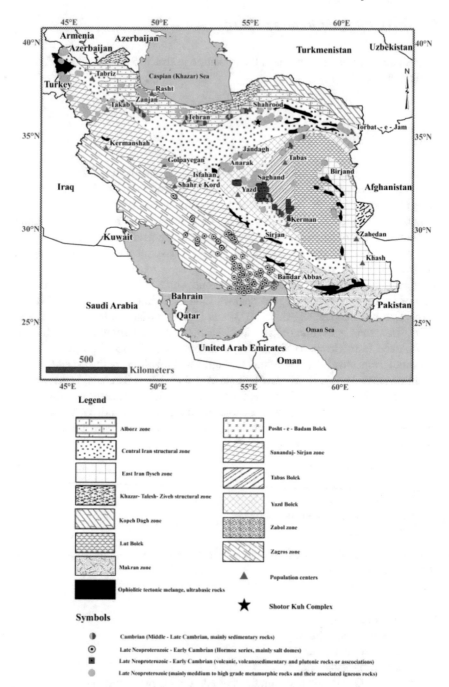

Fig. 8.1 Metamorphic zones of Iran (Sadeghian et al. 2019)

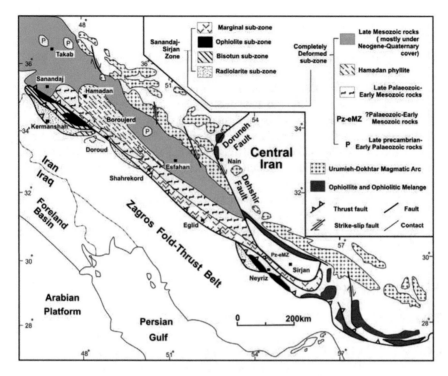

Fig. 8.2 The Sanandaj-Sirjan subzones (Mohajjel and Sahandi 1999)

B.

In addition to the mentioned metamorphic areas that are mostly of late Paleozoic, Triassic, Jurassic and Cretaceous age, the Precambrian metamorphic rocks of Sanandaj-Sirjan zone can be named.

The length and width of the Sanandaj-Sirjan metamorphic belt is 1200 and 300 km, respectively, which includes the Divandarrah, Qorveh, Songhor, Sanandaj, Hamedan, Malayer, Tuyserkan, Boroujerd, Arak, Aligoudarz, Golpayegan, Saman, Bavanat, Dehbid, Neyriz, Esfandaqheh, Sirjan and Hajiabad metamorphic complexes.

Different types of the metamorphic facies and rocks can be found in this zone. The migmatite, gneiss, mica schist, amphibolite and marbles with Precambrian to Cenozoic age are frequency exposed at southern parts of the zone. The green-schist to amphibolite facies are reported from Paleozoic-Cenozoic rock units at northern parts of the zone. The metamorphic rocks can be classified as follows:

1. A low grade metamorphism has occurred in the rocks equivalent to the Kahar Formation (Tashk and Kalmard) during the Precambrian. It can be mostly observed at northwestern part of the Sanandaj-Sirjan zone (i.e. Mahabad area).
2. The metamorphism caused by the Early Cimmerian is widely distributed in this zone. The metamorphic rocks previously attributed to the Precambrian

are actually associated with the Early Cimmerian metamorphic phase. These metamorphic rocks are exposed in NW Divandarah, northern Songhor (Baba-ali to Galali belt in Hamedan-Qorveh area), Aligoudarz, Saman, Kolikosh, Surian, Tutak and Sirjan.

3. The Middle and Late Cimmerian metamorphic phases resulted in the metamorphism of some rock units to slate and phyllite at Hamedan resulted from the subduction of Neo-Tethys. They are traceable in southeast of the Sanandaj-Sirjan zone such as the metamorphic rocks of Hajiabad. The grade of metamorphism caused by Middle and Late Cimmerian is lower than those of the Early Cimmerian. For this reason, it is difficult to detect such metamorphism in rocks older than Jurassic.

4. The varied contact metamorphism such as hornfels formation around the Alvand, Qorveh, Mahabad, Piranshahr intrusions and generally contact metamorphism occurred close to the large magmatic bodies in Sanandaj-Sirjan Zone.

8.2.1.1 Mahabad

The Mahabad metamorphic rocks are attributed to the Precambrian and its protoliths is assumed to be the Kahar Formation (Ghorbani 2013). There are two series of metamorphic rocks in the west of Mahabad geological map quadrangle (1:100,000) which are correlatable to metamorphosed Kahar Formation.

A. The old series of mixed metamorphic rocks including very well distributed schist, gneiss, and amphibolite,

B. A series of mixed marbles and amphibolites that is similar to the Sargaz-Abshoura-Qadagheh series.

The metamorphic rocks of the Mahabad are mostly composed of schists metamorphosed at green-schist facies. Furthermore, there are series of metamorphic rocks in the Sirjan area, green-schist in facies, attributed to the Precambrian. These rocks were accurately attributed to the Precambrian; because, in the Mahabad and Piranshahr, the unmetamorphosed Early Cambrian rocks overlie the Precambrian units. At southeastern parts of Mahabad city, the Mahabad formation (composed of metamorphic rocks) can be observed that overlies the Early Cambrian Barut, Zagun and Lalun formations where they do not show any evidence of metamorphism (NIOC 2017).

The Mahabad metamorphic rocks are mostly inhomogeneous schists composed of different types of rock including green schist, meta-sandstone, calc-schist and marble.

8.2.1.2 Northern Dehbid-Surian

In the Kuh-e Sefid (Tutak) area, a relatively thick sequence of the black mica schists is overlain by the Kuh-e Sefid marble complex. The Paleozoic rocks in the Sanandaj-Sirjan zone are exposed as a set of marble, schist, amphibolite and quartzite. They are widely exposed in two zones, Kuh-e Sefid (Tutak) of Bavanat area and the Koli-Kosh area along the Abadeh-Shiraz road. The metamorphic rocks in terms of age can be divided into three complexes:

1. The Tutak complex which is mostly associated with the marble, black schist, and granite-gneiss.
2. The Surian complex which is composed mainly of schist, quartzite-schist with limestone, basalt and basaltic tuff interlayers.
3. The Kolikosh complex that overlies the Surian Complex. Ankerite-dolomitic carbonate sequence of its upper parts is comparable to the Shishtu Formation (members I and II; Hadizadeh Shirazi 2010).

The Triassic rocks in the Sanandaj-Sirjan zone of the Eghlid geological quadrangle (1:100,000 in scale) are not less widely-distributed. Two major phases of metamorphism affected the Eghlid region (within the Eghlid geological quadrangle) are:

1. The first phase that is traceable up to amphibolite facies and
2. The second phase is a retrograde metamorphism occurred after the first phase and has resulted in the green-schist facies.

The first phase has occurred during the late Middle Triassic to early Late Triassic (Early Cimmerian orogenic phase) and has resulted in an intermediate-pressure, high temperature conditions. The second metamorphism phase has affected the Lower Jurassic rocks and has resulted in an intermediate-temperature and -pressure metamorphism. It is probably related to the Middle Cimmerian orogenic phase.

The Tutak metamorphic zone is described in more details in this part. The Tutak metamorphic rocks in the Kuh-e Sefid area of Bavanat form an anticline that its bigger and smaller diameters are 20 and 10 km, respectively. The anticline located at the 30° 10′ to 30° 35′ northern latitudes and 53° 20′ to 54° 00′ eastern longitudes can be accessed at 250 km of northeastern Shiraz, 80 km of northern Dehbid and northeastern of Mazayjan village. The anticline falls into the Eghlid geological quadrangle (1:250,000 in scale; Houshmandzadeh et al. 1990).

The anticline is of an elongated oval shape located between two major reverse steep righ-lateral faults, the Surian faults in north and Mozajian fault in South. The activity of these two faults has an important role in the formation of the anticline structure (Ahmadi 2004).

Structurally, the area is a part of the Sanandaj-Sirjan zone, where it is located in the southeastern part of the zone which shows similar NW-SE trend. This area due to being located at the Sanandaj-Sirjan zone is structurally related to the different tectonic events and has its own complexities.

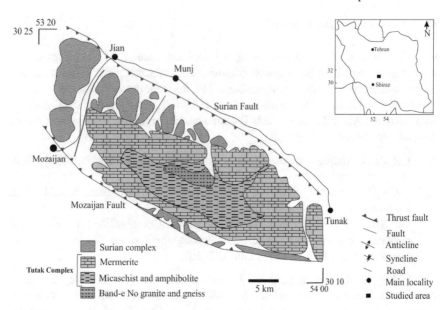

Fig. 8.3 Geological map of the Kuh-e Sefid Tutak anticline (modified after Hosseini et al. 2012)

The protolith of the Tutak complex was sedimentary and magmatic rocks, so, they occur as a sequence of marble, schist and amphibolite (Fig. 8.3). The oldest rock unit exposed in the studied area is a sequence of black schists with sedimentary protoliths. The rocks are thick and very dark in color without any distinct character. Within the schist units, alternations of dark green metamorphosed volcano-sedimentary rocks can be found that are metamorphosed up to the amphibolite facies. However, the existence of these rocks indicates magmatic activities occurred during the deposition of sedimentary sequences. According to Sabzehei (1996), the black schists of the Tutak Complex can be correlated to the slates and black schists of upper part of the Rotshun Complex, Late Ordovician to Early Silurian in age, that can be traced from Bajgan, southeastern parts of the Sanandaj-Sirjan zone, to surroundings of Eghlid at northwestern part of the studied area. A sequence of thick white marbles overlies these units and forms the main part of the Kuh-e Sefid anticline. The marbles that form the highlands of the region are of alternations of marble and micaschist at the lower and upper parts of the sequence. Partoazar identified some fossils in some parts of the Tutak marble complex and assigned the Devonian to Late Carboniferous age for Tutak Complex (Noori Khankahdani 2005). A granitoid pluton (Band-e No; Middle Jurassic in age by U-Pb dating performed by Hosseini et al. 2012) has intruded into the black schists that shows the foliated and gneiss-form textures due to suffering the tectonical events. There are several enclaves of surrounding schists in the granitoid.

According to Houshmandzadeh and Soheili (1990), the Kuh-e Sefid anticline is composed of two metamorphic complexes. The Tutak complex is observed at the anticline core where the Surian complex forms the outer parts. So, the Tutak complex is older than Surian complex. Generally, the main part of the anticline is formed by

Fig. 8.4 The Surian complex overlies the marble-micaschist units of Tutak complex at northern flank of Kuh-e Sefid anticline (view towards SW)

Tutak complex and the Surian complex just can be observed at outer parts (Fig. 8.4). According to present author, these complexes are showing a sequence and they are not metamorphosed simultaneously.

Tutak Complex: The Tutak complex that forms the main part of the Kuh-e Sefid anticline is composed of metamorphosed and deformed sedimentary and basic volcanic rocks. Indeed, this complex is of metamorphosed and deformed granitic rocks located at the central part of the complex. In an overall view, the rock units of this complex can be divided in to three main categories as below.

The black meta-pelite and meta-basite: A relatively thick sequence of micaschists can be observed in Kuh-e Sefid area as an almost homogeneous rock unit showing a dark color. These rocks are in fact the oldest sedimentary rock unit of area (Houshmandzadeh and Soheili 1990) mostly observed as separated rock units.

The alternation of meta-pelites and meta-basites shows that the lavas were flewing simultaneously to the formation of shale and sandstone units of area.

Marble: This complex overlies the meta-pelite unit and is the thickest rock unit in Kuh-e Sefid anticline (Fig. 8.5).

Meta-granitoid: the light-colored granitoids can be observed at the central parts of the Tutak complex. They are of most exposure at Band-e No valley. They are affected by dynamic metamorphism where they can be categorized as deformed granitoids and augen gneisses based on their color and fabric differences.

Generally, their contact with surrounding schist units is normal and no fault is observed. This granitoid intrusion has fragments of schists (several centimeters

Fig. 8.5 Marble overlie the meta-pelite and amphibolite units of Tutak metamorphic complex at the northern flank of Kuh-e Sefid anticline (view towards NE)

to several meters) as enclaves. Indeed, some aplite veins are reported in granitoid intrusion and surrounding schists (Fig. 8.6).

Surian Complex: The Surian complex overlains the uppermost part of marble unit of Tutak complex at northern and southern flanks of anticline with a disconformity. In these areas, this metamorphic complex is showing smooth hills that are resulted from their higher erosionability compared to resistant marble unit of Tutak complex. The rocks of this complex are completely inhomogeneous mostly including sedimentary and basic volcanic rocks showing weak green schist facies.

Based on the recognized rock units of Surian complex, their protolith is mostly the basic volcanic rocks and shales.

Generally, the rocks of this metamorphic complex show evidence of a weaker metamorphic grade compared to what can be observed in Tutak complex.

Finally, the calcareous-clastic rock units of Late Jurassic-Early Cretaceous overlie the Surian complex with a faulty boundary.

8.2.1.3 Northwestern Saman

The area is confined between 32° 30′ to 33° northern latitudes and 50° 35′ to 51° eastern longitudes. It is located in Chadegan geological map (1:100,000 in scale) at northwestern Saman. The Zayandehroud dam is built on these rock units.

Fig. 8.6 The aplite veins hosted by gneiss-granite rocks of Tutak complex at the core of Kuh-e Sefid anticline

The rock units of area show an anticline. The anticline is located at northern Shahr-e Kord, eastern Yan Cheshmeh, south and southwestern Abpouneh village and eastern Zayandehroud dam. It has 20 km of length and 10 km of width with an NW-SE trend that resembles the general trend of Sanandaj-Sirjan zone. In this area, the rock units are composed of a complex of sedimentary and magmatic rocks all of which have been affected by metamorphic and deformation phases. The major rock units of area include meta-pelite, amphibolite, marble and granite-gneiss. The morphology of anticline is totally changed due to tectonics where some parts are completely eliminated. So, tracing the rock units and their changes is difficult. They are most probably recrystallized in Middle-Upper Triassic (Early Cimmerian). The slate and phyllite rock units of Jurassic (Hamedan area) are metamorphosed in Middle Cimmerian. So, the oldest rock units of area are composed of garnet mica-schist and amphibolite that show a smooth topography and are located at the core of anticline. Their age is accurately determined. They are overlain by Permian marble (Zahedi et al. 1993). So they should be of Paleozoic age or older. The Permian rocks are mostly carbonates that are metamorphosed in weaker grades (as marble to recrystallized limestone). According to the field observations of the present author the rock units of the area are as below (as shown in the map):

1. The highly metamorphosed rock units of area which are accompanied by some intrusions and are attributed to Precambrian are studied by Jamali Ashtiani

(2017) where their Proterozoic age is confirmed. They are metamorphosed in Late Precambrian.

2. The equivalent rock units of Routeh Formation in Alborz, Surmaq in Sanandaj-Sirjan and Dalan Formation in Zagros (Permian in age) (Ghorbani 2019).

The marble unit is of a creamy to beige color occasionally is interbedded with sandstone, meta-pelite and amphibolite. Their exposure can be observed mostly in northern flank of anticline close to Abpouneh village.

At the core of anticline, the granite-gneiss can be observed which is of more extend compared to other rock units of area (Fig. 8.7). In some localities, the granitic unit is of schist and amphibolite enclaves up to several meters in diameter. So, it can be concluded that granitic rocks are younger than schist and amphibolite rock units. According to the studies performed by Jamali Ashtiani (2017) the granite is of Neoproterozoic age. In hand-specimen scale, the gneiss shows augen fabric with

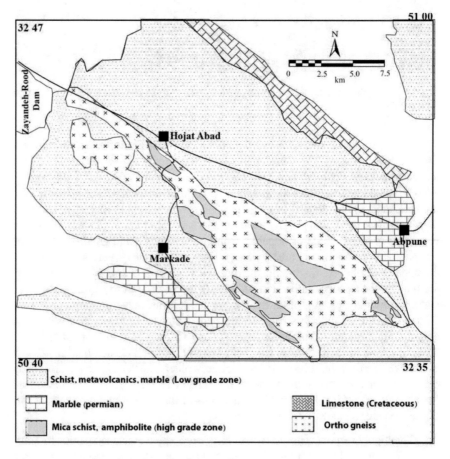

Fig. 8.7 The main rock units of northwestern Saman. Modified after Zahedi et al. (1993). The points refer to samples taken for geochemical analysis

coarse crystals of pink K-feldspar. In some cases, these crystals are forming the majority of rock giving it a pink color. Toward the marginal parts of anticline, the metamorphic grade is obviously weaker than those of anticline core. In these parts, the rock units include Jurassic slate and phyllite occassionaly rich in graphite and organic matter. The protolith of these rocks can be sandstone and shale interbedded with shale layers rich in organic matter affected metamorphosed in low grades. They are of dark gray to black color with a shiny luster showing a well-developed schistosity as cleavage. They are the continuation of Hamedan slate and phyllite developed vastly in northwestern parts of Sanandaj-Sirjan (mostly around Malayer, Arak and Hamedan) toward the southeastern parts of the zone (Upper Triassic to Jurassic in age) (Hosseini 2011). In the area, the relationship of these rock units with those of Palezoic is not recognizable but according to their sudden change in metamorphic grade, most probably thay have a faulty boundary.

8.2.1.4 Shahr-e Kord

The main phases of deformation and metamorphic zones of the Shahr-e kord, Saman and Sanandaj thermodynamically and thermostatically can be summarized in three phases as follows:

1. The first phase in which a high-pressure metamorphism has resulted in a set of marble, amphibolite, schist and gneiss that in the upper parts of it the Permian to Middle Triassic fossils have been found. On the other hand, the age of metamorphism is the Late Triassic to Early Jurassic related to the Early Cimmerian orogeny.
2. The dynamics features of this phase are dominant compared to the role of temperature. The metamorphism has occurred after the Cretaceous and is most probably related to the Laramide orogeny.
3. The third phase is divided into two distinct stages:

   (A) The intrusion gabbroic and dioritic bodies resulted in the contact metamorphism and forming sillimanite, cordierite, andalusite and staurolite.
   (B) The intrusion of granite and granodiorite into the gabbroic and dirotic host rocks as well as their marginal metamorphosed rocks caused that some particles of the host rocks trapped into the acidic intrusions as enclaves. The Alvand granite is one of these intrusions probably emplaced during the Middle Jurassic to Early Paleocene. There are several reports regarding the regional high-grade metamorphic facies in Shahrekord area (the area around the Zayanderud Dam) which is probably related to the shear zones or uplifted basement rocks. The metamorphic facies of Shahrekord can be categorized between the green-schist to amphibolite (Houshmandzadeh and Posht-kouhi 1996).

8.2.1.5 Aligudarz

The studied area is located in the Lorestan province, along the Azna to Isfahan road. The NW-SE Aligudarz metamorphic zone belongs to the Sanandaj-Sirjan zone. It is confined between 33° 20′ to 33° 45′ northern altitudes and 49° 15′ to 49° 45′ eastern longitudes. The metamorphic rocks are composed of a sequence of metamorphosed sedimentary and igneous rocks including various types of schist, marble, quartzite, meta-basalt, meta-andesite and mylonitic granitoid rocks. They have been metamorphosed by regional metamorphism and have suffered deformation phases.

The petrographic and petrofabric studies showed that meta-basic rocks originally were lavas and basic-tuffs metamorphosed in green-schist and beginning of the amphibolite facies. The meta-granitoid rocks are resulted from deformation at the shear zones. The regional metamorphism is low-pressure and -temperature chlorite and biotite facies.

The advanced metamorphism begins from green-schist and ends at the epidote-amphibolite facies (the transition stage between green-schist and amphibolite) in the region and then a retrograde metamorphism has occurred.

The mica-schists form the majority of metamorphic rocks. These rock show a smooth topography as smooth hills (Fig. 8.8). They mostly include sericite schist,

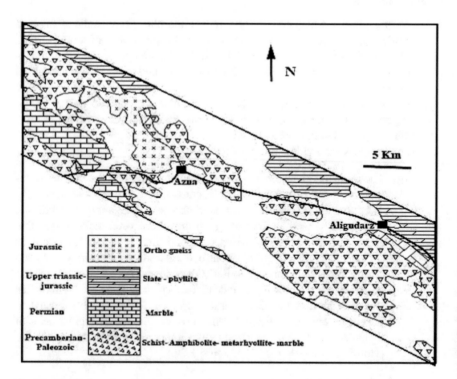

Fig. 8.8 The main rock units of Aligudarz-Azna axis, modified after Soheili et al. (1992)

chlorite-muscovite schist and muscovite-biotite schist. The metamorphic grade is low up to lower green schist. In area, the mica-schists are of notable thickness. Indeed, several changes can be observed in their protoliths. Within this sequence, a large volume of volcanic rocks can be observed introduced as Precambrian-Lower Paleozoic? units in 1:100,000 geological map of Aligudarz (Soheili et al. 1992). The volcanic rocks are attributed to Precambrian in the same map. The protolith of these rock units according to the opinion of the present author and based on the geological evidence is Late Paleozoic-Early Triassic. The complex is dated using U-Pb dating on zircon mineral (Hosseini 2004). So, taking into account that the volcanic rocks are observed within the metamorphosed sedimentary rocks, the age of mica-schist's protolith can be determined.

Northern Sanandaj-Sirjan (Paleozoic-Late Triassic)
These metamorphic rocks that have exposures around Almogholagh, Qorveh, Divan-Darreh and Saqez are metamorphosed up to green schist to amphibolite where their protolith is mostly volcano-sedimentary rocks.

1. Geological background of Babaali-Galali axis (Almogholagh surroundings)
   The sequence is composed of intercalations of metamorphosed magmatic and sedimentary rocks. The magmatic protolith mostly include acidic to intermediate volcanics and pyroclastics, subvolcanics and minor basic volcanics. The sedimentary protolith include limestone, dolomite and pelitic rocks now can be observed as marble, calc-silicates and mica-schist. In conclusion, it should be mentioned that the majority of this metamorphic complex is composed of meta-rhyolite to meta-rhyodacite as well as minor volumes of other rocks. In fact, the exposure of metamorphic rocks around the Almogholagh mount is the result of its upward movement forming dome-shape morphology. The metamorphic grade is up to amphibolite. In this metamorphic complex coarse folded kyanite can be found. Apart from this area, the phyllite and slate of Hamedan overlie the mentioned rock units. Such relationship can be observed in Hamekasi village (close to Hamedan-Qorveh road). The ptotolith of Hamedan pyllites is of Upper Triassic-Lower Jurassic age with defined metamorphic grade. But the underlain rocks are metamorphosed up to amphibolite grade. So, this area is one good example where it can be said that the metamorphic rocks of Babaali-Galali, Divan-Darreh and Aligoudarz have been metamorphosed before Upper Triassic. Their protolith is of Late Paleozoic-Early Triassic age. So, it can be concluded that these rocks have been metamorphosed in Middle to Upper Triassic.

Geological Background of Divan-Darreh (Saqez) Area
The metamorphic rocks of Divan-Darreh area are exposed at northwestern part of Davan-Darreh city where the Zafarabad iron prospect is located. The area is mostly covered with metamorphic rocks that show resemblances to those of Babaali-Galali, Aligudarz and other adjacent areas. These rocks are attributed to Pre-Permian by Alavi-Naini et al. (1982b) in 1:250,000 Takab geological quadrangle.

The lithological composition of the rocks include gneiss, marble, meta-volcanic, green schist and slate. In this complex a serie of intrusive rocks including diorite to granite has intruded that most of them are also metamorphosed (Hosseini 2004).

The iron ore of Zafarabad prospect is as a lens found between the meta-volcanics and schists with a NNE-SSW trend. The host rock shows a westward dip direction and by detail observations it can be seen that the iron lens is parallel to host rock. The ore due to its resistance against the erosion is exposed well within eroded host rocks that have turned into farms.

The host rock of the ore can be divided into two categories:

1. Recrystallized limestones of Routeh Formation that form the hangingwall of ore. So, the upper contact of ore was carbonates that are now metamorphosed.
2. The metamorphosed igneous rocks including meta-basalt, meta-andesite and metamorphosed tuffs forming the footwall of the ore.

The ore is mostly mixed with the meta-andesite unit and the genesis of mineralization is also related to the mentioned igneous rocks. Occassionaly within the marbles, small lenses of iron are observed. The main mineralogy of ore is magnetite and hematite accompanied by some impurities such as carbonates, calc-silicates, sulphides and silica. Based on field observations the volcanic rocks and ore have formed at the same time (Ghorbani 2007).

In fact, this iron-bearing horizon can be traced in northern Sanandaj-Sirjan zone accompanied by metamorphic rocks of Late Paleozoic-Early Triassic. So, this is why the iron ore minerliaztion of area is discussed here.

8.2.1.6 Hamedan

There are several outcrops of schist and micaschist beds along the of Hamedan-Malayer road that are associated with Aliabad Damagh gneiss rock units (Darvishzadeh and Valizadeh 2001).

The Hamedan metamorphic rocks studied by Majidi and Amidi (1980) and Naderi (2012) include:

1. Pre-Jurassic crystallized limestone,
2. Greywacke-schist and
3. Phyllite, schist and meta-sandstone.

The Alvand granitic-granodioritic batholith has intruded into the rocks of Hamedan area which suffered regional metamorphism. At the margin of the intrusion, especially in the eastern part, a vast area of dense black hornfels rocks can be found which consist of cordierite, garnet, biotite and andalusite (Baharifar 2004).

In general, based on the mineralogical and textural changes, the Hamedan metamorphic rocks are classified into two groups as below:

A. Extensive regional metamorphic rocks

Slate: Being very fine in grain size, the texture of the rock due to the occurrence of phyllosilicates as oriented crystals is lepidoblastic. Although, because of its fine grain size it shows slaty texture (Baharifar 2004). The rock forming minerals are:

Quartz, chlorite, mica and opaque minerals (iron oxides or pyrite are sometimes observed in these rocks, especially in vicinity of the Arak city). These metamorphic rocks can be traced from the Hamedan to Golpayegan and Arak, Khansar, Khomein and Malayer.

Phyllite: The phyllites are lepidoblastic and lepidogranoblastic in texture. The main minerals are quartz, biotite, muscovite, chlorite and minor amounts of garnet and opaque minerals. Distinguishing the slate from phyllite is difficult, but westwardly, from Arak to Malayer, Boroujerd and Hamedan, the volume of phyllites is increasing (Goudarzi 1995).

Garnet schist: In Hamedan and Tuyserkan, the metamorphism grade is higher than phyllite facies; that may be resulted from the occurrence of the intrusions coincided with the metamorphism (Baharifar 2004).

The texture of the garnet schist is porphyroblastic, porphyro-lepidoblastic and occasionally poikiloblastic.

Porphyroblasts of these rocks are mainly garnet in composition but occasionally mica and chlorite crystals are observed accompanied by garnet. The main minerals of these rocks are quartz, biotite, muscovite, garnet and chlorite as well as plagioclase and iron oxides as accessory minerals. Garnet occurs in two forms:

1. Inclusion-free Auto morph garnets without pressure shadow structure. The chlorite halo around the garnet grains can be found due to the retrograde metamorphism.
2. Poikiloblasts of garnet containing abundant inclusions of quartz, biotite and opaque minerals.

Garnet-staurolite schist: The texture of the rocks is porphyro-lepidoblastic. The garnet-staurolite schist rock-forming minerals are:

Staurolite, garnet, biotite, muscovite, chlorite, quartz and opaque minerals.

The staurolites show poikiloblastic texture and are of abundant inclusions of quartz and fine-grained opaque minerals. They have grown with garnet simultaneously and surrounded porphyroblasts of garnets (Baharifar 2004). The S_1 and S_2 cleavage systems are visible but S_0 can be identified only at a macroscopic scale.

Meta-sandstone: The texture of the rock is grano-lepidoblastic. The rock-forming minerals include quartz, mica, garnet and opaque minerals. Some portions of the meta-sandstones contain a high percentage of garnet (almandine). Consequently, it can be considered as garnetite. Also, the mica crystals show some orientation resulting in the foliation of the rock (Baharifar 2004).

Amphibole schist: The texture of the rock is nematoblastic, porphyroblastic or porphyro-nematoblastic. The rock-forming minerals are tremolite-actinolite, quartz, plagioclase, garnet, zoisite and opaque minerals. Amphiboles have inclusions of quartz and mica. The inclusions are concentrated more in the central part than the rim of amphiboles. In some cases, a coating of quartz encompasses the amphibole crystal. Zoisite is one of the main minerals of the rock and can be found as fine grains in the matrix of the rock. These rocks are devoid of plagioclase or a very small percentage of fine grain plagioclase can be identified as scattered crystals in the matrix (Baharifar 2004).

Amphibolite: The rocks in field and microscopic studies are similar to Amphibole-schist. The texture of the rock is porphyro-nematoblastic and bouquet shape. The rock-forming minerals are: tremolite-actinolite, hornblende, quartz, plagioclase, garnet, zoisite and opaque minerals (Baharifar 2004).

Carbonate rocks: These rocks include marble and calk-silicates which are not scattered widely in the area.

B.   Contact Metamorphic Rocks

Cordierite hornfels: The texture of hornfelses is porphyro-lepidoblastic. The main mineral is cordierite showing poikiloblastic texture and contains quartz and biotite inclusions. The matrix includes fine grains of biotite, muscovite, quartz and opaque minerals. There are few occurrences of andalusite and sillimanite in some samples. The contact metamorphism is observed around the Alvand batholite where by taking distance from the batholite they show weaker grades of metamorphism (Fig. 8.9).

Andalusite-cordierite hornfels: The main minerals of the rock are: andalusite, cordierite, quartz, alkali-feldspar (microcline), perthite, plagioclase and minor amounts of mica (biotite and muscovite), garnet and opaque minerals. Garnet porphyro-blasts are isomorphous to amorphous and encompass inclusions of mica and quartz. The mica crystals are isomorphous to amorphous scattered in the matrix of the rocks. The texture of the rocks is porphyro-granoblastic (Baharifar 2004).

8.2.1.7   Qorveh

The Qorveh metamorphic area, located at the southern part of the Qorveh quadrangle, is a part of Sanandaj-Sirjan zone which is located at southwestern part of the Hamedan-Qorveh area. The area consists of igneous, metamorphic and sedimentary rocks. The metamorphic rocks are schist, gneiss, marble and hornfels in composition (Abdi 1996).

The metamorphic rocks of the region have been subjected to several types of metamorphism. Based on the petrography and petrofabrics of the rocks, the sequence of metamorphism in the study area is as follows, from the oldest to youngest: regional metamorphism, contact metamorphism and dynamic metamorphism (mylonitic and cataclastic; Abdi 1996).

It can be taken into account, regarding to the mineralogical composition of the metamorphic rocks and metamorphic reactions the grade of metamorphism is greenschist to lower amphibolite facies (Abdi 1996).

The contact metamorphism has resulted in formation of the schists, marbles and quartzites in the region. In these circumstances, hornblende hornfels and pyroxene hornfels facies have been identified.

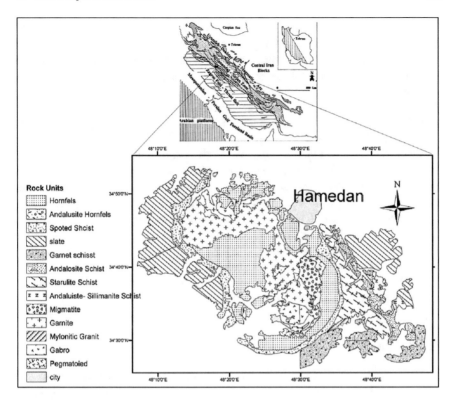

Fig. 8.9 The geological map of Hamedan area (Izadi Kian et al. 2014)

Also, this series of metamorphism was influenced by the dynamic metamorphism. Based on the occurrence of the metamorphic structures the dynamic metamorphism can be divided into three plastic-mylonite, brittle-plastic and cataclastic. In some areas, the metamorphism formed with two different natures (Abdi 1996).

Types of Metamorphic Rocks in the Qorveh Area
Following metamorphic rocks have been recognized in the Qorveh area:

Slate and phyllite: The main part of metamorphic rocks is composed of slates, phyllites and schists which are classified as pelitic and semi-pelitic rocks. The high amount of K_2O, Al_2O_3 and MnO as well as the variable amounts of Fe_2O_3, FeO, Mgo, and CaO (Barker 1990; Hyndman 1985) are the geochemical characteristics of these rocks. According to petrographic studies, the protoliths of the pelitic and semi-pelitic rocks are shale, siltstone, limestone and greywacke that in some cases the amount of CaO increases (calcareous clays) close to the CaO part of the diagram (for instance in the Amirabad garnet-schist).

Quartzite sandstone rock group: The psammite and quartzite rocks fall into the sandstone group. Residual plagioclase, with polysynthetic twins as well as variable amounts of quartz and mica indicate the origin of the rocks to be arkose, arkose-wacke and feldspathic Greywacke, for the psammitic rocks (Sheikh Zakraei 2008).

Calc-silicate Rocks (regardless the para-amphibole): These rocks consist of different types of amphibole schist. The protolith of amphibolites and para-amphiboles is the sequence of calc-silicate rocks. The calc-silicate rocks are found as interlayers of the main sequence and abundant amounts of amphibole has been identified in their composition. In addition, all of the rocks are composed of quartz and graphite (Sheikh Zakraei 2008).

Amphibolites: The Qorveh amphibolites are mainly exposed in the Garmkhani, Garmkhany-Sanduqabad, northern Polusarkan, northwestern Meimanatabad, southern Ghaleh, around Shirvaneh and Tazehabad areas. Generally, the amphibolites have two completely different origins: igneous (Ortho) and sedimentary (Para).

Granite and Aplitic Gneiss: Regarding the field evidence, mineralogy, texture and matrix of Qods Military Base and Sanginabad gneisses, and the aplites, the Jamal and Shanoureh-Moshirabad igneous complex are of igneous origins. The Sarghol, Polusarkan, Minabad and Sartipabad gneisses are of sedimentary origin.

Marble: These rocks generally consist of carbonate minerals and are found where regional and contact metamorphism occurred.

8.2.1.8 Boroujerd

Two types of the metamorphic rocks can be found in the Boroujerd region. The first type is the widely-extended low-grade metamorphic rocks which include metapelite, slate and phyllite well known as the Hamedan slates and phyllites showing a poor bedding. They are metamorphosed in a range of the prehnite-pumpellyite to the beginning of the green-schist facies. The second group is composed of contact metamorphic rocks, medium to high in grade, that are formed around the Boroujerd granitoid (Ahmadi Khalaji 2006). They are generally consisting of spotted schist and hornfels that occur at albite-epidote hornfels, hornblende-hornfels and pyroxene-hornfels facies. The pneumatolitic fluids mostly derived from the intrusions have caused the alteration, metasomatism and retrograde metamorphism which finally resulted in the formation of the low pressure and temperature minerals such as sericite, chlorite and epidote along with andalusite, biotite and plagioclase, respectively. The protolith of the mentioned rocks is determined by geochemical analysis and petrographic studies which is generally consists of shale, marl and greywacke.

According to the indicator minerals such as the biotite, andalusite, cordierite and sillimanite, their zones have been identified in the area.

Southeastern Sanandaj-Sirjan Zone

8.2.1.9 Esfandagheh-Dolatabad

In southeastern Sanandaj-Sirjan zone, Esfandagheh-Dolatabad area, two zones can be distinguished:

1. In Abshor-Sargaz zone, Precambrian-Paleozoic? in age, one major and two minor orogenic phases have occurred.

   - The early Cimmerian orogeny,
   - The late Cimmerian orogeny and
   - The Alpine orogeny.

   The metamorphism of the region is high grade, Barrovian in type, formed during the early Cimmerian orogeny.

2. The colored mélange zone is associated with mobile oceanic subduction zones. Four phases of deformation after the Late Cretaceous have affected this zone. The last phase also has been formed during the main Zagros orogeny. The glaucophane schist probably has been formed during the first phase of deformation by metamorphism of the Colored Mélange sediments (Sabzehei 1974).

The metamorphic phases affected the region and its grades include:

The first phase is the low-grade Barrovian metamorphism, amphibolite facies. Its temperature is estimated to be between 500 and 550 °C accompanied by moderate to high pressures.

The second phase is also Barrovian in type. At the end of the second phase a retrograde metamorphism took place at a temperature between 350 and 400 °C and a pressure of 4 to 5 kilo bars along with an increase in CO_2 pressure. In general, the metamorphism is of the regional metamorphism type which occurred after the Permian and before Early Jurassic (Sabzehei and Berberian 1972).

8.2.2 Central Iran

In fact, the basement of Central Iran has been made of the Neoproterozoic rock. The Precambrian rock units in Central Iran, same as those of the other areas, have exposed where the paleo-highs were formed. Metamorphic rocks and metamorphic complexes in central Iran consist of Precambrian metamorphic formations and sometimes Phanerozoic formations have been metamorphosed. Some particular metamorphosed complexes and formations are described as follows (Fig. 8.10).

8.2.2.1 Chapedony Complex

It is one of the oldest Iranian regional metamorphism. The lower or deeper parts of the complex, in some cases, show anatexis texture. Because of the intensity of metamorphism, the sediments are migmatized and then by increasing the temperature, pressure and melting, the anatexis granite or diorite have been formed. In Iran, the orogeny which have caused the folding of this complex is named Chapedonian Orogeny (Haghipour 1974). The upper boundary of this complex with the upper younger complex called Bone-Shuro is not specified, but it can be likely introduced

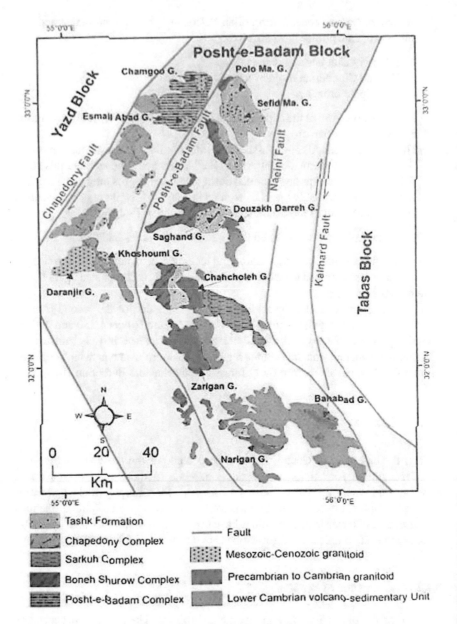

Fig. 8.10 The simplified geological map of Posht-e Badam micro-plate (modified after Haghipour 1977)

as an unconformity. The Chapedoni Complex with a thickness of 4000 m is the oldest known rocks of the central Iran (Haghipour 1974).

There are some ideas that metamorphism of the Chapedoni complex has been occurred during the younger periods (Houshmandzadeh Pers. comm.). However, their protoliths belong to the Precambrian or Neoproterozoic rocks. It seems that the metamorphism has occurred in Late Precambrian.

Apparently, the Chapedoni rocks were affected by two phases of metamorphism. The regional metamorphism is core-complex in type, Late Precambrian in age, which has resulted in the ruining of the evidence that belongs to the older metamorphic phase.

8.2.2.2 Tashk Formation

The Tashk Formation includes Tashk I and Tashk II (Ghorbani 2019). Tashk I which is composed of fine-grained greywacke, sandstone, clay and shale is metamorphosed to slate and phyllite and rarely schist and Tashk II is similar to the Tashk I. Moreover, the thickness of shaly parts of Tashk II is more than Tashk I. However, its metamorphic grade is lower than the Tashk I. So, it would be better to call all the sequence as the Tashk 'Series' and do not divide it into two formations. After the deposition of the Tashk II formation, a major folding phase associated with two major phases of metamorphism occurred in Precambrian. The first phase is Barrovian in type (high temperature and moderate pressure), and the second one is Abukuma in type (very high temperature and low pressure; Haghipour 1974).

The present author has not identified the Abukuma metamorphism type in the Tashk 'Series'. The overall evidence indicates that the Tashk I and II, which have mainly sedimentary protoliths, are folded and metamorphosed in the last phase of Pan-African orogeny. The mentioned Abukuma type can be found at Central Iran, nevertheless, only their protolith belongs to the Tashk Formation. However, the regional metamorphism that occurs in the series of Tashk is formed in late Precambrian. The overlying Precambrian Rizu and Kushk 'Series' are not metamorphosed and overlie the Tashk 'Series'.

8.2.2.3 Posht-e Badam Complex

The rock sequence of Posht-e Badam complex consist of the alternation of marble, meta-basalt and minor metamorphic quartzclastic sediments; at the lower part of the sequence pyroxenite, serpentinite and gneiss can be observed. The age of metamorphic rocks is Precambrian to Paleozoic and the grade of metamorphism is medium (Houshmandzadeh et al. 1990). The metamorphic rocks of the Posht-e Badam Complex have Precambrian protoliths which are probably equivalent to the Tashk 'Series' but metamorphosed at higher facies. In fact, the Paleozoic rocks near the Posht-e Badam Complex are not metamorphosed. What mentioned about the metamorphic rocks of Chapedoni can also be generalized to the Posht-e Badam Complex.

Posht-e Badam Meta-Ophiolites

Posht-e Badam meta-ophiolites are located in NE of Yazd province, SW of Robat-e Posht-e Badam village and near to the Mazrae' Ismailabad village. It includes meta-peridotite, meta-gabrro, amphibolite, listvenite and rodingite. These units are covered by the sedimentary rocks which have been metamorphosed to schist and marble. The listvenite is also traceable along the faults with the general trend of the Posht-e Badam ophiolite (north-south). The small and large bodies of amphibolite form the main parts of rock units. The Ismailabad granite has cut the Posht-e Badam ophiolite where both of them can be observed together in various parts of the area. The Ismailabad granite is overlain by the Cretaceous limestone.

The age of Posht-e Badam ophiolite is attributed to the Palaeozoic, although the age of metamorphism is much younger compared to the age of ophiolitic body. Anyway, the age of initial ophiolite can be correlated to the Anarak and Takab ophiolites; i.e. late Precambrian. The metamorphism age can be attributed to Late Precambrian (Ghorbani 2013).

8.2.2.4 Bone-Shuro Complex

The thickness of Bone-Shuro complex is approximately 1000 m and is composed of gneiss, schist and amphibolite. The protolith of this series is similar to that of Chapedoni and Posht-e Badam formed in a sinking shallow basin. Two important phases of metamorphism have been occurred in the Central Iran (including Abukuma and Barrovian phases) that are distinguishable in the Bone-Shuro complex at amphibolite facies. The lower boundary of the series was not detectable, but the upper boundary is separated by a series of dolomite unit that is metamorphosed to marble (Haghipour and Pelissier 1968). Most likely the Bone-Shuro complex has been metamorphosed during the last stage of the Pan African orogeny. Ramezani and Tucker (2003) have determined the age of amphibolite metamorphism 547 Ma. using U-Pb dating method. The geochemical studies of Farzami et al. (2016) revealed that the chemical composition of Precambrian metamorphic rocks of Bone-Shuro and Tashk areas is mostly within the tholeiitic and calc-alkaline series. The investigations come out from Hf isotopes show that the δHf ratio is positive changing from 84.2 to 94.6. These results propose that the granitic magma is originated from the partial melting of mantle. The U-Pb dating on zircon by the use of SIMS method showed the 560 to520 Ma. time interval that is in agreement with the geological observations. It can be suggested that some parts of the geological events of area have been affected by Cadomian orogeny.

8.2.2.5 Jandaq Meta-Ophiolites

The metamorphic rocks of Jandaq are located at northeastern part of Isfahan province close to Khur and Biabanak county. The metamorphic rocks of area can be divided into two meta-ophiolite and meta-pelite groups. The Meta-ophiolites of Jandaq are among the oldest ophiolite exposures of Iran which is attributed to the Proterozoic and Paleozoic and has suffered numerous metamorphism phases. The ophiolite consists of serpentinite, meta-peridotite, serpentinized meta-peridotite, meta-gabrro, amphibolite, meta-basic dikes, listvenite and rodingite where it is covered by Paleozoic schists and marbles. All the sequences have been cut by the granitic bodies (Torabi 2012).

The $^{40}Ar/^{39}Ar$ dating of hornblendes in amphibolites indicates 157 ± 33 Ma. for the ophiolite complex (Middle Jurassic; Bagheri 2007). A considerable thickness of metamorphic rocks such as schist and marble covers the ophiolitic series. The age determination of muscovites in micaschists using Argon-Argon technique signifies 164 ± 2 Ma. (Middle Jurassic; Bagheri 2007). The last mentioned age is the age of metamorphism, not the age of protolith.

The above ages show the effect of the Middle Cimmerian orogeny, resulted in the metamorphism of shale to schist and basalt to amphibolite. All of the ophiolitic rocks have exprinced different phases of metamorphism. Consequently, the Russian geologists of the Technoexport project have named the Jandaq ophiolite to as meta-ophiolite.

The serpentinized peridotite of Jandaq ophiolite is an identified and developed event. The Jandaq ophiolite complex, during the formation, emplacing and orogenic phases, undergone several static and dynamic serpentinization phases. According to the present author, probably the age of protolith of the Jandaq ophiolite is similar to the age of those of Takab and Anarak ophiolites. Although, the absolute age determination shows Jurassic, Late Precambrian can be suggested due to regional evidence.

The studies performed on Jandaq meta-pelitic rocks (Tabatabaei Manesh and Sharifi 2011; Rajabi and Torabi 2012; Fereidouni et al. 2010, Torabi 2012, 2017; Baluchi et al. 2019) show that these rocks based on their mineralogical content are of members of quartz muscovite schist, quartz muscovite biotite schist, garnet muscovite chlorite schist and garnet staurolite schist groups.

The garnet muscovite chlorite schists of Jandaq show the first occurrence of garnet grains which are mainly 58–76% almandine, 1–18% spessartine and 8–20% grossular. The Mg content increases from the central to marginal parts of zoned garnets from meta-pelites of Jandaq area. This evidence shows that they have been formed during a prograde metamorphism and rapid cooling without being affected by the next metamorphic phases. The occurrence of staurolite in garnet muscovite staurolite schist is the entering gate of amphibolite facies. The thermobarometric studies showed that the meta-pelites are formed in 400–670 °C and 2–6.5 kbars. These results are in agreement with mineral paragenesis and it can be concluded that the pelitic sediments of Jandaq area are metamorphosed in green schist and amphibolite facies.

8.2.2.6 Bayazeh Meta-Ophiolites

The peridotites (mantle-derived) of Bayazeh ophiolite complex are metamorphosed and fully serpentinized. The water absorption and metamorphism processes leading to the conversion of the mantle derived peridotites to the metamorphic peridotites as follows:

– Serpentinite: By developing the serpentinization process, the essential minerals of primary peridotites in ultramafic ophiolite of Bayazeh, has fully changed to serpentine group minerals, and only chrome-spinel mineral is remained as a primary igneous mineral in these rocks (Torabi 2012).
– Listvenite: The listvenite of Bayazeh area is exposed next to the serpentinite rocks. The co-occurrence of these two rock types could be related to the hydrothermal alteration of serpentinized peridotites and serpentines of the ophiolite series. The listvenite (after forming) has been affected by regional metamorphism, as a result the meta-listvenite has been formed by hydrothermal alteration of serpentinized peridotites and serpentinite in the region (Torabi 2012).

According to Torabi (2012) based on the field studies, the meta-listvenites are exposed along the shear zones that show a rough morphology. The meta-listvenites are of two different colors, which are related to the mineralogy of the rock. According to the mineralogy of the rocks, they can be divided into two groups; carbonate meta-lisvenite and carbonate meta-lisvenite resulted from the metamorphism of carbonate serpentanite. The carbonate meta-lisvenite is cream to dark yellow in color. The carbonate serpentanite rocks have a cream color to dark yellow. They have stratiform structure and are found close to the serpentinite occurrences.

The meta-lisvenite resulted from the metamorphism of carbonate serpentanite is dark yellow, brown and green in color and can mostly be observed as massive bodies. Needle-shaped amphibole crystals are observed within the meta-lisvenite in hand sample. The carbonate meta-lisvenite is the dominant type. During the lisvenitization phenomenon, the released SiO_2 will form the amorphous (chalcedony) silica veins in hand sample.

– Meta-gabbro: The meta-gabbro in Bayazeh area is a part of the ophiolite sequence. The rocks are melanocratic and have a massive and dense structure, mostly medium- to coarse-grained in hand samples. Meta-gabbro occurrences of Bayazeh ophiolite are widely altered and metamorphosed.
– Meta-picrite: Meta-picrites form part of Bayazeh ophiolite sequence. Field studies indicate that these rocks are of limited extent outcropped in the eastern part of the area. The texture of the rock is massive, dense and dark green in hand sample (Torabi 2012).
– Metamorphic ultrabasic dikes: Field studies of the dikes signify a limited volume, dense texture, and green to yellowish color in hand sample. These rocks contain some minerals such as pumpellyite, garnet, chlorite, titanite and magnetite (Torabi 2012).

The metamorphism age of the Bayazeh meta-ophiolites is probably similar to that of Jandaq area and has likely been endured several metamorphism phases. The last metamorphism has been occurred during the Middle Jurassic, but likelihood this meta-ophiolitic sequence have endured different phases. The protolith of meta-ophiolites is similar to those of Anarak and Takab areas.

8.2.2.7 Sarkuh

The Sarkuh metamorphic complex located in the Saqand area, close to the Chadormalu iron mine, can be divided into three zones as follows:

1. Garnet, andalusite zone,
2. Garnet, staurolite zone and
3. Garnet, sillimanite zone.

The Sarkuh protoliths are mainly sedimentary in origin (pelitic, greywacke and sandstone) metamorphosed in the amphibolite facies. The original rocks of Sarkuh belong to Early Precambrian. The metamorphism of Sarkuh is equivalent to the Posht-e badam and Bone-Shuro complexes, although some researchers have attributed Sarkuh metamorphism to Early Cimmerian (Houshmandzadeh pers. comm.). Some other researchers, due to the occurrence of non-metamorphosed rocks of Rizu series, have attributed it to 180 Ma. (Late Precambrian). Nevertheless, the geological evidence indicates a probably old metamorphic phase which had been already occurred. Although the phenomenon of the new core complex metamorphism has causes disappearing of the primary metamorphic evidence. Generally, the Sarkuh complex is of kyanite, sillimanite, garnet, andalusite-bearing mica-schists occassionaly interbedded with marble and scapolite.

8.2.2.8 Northwestern Mashhad

The main part of the northwestern Mashhad metamorphic rocks can be observed in Dehno area. Dehno is located within the Binaloud structural zone. In Mashhad region, Torbat-e Jam Plain, a wide range of intrusive and metamorphic rocks is exposed of which the metamorphic rocks are dominant. Most of the intrusive rocks are related to the early Mesozoic or the Late Paleozoic (see the table of intrusive rocks of Chap. 1).

The metamorphism evidence such as the low pressure and high temperature facies indicates that the intrusive bodies have been involved in the metamorphic event. On the other hand, there is not too much differences between the age of metamorphism and age of intrusion emplacing. Actually they have been simultaneously formed (late Paleozoic to early Mesozoic). Majidi (1978) introduced three different phases of metamorphism including contact, dynamic, and regional around Mashhad as follows:

1. The regional metamorphism which causes the formation of slate, phyllite, schist, marble and quartzite,

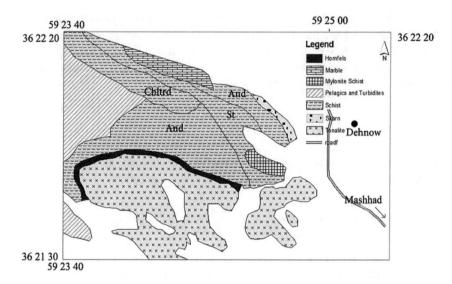

Fig. 8.11 Geological Map of the main structural units of northeastern Iran (modified after Alavi 1991)

2. Dynamic metamorphism to form mylonite and cataclastic rocks, and
3. The contact metamorphism due to the influence of the granitoids, resulted in the contact metamorphic rocks (hornfels) and skarn.

Generally, the metamorphic rocks of the whole region can be divided into two groups:

1. The regional metamorphic rocks, including schists (garnet schist and chloritoid garnet schist) and
2. The contact metamorphic rocks, including hornfelses (Fig. 8.11).

8.2.2.9   Khajehmorad (Mashhad)

The Khajehmorad area is a part of Binaloud sub-zone in NE Iran. The rocks of this region constitute a complex of igneous, metamorphic and sedimentary rocks. The igneous and metamorphic complexes are related to the Paleo-Tethys (Tethys II) closure in Iran. The metamorphic and igneous complex in Khajehmorad, Nurabad (Virani), Shandiz, Vakilabad and Khalaj include the following rocks (Aghanabati and Shahrabi 1987):

Meta-ophiolite: It consists of schist (volcanics of the ophiolitic sequence as protolith) and hornblende-gabbro (the gabbro of the ophiolitic sequence as protolith). The protolith of the rocks dates back to the Permian which is approximately 287–281 Ma. by Ar^{40}-Ar^{39} method (Ghazi et al. 2001). Probably the metamorphism age is Triassic in Mashhad area, like other metamorphic rocks of the region formed during the closure of the Paleo-Tethys.

– Meta-sedimentary rocks: These rocks include slate, quartzite, phyllite and marble resulted from the metamorphism of the primary sedimentary rocks associated to the ophiolitic series. The metamorphic characteristics of these rocks show a low pressure and medium temperature regional metamorphism. Karimpour et al. (2011) classified the meta-sedimentary rocks into two groups such as meta-phyllite and meta-sediments (Fig. 8.12).

The Khajehmorad intrusive body has intruded into the meta-sedimentary rocks. Based on the researches performed by Karimpour et al. (2011) the zircon's age has been detected to be about 205 ± 4.1 Ma. based on the U-Pb method.

It seems that the age of all metamorphic rocks with different protoliths is similar to the age of the Khajehmorad granitoid resulting from the closure of Paleo-Tethys (Tethys II).

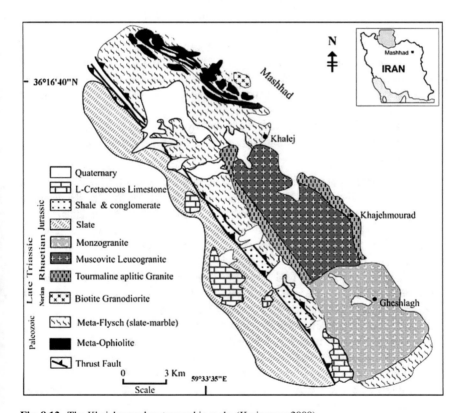

Fig. 8.12 The Khajehmorad metamorphic rocks (Karimpour 2009)

8.2.2.10   Torud

Two types of metamorphism can be found in the Torud area. The first in the middle parts of the zone that is metamorphosed up to the green-schist facies and is related to the early Cimmerian movements; and the second in the eastern zone that in terms of lithology and alteration intensity is completely different from the first one.

The Precambrian? rocks are of a higher metamorphism grade than the those of the Paleozoic to Mesozoic and include schist, gneiss and amphibolite. These rocks are exposed in the NE Torud.

The grade of metamorphic sediments of Paleozoic and Mesozoic are weak and do not exceed the green schist facies, although depending on the tectonic situation, the intensity and weakness of metamorphism varies.

Metamorphic rocks in Chah-e Shirin and Reshm mountains include quartz-schist, calc-schist, quartzite, metamorphosed sandstone, marble, and medium to basic metamorphosed igneous rocks (amphibolite).

The Reshm-Chah-e Shirin metamorphic rocks are observed in northwestern Moaleman, especially the Reshm valley and Chah-e Shirin that the metamorphism age and their protolith age is not well defined. Hereby, the age of rocks is discussed and then the opinions of the present author which are based on his field observations will be mentioned.

Generally, according to the studies performed by the authors of 1:250,000 geological map of Torud, the metamorphic rocks of the Chah-e Shirin-Reshm Mountains are as below:

– Quartz schist: they are mostly found as interbedded layers within Niur and Padeha formations showing gray to green colors. The schistosity is defined by the cleavage surfaces of muscovite and chlorite that are interrupted by quartz crystals.
– Calc-schist: the limy-clayey green thin-bedded rocks of this group are found in Niur, Padeha and Sibzar formations as well as the basal part of Permian-Triassic sedimentary rocks.
– Quartzite and metamorphosed quartzitic sandstone: The majority of this group can be found in Niur, Padeha, Sardar and those sediments of Lias. The observed colors of these rocks include gray, green, black and brown.
– Marble: They are found within the Niur Fm. Showing a dirty beige color. Usually they can be observed in Padeha Fm. with brown color. They are of white or gray colors and granular fabric in Sibzar and Bahram formations. The marble units of Permian-Triassic show different ranges of grain size and colors.
– The intermediate to basic igneous rocks (amphibolite): in the areas confined between Chah-e Shirin and Reshm (mostly the western flank of Kalleh-Kajou mount granitic to grano-dioritic stocks are exposed that increase the metamorphism grade to higher grades where coars grains of muscovite, chloritoid and biotite are formed occassionaly show pegmatitic fabric. These stocks are most probably the evidence of bigger intrusions responsible for the increasement of geothermal gradient resulted in vast metamorphism of higher grades. They can also affect the thermo-dynamic phenomena of area.

The present author has visited and sampled the Niur, Padeha, Bahram and Jamal formations of Reshm valley and southern Damghan (Qusheh, Iravan and some other localities) during the Paleozoic project (supervised by NIOC). The geological evidence shows that none of the mentioned formations is metamorphosed. According to the regional geology of area and the metamorphism age of rocks of Shotor Kuh, all metamorphic rocks of Reshm-Chah Shirin (as described above) belong to Late Precambrian and the mentioned formations (Niur, Padeha, Bahram and Jamal) are not metamorphosed. The protolith of all metamorphic rocks of area had sedimentary to volcano-sedimentary origin.

The Kuh Kaftari (Shotor Kuh) metamorphic complex: Shekari et al. (2017) and Hame Rezaei et al. (2015) have mentioned that NE-SW the Kuh Kaftari (Shotor Kuh) metamorphic complex (22 sq km in 1:100,000 geological map of Rezveh) is located at 80 km of shouthern Shahrud county, northeastern Torud, Rezveh area, northern Sahl village (Fig. 8.13).

Studies show that the Kuh Kaftari complex is composed of gneiss, amphibolite, micaschist, phyllite and slate. The mineralogical studies indicate the intermediate green schist to intermediate amphibolite metamorphic faceis. Based on the mentioned studies and field observations, the protolith of these rocks were composed of an intercalation of shale and carbonate deposits occassionaly accompanied by sandstone and conglomerate. The amphibolite can be the originated from the primary meta-basic rocks of area.

The main rock units of area include gneiss, schist and amphibolite where the micaschists encompass biotite schist and garnet schist. The micaschists are mainly composed of quartz, alkaline feldspar, plagioclase, biotite, garnet and occassionaly muscovite. They are formed from the metamorphism of pellitic rocks up to intermediate amphibolite facies.

The Shotor Kuh metamorphic-igneous complex is unique due to its huge basic rocks that are metamorphosed up to forming garnet amphibolite and even advanced basic migmatites.

The metamorphic zones include chlorite, biotite, garnet, staurolite, kyanite and sillimanite.

Based on the mineral paragenesis of metamorphic rocks of area and their deformation mechanism it can be mentioned that the metamorphism grade is up to intermediate green schist and intermediate amphibolite. The microscopic investigations show that the studied rocks of area have affected by a regional deformation and metamorphism in a P-T window of 400–900 °C and 208 kbars.

Thermo-barometric studied performed on garnet, amphibole and plagioclase of meta-basite showed 602–711 °C and 9–11 kbars for the formation, exchange and the final equilibrium indicating the amphibolite and upper amphibolite facies.

Fig. 8.13 Metamorphic units of Shotorkuh (taken by Zadsaleh 2019)

According to the U-Pb dating performed on zircon mineral of meta-basite unit, the metamorphism age is assigned to 526–577 Ma. (correlated to the final stages of Neoproterozoic and Cadomian orogeny in Gondwanian territories).

The Metamorphic Rocks of Biarjomand (Delbar)

The metamorphic rocks of Biarjomand (Delbar) are located at 130 km of southeastern Shahrud where they include schist, micaschist, gneiss, marble and meta-basite (Balaghi et al. 2015).

This complex is of sedimentary protolith such as pellitic sandstone, carbonate rocks as well as igneous rocks including basic dikes, volcanic rocks and small intrusions all metamorphosed up to green schist to amphibolite faicies.

According to studies on different calibrations of garnet-biotite for acquiring the temperature windows and estimated pressures performed by Balaghi et al. (2015), Ghasemi and Asyabasha (2006) and Malekpour et al. (2005), the mica schists and garnet-bearing gneisses have formed in 467–498 and 640–706 °C respectively and 4, 2–6, 7, 4, 2–8 and 13 kbars of pressure (green schist and upper amphibolite) facies.

The temperature and pressure increase in formation of the mentioned rocks indicate a progressive metamorphism continued up to mekting point and partial melting. The resulted felsic magma has caused varied migmatitic fabrics in meta-pellitic rocks now observed as granite and anatexic leucogranite.

The U-Pb dating performed on gneiss and granite of this complex showed 541–547 Ma. (final stages of Neoproterozoic-Early Cambrian) and is most probably related to Pan-African orogeny.

According to Balaghi et al. (2015), Ghasemi and Asyabasha (2006) and Malekpour et al. (2005) the formation of mylonitic foliation and extensional lineament of metamorphic rocks of Biarjomand area is a strong evidence of their formation in a shear zone. So, based on the mineralogical composition and observed micro-structures it can be noted that these rocks have been metamorphosed in a P-T condition of intermediate amphibolite sub-facies. According to the micro-structires and the mineral composition the 650–700 °C of temperature is assumed for their formation that points to an active metamorphism up to 20–23 km of depth in crust affecting the median crust. It indicates an important extensional environment in median crust of Central Iran at most probably Late Cretaceous-Paleocene.

8.2.2.11 Takab

The metamorphic rocks are widely exposed in the Takab area (Fig. 8.14). Most of rocks located in the C and I zones of the Takab region (especially those of late Precambrian-early Cambrian) excluding the Cenozoic formations are metamorphosed. The Grade of metamorphism of Takab which are shown in the Takab Quadrangle Map (1:100,000) varies from low to high grades.

Several researchers like Pelissier and Bolourchi (1967), Alavi-Naini et al. (1982a, b), Ghazanfari (1991) and Ghorbani (1999) have studied the regional metamorphic

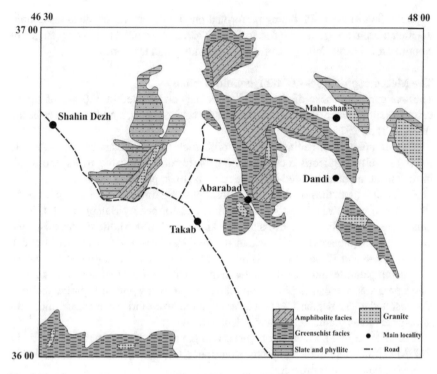

Fig. 8.14 Metamorphic rocks scattered in the Takab area (Ghorbani 1999)

rocks of the Takab. The metamorphic rocks in the Geological Map of the Takab are divided into five complexes as below:

1. The Sorsat mountain complex in eastern Shahindezh includes mica schist, gneiss, granitic gneiss migmatites and small granite bodies.
2. Mahneshan Complex in NE Takab includes Aghkand gneiss, Poshtkuh schist and white marbles of Almalu.
3. Kheyrabad Complex consists of the following units:

   (a) Daveh Yataghi Gneiss including gneiss, almandine-gneiss with schist interlayers.
   (b) Alamkandi Amphibolite including amphibolite with interlayers of gneiss and marble.
   (c) Anguran Schist including a variety of schists, gneiss, marble and Meta-volcanics.

4. Amirabad Complex in southern Dandi, south and SE Anguran Mine consist of mica-schist, amphibolite schist, marble, gneiss and migmatite, which Pelissier and Bolourchi (1967) classified their metamorphism grade as green-schist facies.

5. The metamorphic rocks in the southwestern basin of the Takab including amphibolite, gneiss, mica-schist and marble which have been assigned to the Permian and older by Alavi-Naini et al. (1982b). This unit has different age and facies compared to the abovementioned metamorphic complexes.

Ghazanfari (1991) evaluated in detail the metamorphic rocks of NE Takab, so he classified the metamorphic and igneous rocks in three categories.

1. A metamorphic complex consisting of low, medium and high grades (according to the classification of Winkler 1979), and green-schist to amphibolite facies (according to Turner 1981).
2. The low to very low grade metamorphic Kahar Formation (Winkler 1979) and low-grade green-schist facies (Turner 1981).
3. Granodiorite intrusions with unknown age that have been attributed to the Precambrian by Alavi-Naini et al. (1982a, b). Pelissier and Bolourchi (1967) and Alavi-Naini et al. (1982a), based on field observations and petrographic studies believe that the higher grades of metamorphism mean the older rock and vice versa.

Studies of Ghazanfari (1991) is based on the realistic data, such as petrological, petrographic, geochemical and field evidence and is more useful in this regard. The studies have focused in Anguran area and Belghais Mountain.

Generally, the metamorphic and igneous rocks that are merged with metamorphics can be classified in three groups based on the Takab Geological Map (1:100,000) as follows:

1. The giant set of metamorphic rocks of Takab is the main metamorphic sequence at the middle part of the Takab area that includes metamorphic complexes of Mahneshan, Amirabad, Kheyrabad and Sorsat Mountain areas. All abovementioned rocks are metamorphosed at Green-schist and amphibolite facies; they include many types of rocks, such as schist, amphibolite, gneiss, migmatite and anatexis granites. Regarding their geological situation within Takab area and petrographic rock's composition, they can be divided into two categories:

   – The large complex of metamorphic rocks at northeastern Takab that includes all metamorphic rocks of southeast of Dandi, Amirabad, Belghais mountain, Alamkandi, northern Kheirabad and Mahneshan areas. The complex is consisting of different rock types including the metamorphosed ultramafic, mafic, mafic and acidic volcanic rocks along with tuffs, carbonate and pelitic rocks (Ahmadi Rohani 1999).
   – The NW Takab complex (the Sorsat Complex) includes all metamorphic rocks of Zeydkandi, Khazaei, Khan Qoli, Ouch Darreh and the Maein Bolagh. The lithological variation of this complex due to the lack of diversity of source rock is not worth mentioning. Most of the protoliths are pelitic in type which sometimes have been associated with volcanic and carbonate rocks.

The difference between these two groups of rocks is not based on the grade of metamorphism, but is related to the composition and diversity of the protoliths. Therefore, the given name is the "Takab Giant Metamorphic Complex".

2. The rocks with weak metamorphic grade that are mostly at the lower parts of the green-schist facies. The main protolith of these rocks are tuffaceous carbonate and clastic rocks named Kahar, Soltanieh and Barut formations. These rocks are outcropped in eastern and western parts of the Takab Quadrangle. The grade of metamorphism decreases by getting away from the middle part of the Takab Quadrangle. So, they can be categorized as non-metamorphic rocks in the marginal area. Similar to these metamorphic rocks can be found in the Kahar formation of the Alborz zone.

3. Metamorphic rocks of southwestern Takab Map which have been attributed to the Permian and older by Alavi-Naini et al. (1982b) and include schist, marble, limestone, recrystallized limestone, meta-volcanic rocks and spilite. But, according to the studies conducted in other parts of the Sanandaj-Sirjan zone, these rocks are equal to the Paleozoic volcano-sedimentary rocks in other areas of the later zone. Therefore, they belong to the Paleozoic to Early Triassic and are likely metamorphosed in the Early Cimmerian phase. They show similarities to metamorphic types of northern Sanandaj-Sirjan zone and are similar to Divandareh and Qorveh metamorphic rocks; in fact, they are different in terms of metamorphic facies and age compared to the Takab metamorphic complex (Fig. 8.15).

8.2.2.12 Anarak-Khur

The metamorphic rocks of the Anarak-Khur area are same as the Takab area metamorphosed during the Late Proterozoic. They can be categorized as follows.

For the first time, Davoudzadeh et al. (1969) called the metamorphic rocks of the area the "Anarak Metamorphic Rocks". Afterwards, the Russian geologists (Techno Export, Co) have widely applied the term, for the metamorphic rocks of the Anarak and Khur areas. According to studies conducted by Russian geologists the Anarak metamorphic rocks can be divided into five complexes in Anarak Geological Map, 1:250,000 as follows (Table 8.1; Fig. 8.16):

1. Chah Gorbeh,
2. Morghab,
3. Patyar,
4. Mohammadabad and
5. Doshakh Complexes.

Also in the Khur area equivalent rocks (Aistov et al. 1984) to the Anarak metamorphic complexes include (Table 8.2):

1. Chah Gorbeh,
2. Patyar,
3. Kabudan,

Fig. 8.15 The relationship between metamorphic and igneous rocks in Mahneshan, Takab and Shahin Dej regions (A to F) (Ghorbani 1999). **a** Metamorphic rocks of the Mahneshan area, along the Incheh-Aghkand road. **b** Metamorphic rocks of the Mahneshan area, along the Kheyrabad-Qelichkhan road. **c** Metamorphic rocks of the Mahneshan area, along the Pari-Alamkandi road. **d** Metamorphic rocks of the Khazaei area, Shahin-Dezh. **e** Metamorphic rocks of the Zeydkandi area, along the Shahin Dezh-Takabroad. **f** Metamorphic rocks of the Shakh-Shakh area, Takab

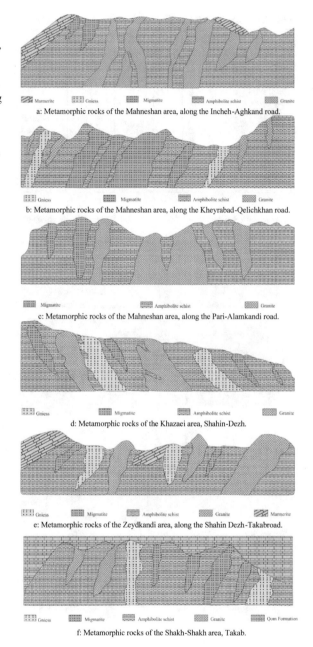

a: Metamorphic rocks of the Mahneshan area, along the Incheh-Aghkand road.

b: Metamorphic rocks of the Mahneshan area, along the Kheyrabad-Qelichkhan road.

c: Metamorphic rocks of the Mahneshan area, along the Pari-Alamkandi road.

d: Metamorphic rocks of the Khazaei area, Shahin-Dezh.

e: Metamorphic rocks of the Zeydkandi area, along the Shahin Dezh-Takabroad.

f: Metamorphic rocks of the Shakh-Shakh area, Takab.

Table 8.1 Anarak metamorphic rocks modified after Sharkovski et al. (1984)

Anarak metamorphic rocks	Lakh Marble	Doshakh complex	600 m	1000 m	Dolomite, marble
	Patyar Complex		1800 2000		chlorite-muscovite schist, chlorite-actinolite-epidote schist, quartzite, marble and dolomite
	Morghab Complex	Mohammadabad Complex	3100 3800	4100 m	Muscovite , muscovite-chlorite, and muscovite- epidote-chlorite schist, quartzite and marble lenses
	Chah Gorbeh Complex		1600 2100		Muscovite- chlorite, epidote-chlorite, epidote-actinolite, muscovite-carbonate schist, and marble
Pol-e Khavand			950 m		Muscovite granite, granite gneiss, amphibolite, mica gneiss and mica schist

(Right vertical label: Upper Proterozoic)

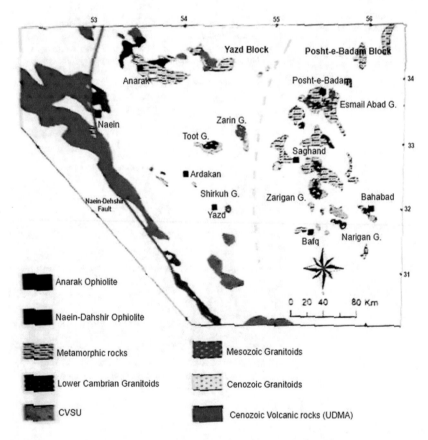

Fig. 8.16 A simplified geological map of a part of central Iran to better understand the position of Anarak ophiolite relative to the Posht Badam block

Table 8.2 Metamorphic rocks in Khur, with modifications (Aistov et al. 1984)

Anarak metamorphic rocks	Patyar complex	Kabudan complex	Doshakh complex	Posht-e Badam complex		Late Proterozoic
					Meta-sandstone, phyllite, dolomite, schist and metamorphosed volcanic rocks	
					Metamorphosed dolomite, limestone and sandstone	
	Chah Gorbeh complex				Schist and marble	
Jandaq complex						
Chapedoni complex					Gneiss-migmatites, anatexis granite and amphibolite	

4. Doshakh and
5. Posht-e Badam complexes.

Thus, the metamorphic rocks in the Anarak and Khur areas are composed of seven complexes. The Chah Gorbeh Complex entirely is equivalent to the Mohammadabad Complex. Also, the Doshakh complex is equivalent to some parts of Patyar Complex.

Kabudan Complex

According to the Russian geologists, the Kabudan complex includes four members; schist, marble, granite and quartzite from bottom to top, respectively.

The schist unit of the complex is equivalent to the upper part of the Chah Gorbeh complex, and the Marble unit of the Kabudan complex is equivalent to the Lakh Marble of the Anarak complex. Also marble-schist and quartzite units are equivalent to the parts of the Anarak and Patyar complexes. So, the Chah Gorbeh complex is the lowest part of the Anarak complex, and in fact, the Kabudan complex is equivalent to the upper parts of the Chah Gorbeh complex and to the Lakh Marble as well as the Patyar complexes of the Anarak metamorphic rocks. The Doshakh complex is generally equivalent to different parts of the abovementioned complexes.

Chah Gorbeh Complex

The Chah Gorbeh Complex consists of 5 members, from bottom to the top, respectively:

1. Se-barz Schist Member,
2. Lower Marble Member,
3. Lower Schist Member,
4. Middle Marble Member and
5. Upper Schist Member.

Se-barz Schist Member: The lowest member of the Chah Gorbeh Complex is mainly composed of light-colored muscovite-schist (very light green). Its age is Upper Proterozoic and has an average thickness of 1000 m.

Lower Marble Member: It is generally including an alternation of the light and dark gray marbles and mica schist. The marbles are characterized by folding structures (slump folds) where often observed as recumbent and parallel folds. This unit (700 m-thick) is attributed to the late Proterozoic.

Lower Schist Member: It is mainly composed of muscovite, chlorite, biotite, epidote and carbonate schists. Moreover, there are some interlayers of quartzite and marble with a thickness which is between 500 and 600 meters and its age is attributed late Proterozoic.

Middle Marble Member: Generally, includes an alternation of dark gray to light marbles and schists. The thickness is 800 meters and an early Cambrian age is considered for this member.

Upper Schist Member: It includes epidote, chlorite, muscovite, biotite and albite schist. Also marble and quartzite can be observed as intercalation. Its thickness, at the middle parts of the Lakh Mountain is about 1500 meters.

Lakh Unit

The Lakh Unit which overlies the Chah Gorbeh complex is well known as a prominent lithological unit in Central Iran. This unit is equivalent to the Anguran Marble and Chaldaq Marble in the Zarshuran Mine, Takab area.

Patyar Complex

The best type section of this complex is exposed in the eastern Patyar abandoned antimony mine. This complex mainly consists of the chlorite schists, epidote schists, and glaucophane schists and mica schists accompanied by quartzite, dolomite and thin layers of chert in the dolomite.

Morghab Complex

The Morghab complex is the uppermost part of the Anarak complex. The best outcrop is in the northern part of the Moala village. This complex is mainly composed of chlorite, epidote and mica schists, relatively abundant quartz veins can be observed. The Morghab complex is located on the most upper part of the Patyar carbonate part.

8.2.3   Northern Iran (Shanderman-Gasht)

8.2.3.1   Northern Iran

The metamorphism of northern Iran is located in the Masal-Shanderman area (55 km of southwestern Bandar-Anzali). The metamorphic complex of Shanderman, consists of a wide range of metamorphic rocks including slate, schist, calc-schist, marble, serpentinite, eclogite? And meta-gabbro.

Clark et al. (1975) attribute its age to Precambrian, but Eftekharnezhad et al. (1992) refer these rocks to the Hercynian orogeny. Ghorbani (1999) also consider these metamorphic rocks to late Paleozoic to Triassic.

Based on the studies of Zanchetta et al. (2009), the Shanderman complex is an obducted series of the southern Caucasus region which was replaced during the regional deformation of the Variscan event. According to Zanchetta's idea, the thrust sheet is moved towards the south during the Cimmerian orogeny. As a result, it is translated eastwards during the Permian along a dextral mega shear zone taking from a Pangea-B to a Pangea-A plate configuration. The age of the rocks is early Carboniferous to Late Cretaceous. The Paleozoic rocks have been metamorphosed. A set of mylonitic rocks are exposed in different parts of the Masal-Shanderman metamorphic zone. The dynamic type of metamorphism is approved by the evidence of the mylonitic rocks in some parts of the region, especially in the vicinity of the shear zones. Due to the fact that shears zones accompany the narrow strips of severe deformation, thus the shear zones are formed during the regional metamorphism. Thus, probably by closing the Paleo-Tethys Ocean in the Masal-Shanderman region, a regional metamorphism have occurred and shear zones have been simultaneously formed to develop a dynamic metamorphism in some parts of the metamorphic complex.

A series of metamorphic zones in the pelitic, basic, quartz-feldspar and calc-silicate rocks is detectable. The type of metamorphism is low to high temperature and pressure in the Masal-Shanderman zone.

According to some researchers such as Omrani et al. (2008) and Zanchetta et al. (2009) by applying $^{40}Ar/^{39}Ar$ on mica mineral of metamorphic rocks indicated that the subduction of the Paleo-Tethys Ocean has been started during the Late Carboniferous to Triassic, simultaneously with the development and opening of the Neo-Tethys ocean, where Paleo-Tethys Ocean were fully closed. The age of the metamorphic rocks is actually younger than Carboniferous. So, the subduction cannot be started in Carboniferous. The Paleo-Tethys evidence (Tethys II) shows that the subduction should be started in the Permian.

It seems that during the middle Paleozoic, the continental crust of Alborz began to fracture, develop and several magmatic complexes emplaced. These activities have been extended up to the Middle Devonian period. In Northern Iran, the magmatic activities stopped and the continental shelf extended during the Permian to Early Triassic. The Intrusions bodies cut the continental crust after the closure. The carbonates and shallow clastic sediments covered the intrusive rocks. The continental shelf developed by the collision of the continental lithosphere of Alborz and Turan plates, during the Late Triassic subduction. As a result of the collision of continental plate margin of Turan and passive margin of Alborz (Cimmerian orogeny), the oceanic sequence and continental crust rocks are metamorphosed.

8.2.3.2 Gorgan Schists

Ruttner (1984) reported several Triassic rocks in the northern and southern parts of Aghdarband that much of them consist of a metamorphosed sedimentary sequence of shale, limestone and volcano-sedimentary rocks.

Ruttner (1984) attributed the marble and high-grade metamorphosed limestone to the reversed sedimentary sequence that underlies the Devonian sediments. He believed that this sequence belongs to the Proterozoic. If fact, the metamorphism has been formed during the Precambrian or Caledonian orogeny (There is no metamorphism in the Caledonian). The upper parts of the marbles are massive and are of a light gray color. Thus, the marbles found beneath the Late Devonian-Carboniferous sequence can be placed at the upper part of the Devonian-Carboniferous, in a normal sequence. Also, related to the age of marbles, metamorphism and deformation phases, they cannot be attributed to the Precambrian and Caledonian of the Turan plate. The metamorphism and deformation is a localized phenomenon that the intensity of crystallization of Carboniferous limestone is higher than the Devonian (at southern Kashafrud), and probably a younger metamorphism has severely affected the Early Triassic limestone in eastern Aghdarband.

The Hercynian orogenic deformation and metamorphic phase seems to be a major factor in eastern part along the Torbat-e Jam and Fariman regions during the late Paleozoic to late Early Triassic. The consequence of this movement has caused a hydrothermal circulation, synonymous with emplacement of granite bodies and a dynamic metamorphism. The Rhaetian-Liassic rocks (Shemshak Group) are the youngest unit that has been influenced by this metamorphism in the Binaloud Mountains (eastern Mashhad; Aghanabati and Shahrabi 1987). The major metamorphism is likely occurred during the closure of the Paleo-Tethys (Tethys II) at the end of the Paleozoic to Late Triassic in the Kopet Dagh basement, and generally northeastern parts of Iran.

8.2.3.3   The Azerbaijan (Western Central Iran)

In Azerbaijan area, similar to other localities of Central Iran, two main metamorphic phase can be mentioned.

– Dynamothermal metamorphism of Late Precambrian that can be observed in green schist to amphibolite facies. The main part of these rocks is of green schist facies exposed in Maku, Siah Cheshmeh, northern Urmia, Mahabad and Zanjan (Kahar Fm., Fig. 8.17) areas.
– The metamorphic rocks of Late Triassic to Cretaceous mostly accompanied by ophiolitic complexes occurred simultaneously to the subduction and collision of the third Neo-tethys.

Maku

At northern Maku, a series of metamorphic rock is exposed composed of schist, chlorite schist, slate, phyllite and meta-volcanics (equivalent to Kahar Fm. of Alborz-Azerbaijan area) metamorphosed in lower green schist facies.

The metamorphism age of these rocks is assumed to be Late Precambrian. Although some of the researchers (Berberian and Alavi-Tehrani 1977; Poshtkoohi 2009) have attributed them to Paleozoic, according to the field observations of the present author, far from the ophiolitic complexes, the Lower Cambrian rocks of northwestern part of country (i.e. Barut, Zaigun, Lalun and Mila formations) are not metamorphosed and the metamorphic rocks of area are older than Early Cambrian.

Fig. 8.17 A view of the metamorphic Kahar Formation at the southwestern Zanjan

Chaldoran (Siah Cheshmeh)

In this area most probably the metamorphic rocks of Late Precambrian and Mesozoic can be observed. The Precambrian metamorphic rocks include amphibolite, green schist, marble, micaschist and meta-pellite. The meta-basics, actinolites and epidote are probably related to ophiolitic complexes.

The Precambrian metamorphic rocks are overlain by carbonate Permian rocks (equivalent of Surmaq and Ruteh formations). Generally, these rocks show an angular uncomformity with those of upper Paleozoic.

Here, we will count some notes by which one would be able to distinguish the Precambrian metamorphic rocks from those of Mesozoic which are related to ophiolitic complexes:

1. The metamorphic rocks of Precambrian are mostly of sedimentary protolith. The volcanic-tuffy protolith is also observed. On the other hand, the metamorphic rocks of Mesozoic are mostly meta-basics.
2. The metamorphic rocks of Late Precambrian are mostly overlain by non-metamorphosed younger Paleozoic rocks. There is no regional metamorphism in Paleozoic and Mesozoic except those metamorphic rocks which are related to ophiolitic complexes. The origin of metamorphic rocks of Siah Cheshmeh area is a subject of debate (Aminiazar and Abbasi 2003; Hajialioghli et al. 2016). Some of them including the authors of 1:250,000 geological map of Maku (Alavi and Bolourchi 1973) have attributed these rocks to Precambrian while some others

relate them to ophiolitic complexes. Anyway, considering two mentioned notes can clarify the situation.

Ophiolite-Related Metamorphic Rocks

At Azerbaijan area (western Central Iran) the ophiolites are exposed in several localities including Chaldoran, Khoy, Sarv, etc.

Accompanied by the ophiolitic complexes, the metamorphic rocks are observed formed simultaneously to the Third Neo-Tethys collision (Jurassic to Late Cretaceous). The most important part of the metamorphic rocks is exposed in Khoy area.

Metamorphic rocks of Khoy: Geologically, the rocks of northwestern Khoy can be classified in three main parts (Azizi and Mohajjel 2007) including ophiolitic complex, metamorphic complex and non-metamorphosed sedimentary rocks (Figs. 8.18 and 8.19).

The metamorphic complex of northern Khoy is composed of two main types of rock (Azizi and Mohajjel 2007) including meta-basic (metamorphosed volcanic rocks, tuffs and basic intrusions) and meta-sedimentary (different types of schist, quartzite, meta-arkose and marble) rocks metamorphosed in green schist to amphibolite facies.

In field observations, these two types of rock are occassionaly mixed together where meta-sedimentary rocks are observed as intercalations in meta-basics.

Meta-gabbro lenses can be found in meta-volcanics and meta-sedimentary rocks (up to 20 in number) (Azizi and Mohajjel 2007).

According to Azizi et al. (2001) the granitic intrusions have penetrated meta-sedimentary units and have affected the ophiolitic complex.

These rocks are formed in 4.5–7 kbars and 400–700 °C based on the studies performed by Azizi et al. (2002).

The metamorphic rocks of northwestern Khoy are of to S_1 and S_2 foliations and S_M mylonitic foliation formed in a three stage deformation phenomena. Since such deformation is permanent, the compression of area has occurred in Late Cretaceous after the collision of Arabian plate and Azerbaijan block.

Marand (Mishu Mountains)

The oldest rocks are metamorphic rocks with a thickness of approximately 1000 m exposed in central part of the Mishu Mountains. These rocks are considered to be metamorphosed at two stages: the first at the green-schist facies and regional metamorphism and then the contact metamorphism at garnet-cordierite-biotite hornfels facies.

This complex contains a thick sequence of schists and micaceous shale, hornfels, slate, spotted hornfels, chlorite schist, spotted schist, phyllite and meta-sandstone. The protolith of these rocks actually is the Kahar Formation. The Kahar Formation has endured two types of metamorphism in the region. The first one is the low-temperature regional metamorphism in green-schist facies and the second is contact metamorphism because of granite intrusion resulted in the formation of hornfels and

Legend

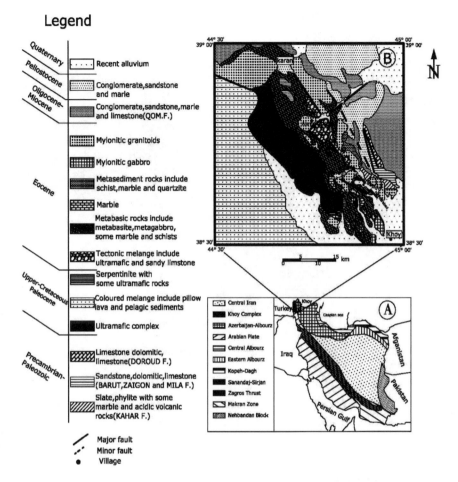

Fig. 8.18 A: The structural zones of Iran (Stocklin and Nabavi 1973); **B**: The geological map of Khoy (Azizi 2001)

spotted hornfels. The lower boundary of the Kahar Formation is unknown and the upper boundary is isocline with the Barut Formation. While Kahar Formation has a tectonically contact with the intrusive body (Shah Zeidi 2013). Also, there are some volcanic rocks in the western part of the area attributed to the Precambrian. This unit consists of small conical shape outcrops of volcanic rocks associated with the Kahar Formation. The composition of these rocks is acidic, and in some cases rhyolitic tuff combined with the acidic rocks. Their texture is porphyritic with a glassy to microlitic matrix.

These rocks are probably equivalent to Qareh-dash rhyolite. Shah Zeidi (2013) believe that the age of metamorphic rocks is Paleozoic and the contact metamorphism is Precambrian in age. It seems that two complexes have been almost metamorphosed

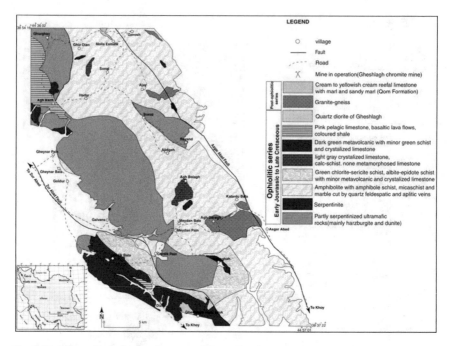

Fig. 8.19 The geological map of metamorphic rocks of eastern Khoy (modified after Radfar and Amini 1999)

simultaneously. If Mishu granitoid of Precambrian is equivalent to the Doran Granite, thus all metamorphic units are the same age as Precambrian.

8.2.4 Eastern and Southeastern Iran

8.2.4.1 Makran Subduction Zone

In the Nikshahr area, the pre-Eocene rocks consist of a sequence of the ophiolitic rocks including serpentinite, peridotite, gabbro, diabase, spilite and plagio-granite. The pelagic sediments associated with the ophiolite consist of radiolarite shales and siltstone accompanied by the Late Cretaceous *Globotruncana* limestone (Santonian-Maastrichtian). The ophiolite complex is a combination of ophiolitic rocks and the pelagic sediments.

At the western end of the Makran zone, a series of abnormal mixed rocks can be observed in a small area which is the continuation of the Azava complex (Fonuj Geological Map, 1:100,000). These rocks are in fact, the extension of the Dorkan complex (Island arc) in Minab, Fonuj, Taherui maps which consists of schist, phyllite and white crystalline limestone resulted from a weak metamorphism and also meta-diabase that are surrounded by serpentine.

The Azava complex in the former geological map is surrounded by faults, and forms the eastern basin of Dorkan-Bajgan continental crust. This area was a deep marine basin characterized by a continental rift zone developed by extensional movements of the continental crust.

In fact, the main metamorphic rocks of the Makran region are reported from the northwestern Makran zone and Southern Jazmurian that is often related to the ophiolitic rocks. In this area, accompanied by ophiolitic rocks, high-pressure low-temperature metamorphic rocks (beautiful glaucophane schists) are found that till now no comprehensive study is performed on them.

In addition to the previous metamorphism in Makran, there is a core complex metamorphism in the Hemunt mount, south of Iranshahr, which shows a real sense of the core complex metamorphism at green-schist and amphibolite facies (the meta-morphism age is assumed to be the Cenozoic). At the core of the Hemunt mount there is an igneous intrusion intruded in ophiolites and has metamorphosed its surrounding rocks.

8.2.4.2 Eastern Iran Metamorphism (Lut Block and Flysch Sub Zone)

There was a separated continent in the western Afghanistan and Eastern Iran by the early Cretaceous. At the mid-Cretaceous a N-S trending rift zone has been developed in the Zahedan (Sistan-Baluchistan province), and divided the continent into two parts, eastern and western of which the eastern and the western parts are called Afghan and Lut blocks, respectively. These two segments have gradually been separated, thus a narrow and small ocean was formed. The new ocean was subsequently closed at the Late Cretaceous to Eocene leading to the collision of the Afghanistan and Lut blocks. Consequently, the colored mélanges have been emplaced along the faults, thrust faults and the metamorphosed flysch zone. The Sistan Ocean closure leads to the formation of the Sistan suture zone. Colliding the Lut and Afghan blocks has gradually caused developments of folds and fractures, uplifting and metamorphism in the area during the Cenozoic. Therefore, the folded belts of the flysch zone are associated with the collision of the micro-continents of Helmand and Lut blocks after the closure of the new ocean. The position of flysch folded zone demonstrates ascending oceanic lithosphere, a series of ophiolites; and cluster-shape of the mafic and ultramafic complexes.

Geology and Metamorphic Rocks of Eastern Iran

The Sistan suture zone in the eastern Iran is extended in an area of 500×100 km and the north-south Cretaceous-Cenozoic orogenic belt is located along the Afghan and Lut blocks.

The Sistan suture zone (Eastern Iran flysch zone) consists of three main units as follows:

1. The Ratuk Complex, the oldest part of the Sistan suture zone represents the initial phase of the closure of the Sistan Ocean. It is signified by the subduction zone, trending to the east (Afghan Block). The area includes the Cretaceous Flysch, metamorphic ophiolite melange along with concentrated fault zones (15 km in length). Within the Ratuk metamorphic complex, meta-chert, meta-spilite and meta-basalt metamorphosed at low temperature and high pressure have been occurred. The rough faulted lenses (measuring approximately 200 × 100 m) are surrounded by low grade chlorite, talc and mica schists. Meta-basic rocks including eclogite and blue schist represent the lithological characteristics of a subduction zone, remnants of subducted oceanic lithosphere in Sistan (Fotohi Rad and Amini 2008). The main outcrops of metamorphic melange zone in the north of the Sistan suture zone include Qazik, Gourchang and Solabest areas.
2. Neh complex is located where the subduction zone relocated towards the western part of the basin and consists of the Late Cretaceous Flysch and less ophiolitic rocks.
3. Sefidabeh basin covers the Neh and Ratuk complexes by an angular unconformity and includes the Maastrichtian to Eocene foreland arc non-metamorphosed sediments.

In other words, the eastern Iran metamorphism is caused by the western movement of the Afghan Block toward the Lut Block. Consequently, a metamorphism and folding occurred in the Eastern Iran. It is clear that the where the collision is occurred sooner the metamorphism grade is higher. So, a variety of metamorphism grades from the eclogite to slate facies can be resulted from the collision.

8.2.4.3   Eastern Lut

The oldest rocks in the Naibandan area are only located on the northern side of Nayband mountain. These folded rocks are as follows:

The Shale, siltstone and sandstones metamorphosed into phyllite. They show too much similarities compared to the Morad 'Series' in the Kerman region. These metamorphic rocks could be ascribed to the Late Precambrian, but the exact age is unknown. The Permian rocks cover the metamorphic rocks of eastern Iran by an angular unconformity. In two different parts of the Lut Geological map some metamorphic rocks are found that are of Triassic age. They are in the south of Dig-e Rostam area including.

The hornfels formed due to the diorite emplacement into the Nayband Formation, and Eastern Garmab Mountain area where the metamorphism is more severe which includes schist and less amphibolite relocated by a main over-thrust fault zone.

The Cretaceous metamorphic rocks of the area consist of phyllites and amphibole-bearing mica schist. There is evidence that the age of the flysch zone is the Early Cretaceous (Neocomian), but it is also possible to be the Late Cretaceous (Alavi-Naini and Griffis 1981).

8.2.4.4 Deh-Salm (Lut Block)

The metamorphic rocks, granite intrusions and pegmatite veins of Deh-Salm meta-morphic complex are exposed near Deh-Salm village in the eastern Lot Block (Fig. 8.20). The Deh-Salm metamorphic complex (100 km length and 8 km width,

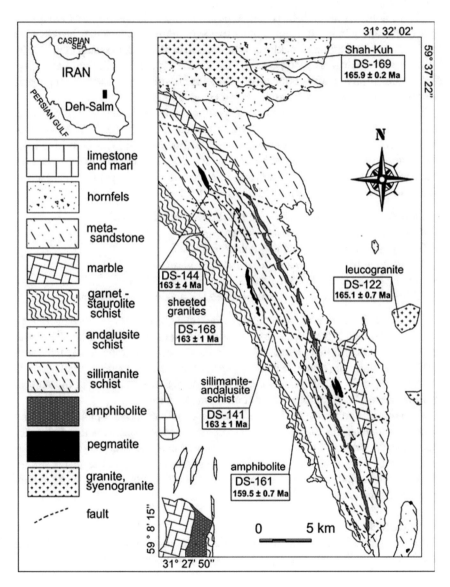

Fig. 8.20 Metamorphic rocks in the Deh-Salm area (Mahmoudi et al. 2010)

latitude 31° 32′ to 31° 13′ longitude 59° 40′ to 59°23′) is located in 60 km west of the Naibandan City, southeast of South Khorasan Province.

The NW-SE Deh-Salm metamorphic complex is the main metamorphic unit of the Lut Block. It has been formed owing to tectonic development of the Eastern Iran.

The general geology of Deh-Salm Metamorphic Complex has been discussed by many geologists.

The protolith of the Deh-Salm metamorphic complex is composed of unknown-aged pelitic and semi-pelitic sedimentary rocks. Nevertheless, they may be related to the Shemshak Formation (Stocklin 1968; Stocklin et al. 1972; Masoudi et al. 2006). The age of the Shemshak group is Late Triassic to Early Jurassic (Besse et al. 1998). Nevertheless, the age of a clastic zircon mineral of northern Iran actually indicates younger ages than the Late Triassic (Horton et al. 2008).

The meta-pelitic rocks are mostly slate, phyllite, andalusite schist, staurolite garnet schist and sillimanite schist. These rocks represent a regular SSE-NNW foliation deeping towards NE (60°). Various stages of deformation in Deh-Salm Metamorphic Complex have been detected by Mahmoudi (2003). According to the metamorphic rock compositions, six main groups are signified in meta-pelitic rocks rich in aluminum (Al_2O_3 content is more than 15% wt%):

- Chlorite + quartz + phengite +muscovite + albite ± calcite
- Biotite + chlorite + muscovite + quartz + albite ± calcite
- Garnet + biotite + muscovite + quartz ± albite
- Staurolite + biotite + muscovite + quartz ± garnet
- Andalusite + biotite + muscovite + quartz + plagioclase
- Sillimanite + biotite + muscovite + quartz ± plagioclase ± garnet

The Deh-Salm metamorphic complex formed at the temperatures less than 350 °C, green-schist facies, in the western part of the area and at higher than 3.5 kb of pressure and 680 °C, at the mid-amphibolite facies in the central part of the area (Mahmoudi 2003). The high temperature and low pressure sillimanite facies is equivalent to the facies series Type III (Pattison and Tracy 1991). The evidence of the highest temperature has been found near the central granite sheets.

The amphibolite is occurred as the local intercalations in meta-pelitic rocks and meta-sandstones of the Deh-Salm metamorphic complex. These meta-sedimentary rocks are characterized by weak foliation composed of amphibole (30–45%), plagioclase (25–40%) and biotite (5–10%). The amphiboles are of the euhedral and subhedral shape and radial structure. Opaque minerals, epidote, titanite, zircon and apatite are amongst the accessory minerals.

Plutonic Rocks

The granitic intrusive rocks of Deh-Salm Metamorphic Complex have been divided based on the distribution, petrology, topography and shape as the following types:

1. Shahkuh body and its halos,
2. Some leucogranite bodies characterized by low topography and high alteration in the eastern Deh-Salm Metamorphic complex,

3. Granite sheets have intruded at high temperatures parallel to andalusite, and sillimanite schists and
4. Metamorphosed pegmatite veins.

The U-Pb dating on zircon, monazite, xenotime and titanite by ID-TIMS shows that the granitic rocks are emplaced at 166–163 Ma., which confirms that high metamorphism's temperatures coincide with the intrusive activities and the region is rapidly cooled. The next post-magmatic hydrothermal activities at 159 Ma. (confirmed by dating on zircon and titanite in an amphibolite and monzonite in a granite) has probably resulted in the later crystallization. Therefore, it seems that the age of the Deh-Salm metamorphism is the same or close to the age of the intrusive bodies and has been probably occurred as a result of the mid-Cimmerian orogeny.

References

Abdi G (1996) Petrological study of northern Qorveh (Kurdistan) volcanic rocks. M.Sc. thesis, Shahid Beheshti University

Aghanabati SA, Shahrabi M (1987) Geological quadrangle map K 4, Mashhad. Geol Surv Iran, Tehran

Ahmadi Khalaji A (2006). Petrology of the granitoid rocks of the Boroujerd area. Ph.D. thesis, University of Tehran

Ahmadi Rohani R (1999) Study of the metamorphic and igneous rocks of the southeastern Dandi. M.Sc. thesis, Shahid Beheshti University

Ahmadi AR (2004) Petrography, petrogenesis and geochemistry of the metamorphic rocks of Tutak complex. M.Sc. thesis, Shahid Beheshti University

Aistov L, Melnikov B, Krivyakin B, Morozov L (1984) Geology of the Khur area (Central Iran). Explanatory text of the Khur quadrangle map 1:250,000

Alavi M (1991) Sedimentary and structural characteristics of the Paleo-Tethys remnants in northeastern Iran. Geol Soc Am Bull 103:983–992

Alavi M, Bolourchi MH (1973) Explanatory text of the Maku Quadrangle Map, 1:250,000. Geol Surv Iran, Tehran

Alavi-Naini M, Griffis R (1981) Geological map of Naybandan, Quadrangle J8, scale 1:250,000. Geol Surv Iran, Tehran

Alavi-Naini M, Hajian J, Amidi M, Bolourchi H (1982a) Explanatory text of the Takab quadrangle map, 1:250,000. Geol Surv Iran, F8, 99 pp

Alavi-Naini M, Hajian J, Amidi M, Bolurchi H (1982b). Geology of Takab-Saein-Qaleh Quadrangle. Geol Surv Iran, Tehran

Aminiazar R, Abbasi S (2003) Geological map of maku, scale 1:100,000. Geol Surv Iran

Azizi H (2001) Petrography, petrology and geochemistry of metamorphic rocks of Khoy. Ph.D. thesis, University of Tarbiat Moalem, Tehran, Iran, 253 pp (in Persian with English abstract)

Azizi H, Mohajjel M (2007) Deformation in the northwestern tectonites of Khoy. J Tehran Univ Sci 33(1):65–73

Azizi H, Moinevaziri H, Noghreayan M (2002) Geochemistry of metabasites rocks in the north of Khoy. J Sci (in Farsi), University of Isfahan, Iran 15:1–20

Azizi H, Moinevaziri H, Yaghobpoor A, Mohajjel M (2001) Mylonitic granitoids in north Khoy. J Sci (in Farsi), University of Tehran 27:81–101

Bagheri S (2007) The exotic Paleo-Tethys terrane in Central Iran: new geological data from Anarak, Jandaq and Posht-e Badam areas. Ph.D. thesis, University of Lausanne, 208 p

Baharifar AA (2004) Petrology of metamorphic rocks in the Hamedan area. Ph.D. thesis, Kharazmi
    University, Tehran, Iran, 218 p
Balaghi M, Sadeghian M, Ghasemi H (2015) Mineralogy, geochemistry and thermobarometry of
    garnet-amphibolites in Delbar metamorphic complex, Biarjmand (Southeast of Shahrood). Iran
    J Crystallogr Mineral 23(3):479–494
Baluchi S, Sadeghian M, Ghasemi H, Minggou Z, Chioli L, Yanbin Z (2019) Mineral chemistry,
    geochemistry and isotope geochronology of kalateh region (NW of Khur): implication for Late
    Triassic magmatism of central Iran zone. Iran J Crystallogr Mineral 26(4):827–844
Barker AJ (1990) Introduction to metamorphic textures and microstructures. Blackie, Glasgow and
    London, 162 pp
Berberian M, Alavi-Tehrani N (1977) Structural analyses of Hamadan Metamorphic Tectonites; A
    Paleotectonic Discussion, Material for the Study of Seismotectonics of Iran, III, 1st edn, Berberian
    M (ed). Geol Surv Iran, 40, pp 263–280
Besse J, Torcq F, Gallet Y, Ricou LE, Krystyn L, Saidi A (1998) Late Permian to Late Triassic
    palaeomagnetic data from Iran: constraints on the migration of the Iranian block through the
    Tethyan Ocean and initial destruction of Pangaea. Geophys J Int 135(1):77–92
Clark LD, Cannon WF, Klasner JS (1975) Bedrock geologic map of the Negaunee SW Quadrangle,
    Marquette Co., Michigan: U.S. Geol Surv Geol Quad Map, G-Q-1206
Darvishzadeh A, Valizadeh A (2001) Introduction of Hamedan Regional metamorphic type based
    on the Zoning of metamorphic minerals in the Northwest of Alvand. In: 5th Iranian Geological
    Survey Symposium
Davoudzadeh M, Seyed-Emami K, Amidi M (1969) Preliminary note on a newly discovered Triassic
    section in the northeast of Anarak (Central Iran). Geol Surv Iran, Geol Note 51, 23 p
Eftekharnezhad J, Asadian A, Mirzaei AR (1992). Age of Shanderman Asalem metamorphic and
    ophiolitic complex and relationship with Paleotethys and sub oceanic crust of Caspian Sea.
    Geosciences, 3, Geol Surv Iran
Farzami F, Ghorbani Q, Shafaei Moghaddam H (2016) Age determination of granite gneiss of Bone-
    Shuro complex (Saghand). Crystallography and Mineralogy of Iran Conference, 23th, Damghan
Fereidouni M, Emami MH, Nasr-e Esfahani AK, Hojjati H (2010) Determination of Tectono-
    magmatic environment in Jandaq Amphibolite, NE Isfahan. Applied Petrology Conference
    2010
Fotohi Rad G, Amini S (2008) Zoning and chemistry of garnets in eclogites and blueschists in
    ophiolitic complex of eastern Birjand: an evidence for subduction process in eastern Iran. Iranian
    Journal of Crystallography and Mineralogy 16(1):141–158
Ghasemi H, Asyabasha A (2006) Introduction and separation of transformation events in Delbar
    Region, Southeast Biarjamand, Central Iran. Isfahan Univ Basic Sci Res J 23(1):231–248
Ghazanfari F (1991) Petrogenesis of Metamorphic Rocks in the northwest of Takab, with an Attitude
    toward Mining of Lead and Zinc Anguran. M.Sc. thesis, University of Tehran
Ghazi AM, Hassanipak AA, Tucker PJ, Mobasher K, Duncan RA (2001) Geochemistry and 40Ar-
    39Ar ages of the Mashhad Ophiolite, NE Iran: a rare occurrence of a 300 Ma (Paleo-Tethys)
    Oceanic Crust. American Geophysical Union, Fall Meeting 2001
Ghorbani M (1999) Petrological investigations of Tertiary-Quaternary magmatic rocks and their
    metallogeny in Takab area. Ph.D. thesis, Shahid Beheshti University
Ghorbani M (2007) Economic geology of natural and mineral resources of Iran. Pars Geological
    Research Center (arianzamin), 492 p
Ghorbani M (2013) The economic geology of Iran: mineral deposits and natural resources. In:
    Ghorbani M (ed) The economic geology of Iran: mineral deposits and natural resources. Springer,
    p 569
Ghorbani M (2019) Lithostratigraphy of Iran. Springer
Goudarzi H (1995) Metamorphism and magmatism in Malayer-Boroujerd region. Master's thesis,
    Kharazmi University, GSI, No G8

Hadizadeh Shirazi M (2010) Petrography and petrogenesis of metamorphic-magmatic rocks in the Kulyakesh Surian region and their links with Mineralogy. M.Sc. thesis, Shahid Beheshti University

Haghipour A (1974) Etude géologique de la région de Biabanak-Bafgh (Iran central), Pétrologie et tectonique du socle precambrien et de sa couverture, France. Ph.D. thesis, Sci Nat, Gronoble University, 403

Haghipour A (1977) Etude geologique de la région de Biabanak Bafgh (Iran central), with coloured map. Geol Surv Iran, Report No 34

Haghipour A, Pelissier G (1968) Geology of the Posht-e Badam-Saghand area (East-central Iran). Geol Surv Iran, Note No 48, 144, 51 Figs, 3 Pls, Map

Hajialioghli R, Fakharinezhad H, Moazzen M (2016) Petrology and geochemistry of amphibolites from southeast of Siyah-Cheshmeh, NW Iran. Sci Quat J, Geosci 25(99)

Hame Rezaei N, Kolahy Azar AP, Eslami SSR (2015) Petrofabric Study of metamorphic Complex of Kaftari Mountain (South of Shahroud). 19th Annual Conference of the Iranian Geological Society and Ninth National Geological Conference of Payame Noor University, Tehran

Horton BK, Hassanzadeh J, Stockli DF, Axen GJ, Gillis RJ, Guest B, Amini A, Fakhari MD, Zamanzadeh SM, Grove M (2008) Detrital zircon provenance of Neoproterozoic to Cenozoic deposits in Iran: implications for chronostratigraphy and collisional tectonics. Tectonophysics 451(1):97–122

Hosseini B (2004) Petrology and geochemistry of metamorphic rocks in Divan-Darreh region (Sanandaj, Sirjan zone). M.Sc. thesis, Faculty of Earth Sciences, Shahid Beheshti University

Hosseini B (2011) Petrology and petrogenesis of Paleozoic metamorphic rocks in Sanandaj-Sirjan zone and their relationship with iron mineralization. Ph.D. thesis, Shahid Beheshti University

Hosseini B, Ghorbani M, Pourmoafi SM, Ahmadi AR (2012) Identification of two different phases of metamorphosed granitoid masses in Kuh-Sefid Tutak anticline based on the U-Pb age dating. Earth Sci J 21(84):57–66

Houshmandzadeh A (1977) Ophiolites of south Iran and their genesis problems (Unpublished). Geol Surv Iran

Houshmandzadeh A, Posht-kouhi M (1996) Metamorphic map of Iran 1:1,000,000. Treatise on the Geology of Iran

Houshmandzadeh A, Soheili M (1990) Explanatory text of the Eqhlid quadrangle map, scale of 1:250,000. Geol Surv Iran, 157 p

Houshmandzadeh AR, Ohanian T, Sahandi MR, Taraz H, Aganabati A, Soheili M, Azarm F, Hamdi B (1990) Geological map of Eglid Quadrangle G10, 1:250,000. Geol Surv Iran, Tehran

Hyndman DW (1985) Petrology of igneous and metamorphic rocks. McGrow-Hill, New York

Izadi kian L, Mohajjel M, Alavi SA (2014) Deformation stages of the metamorphic rocks in Hamedan area and their relationship with Alvand Intrusive Pluton. Sci Q J, Geosci 23(92):187–198

Jamali Ashtiani R (2017) High grade metamorphic rocks from north of Shahrekord, Sanandaj-Sirjan Zone: Age, Fabric, geneses. Ph.D thesis, Shahid Beheshti University

Karimpour MH (2009) Rb-Sr and Sm-Nd isotopic composition, U-Pb-Th (zircon) geochronology and petrogenesis of Mashhad Paleo-Tethys granitoids, Ferdowsi University of Mashhad, Iran (grant P/742-87/7/14)

Karimpour MH, Lang FJ, Chuck A (2011) Geochemistry of radioisotopes Rb-Sr and Sm-Nd, age dating of zircon U-Pb and determinating the origin of leucogranites across Khajeh-Morad, Mashhad, Iran. Geoscience 20(80):171–182

Mahmoudi D (2003) Metamorphic petrogenesis and evolution of Deh-salm metamorphic complex. Ph.D. thesis, Shahid Beheshti University

Mahmoudi S, Masoudi F, Corfu F, Mehrabi B (2010) Magmatic and metamorphic history of the Deh-Salm metamorphic Complex, Eastern Lut block (Eastern Iran), from U-Pb geochronology. Int J Earth Sci 99(6):1153–1165

Majidi B (1978) Etude pétrostructural de la région du Mashhad (Iran), Les problemes des méta-morphites, serpentinites et granitoides Hercyniens. Thése Universite Scientifique et Medical de Granobel France, 277 p

Majidi B, Amidi M (1980) Explanatory text of Hamedan quadrangle map, scale 1:250,000. Geog Surv Iran

Malekpour A, Hassanzadeh J, Mohajjel M, Babaii H (2005) Petrofabric of the Biarjomand meta-morphic rocks, Signs of Tectonic Stretching in Continental crust of Central Iran. Symposium of Geological Society of Iran, Tehran, pp 484–493

Masoudi F, Mehrabi B, Mahmoudi S (2006) Garnet (Almandine-Spessartine) growth Zoning and its application to constrain metamorphic history in Dehsalm Complex, Iran. J Sci, Islam Repub Iran 17(3):235–244

Mohajjel M, Sahandi MR (1999) Tectonic evolution of the Sanandaj-Sirjan zone in northwest Iran, introducing the sub-zones. Sci Q J, Geosci 31–32:28–49

Nabavi M (1976) An introduction to geology of Iran. Geol Surv Iran, Tehran

Naderi F (2012) Petrofabric and petrogenesis of metamorphic rocks in the East Tuyserkan. M.Sc. thesis, Shahid Beheshti University

NIOC (2017) Paleontology and biostratigraphy of Paleozoic rocks of Zagros and Central Iran Basins, Pars Geological research center (Client: Exploration Directorate of National Iranian Oil Company)

Noori Khankahdani K (2005) Investigation of genetic and constructional relationship of Bon-do-no gneiss whit Tootak metamorphic complex. Ph.D. thesis, Islamic Azad University, Science and Research Branch of Tehran

Omrani J, Agard P, Whitechurch H, Benoit M, Prouteau G, Jolivet L (2008) Arc-magmatism and subduction history beneath the Zagros Mountains, Iran: a new report of adakites and geodynamic consequences. Lithos 106:380–398

Pattison DRM, Tracy RJ (1991) Phase equilibria and thermobarometry of metapelites. In: Contact metamorphism. Mineralogical society of America. Reviews in Mineralogy, vol 26, pp 105–182

Pelissier G, Bolourchi MH (1967) East Takab metamorphic complex. Geol Surv Iran, Tehran (Unpublished)

Poshtkoohi M (2009) Poly phase metamorphism of pelitic rocks of Hamedan Area, West Iran based on petrography evidences. Acta Geosci Sinica 30:50–60

Radfar J, Amini B (1999) Geological map of Khoy, scale 1:100,000. Geol Surv Iran

Rajabi S, Torabi G (2012) Mineralogy and geochemistry of xenoliths in the Eocene volcanic rocks, southwest of Jandaq. J Econ Geol 5(1):65–82

Ramezani J, Tucker R (2003) The Saghand region, central Iran: U-Pb geochronology, petrogenesis and implications for Gondwana tectonics. Am J Sci 303(7):622–665

Ruttner AW (1984) The Pre-Liassic basement of the Eastern Kopeh-DaghRange. N Jb Geol Paläont Abh 168(2–3):256–268

Sabzehei M (1974) Les Mélanges ophiolitiques de la région d'Esfandagheh (Iran méridional): étude pétrologique et structurale, interprétation dans le cadre iranien. Université Scientifique et Médicale de Grenoble

Sabzehei M (1996) An introduction to general geological features of metamorphic complexes in southern Sanandaj-Sirjan zone. Geol Surv Iran (Unpublished)

Sabzehei M, Berberian M (1972) Preliminary note on the structural and metamorphic history of the area between Dowlatabad and Esfandagheh, South-East Central Iran. Geol Surv Iran, Int Rep 30; and 1st Iranian Geol Symp

Sadeghian M, Ghasemi H, Shekari S, Zhai M (2019) Petrogenesis and U-Pb dating of the Late Neoprotrozoic Metarhyolites of the Majerad Igneous-Metamorphic Complex (SE Shahrood): an implication to the formation and development. Kharazmi J Earth Sci 4:241–262

Shah Zeidi M (2013) Geochemistry and petrology of granitoids southwest of Marand (south of Eish Abad and Pirbala villages) Northwest of Iran. PhD thesis, University of Tabriz, Faculty of Science

Sharkovski M, Susov M, Krivyakin B (1984) Geology of the Anarak area (central Iran). Explanatory text of the Anarak quadrangle map 1(250)

Sheikh Zakraei J (2008) Petrography and petrology of magmatic stones in Qorveh Region. M.Sc. thesis, North Tehran Branch

Shekari S, Sadeghian M, Zhai M, Ghasemi H, Zou Y (2017) Mineral chemistry and petrogenesis of metabasites of metamorphic—igneous Shotor-Kuh complex (SE Shahrood) an indicator for evolution of intracontinental extensional basins of late Neoproterozoic. J Geosci 27(105):167–182

Soheili M, Jafarian MB, Abdollahi MR (1992) Geological map of Aligudarz, 100000 series, sheet 5956. Geol Surv Iran

Stocklin J (1968) Structural history and tectonics of Iran; a review. Amer Associ of Petr Geol Bulletin 52(7):1229–1258

Stocklin J, Eftekharnezhad J, Hushmandzadeh A (1972) Central Lut reconnaissance, East Iran. Geol Surv Iran, Rept 22:62

Stocklin J, Nabavi MH (1973) Tectonic map of Iran, 1:2,500,000. Geol Surv Iran, Tehran Offset Press, Rep 31, 100 p

Tabatabaei Manesh M, Sharifi M (2011) Evaluation of thermodynamic conditions (P-T) in formation of Jandaq metapelitic schists (Northeast of Isfahan province). J Petrol 2(5):81–92

Takin M (1971) Geological history and tectonics of Iran, a discussion of continental drift in the Middle East since the Early Mesozoic. Geol Surv Iran, Internal Report

Tillman JE, Poosti A, Rossello S, Eckert A (1981) Structural evolution of Sanandaj-Sirjan ranges near Esfahan, Iran. AAPG Bulletin 65(4):674–687

Torabi Gh (2012) Chromitite absence, presence and chemical variety in ophiolites of the Central Iran (Naein, Ashin, Anarak and Jandaq). N Jb Geol Paläont Abh 267(2):171–192. https://doi.org/10.1127/0077-7749/2013/0303

Torabi Gh (2017) Central Iran Ophiolites, Jahad Daneshgahi Publication, Isfahan, Iran, 450 p

Turner FJ (1981) Metamorphic petrology: mineralogical, field and tectonic aspects, 2nd edn. McGraw Hill, Washington, New York, and London, 524 p

Winkler HGF (1979) Petrogenesis of metamorphic rocks. 1st and 5th edn. Spring-Verlag, New York

Zadsaleh (2019) Research seminar on "Petrogenesis of metamorphic rocks of Iran" for PhD students, Shahid Beheshti University

Zahedi M, Vaezipour J, Rahmati Ilkhchi M (1993) Geological map of Shahrekord, No. E8, 1:250000. Geol Surv Iran

Zanchetta S, Zanchi A, Villa I, Poli S, Muttoni G (2009) The Shanderman eclogites: a Late Carboniferous high-pressure event in the NW Talesh Mountains (NW Iran). Geol Soc, London, Special Publications 312(1):57–78

Printed in the United States
by Baker & Taylor Publisher Services